Boris Z. Katsenelenbaum

Electromagnetic Fields – Restrictions and Approximation

Author:

Prof. Dr. Boris Z. Katsenelenbaum
Ha'alya str. 20, Apr. 8, 22383 Nahariya, Israel

Coverfigure:
Zero lines of the auxiliary field for rectangular object reconstruction

Library of Congress Card No. applied for

British Library Cataloguing-in-Publication Data
A catalogue record for this book is available from the British Library.

Bibliographic information published by

Die Deutsche Bibliothek
Die Deutsche Bibliothek lists this publication in the Deutsche Nationalbibliografie; detailed bibliographic data is available in the Internet at <http://dnb.ddb.de>.

Printed in the Federal Republic of Germany

Printed on acid-free paper.

Printing: betz-druck GmbH, Darmstadt
Bookbinding: Großbuchbinderei J. Schäffer GmbH & Co. KG, Grünstadt

ISBN 3-527-40388-4

Preface

The subject of this book falls within the scope of three disciplines – the antennas theory, high-frequency electrodynamics and mathematical physics. There are problems in which these disciplines are fundamentally connected. They are the inverse problems of the high-frequency field theory, more precisely – the connection between the shape of the domain in which the monochromatic currents are located, and the possibility of approximating any given field by the fields of these currents. This connection gives rise to a number of profound questions. Many of them are formulated, and some are even solved in this book.

The following problems are considered:

1. *Antenna synthesis.* Surfaces exist, which have the defect that the fields generated by the current located on these surfaces cannot even approximate many given fields (or their patterns). There are a lot of these surfaces. As a rule, the antenna surface should not be close to them.

2. *Optimal current synthesis.* The case is considered when the fields generated by currents, located on the given surface, do not approximate a given field, or else this approximation requires a very large current. One should then find the field which is (if possible) close to a given one, and so can be realized by the current of the smallest possible norm.

3. *Solvability of the first-kind integral equations.* The existence of solutions to inverse electrodynamic problems depends on the integration domain. This dependence is investigated with different definitions of the term "solution".

4. *Noncompleteness of the set of functions generated by the operator acting over the complete set of functions.* If the operator which transforms the current into the field, acts on the current located on the above-mentioned *defect* surfaces (throughout the book we will refer to such surfaces or lines as specific ones), then the kernel of the conjugate operator is not empty. This means that the generated field is orthogonal (in some metrics) to the so-called orthogonal complement functions. The set of such functions may be highly rich and the actual degree of the surface defectiveness essentially depends on its richness.

5. *Construction of a real solution to the Helmholtz equation* (or the Maxwell equations) *by its zero line* (zero surface). The defect lines (surfaces) are zero lines (surfaces) of some real solutions to these equations; the asymptotics of the solutions at infinity contains the function of the orthogonal complement. The investigation of such solutions is an efficient method of studying a defect line.

6. *Investigation of pseudo-solutions to the first-kind equations* originating from the inverse problems for the fields obtained by the two-dimensional Fourier transformation in arbitrary domains.

The emphases in the book are approximately in the above order. The mathematical technique is not more difficult than that used in diffraction theory problems. The majority of the intermediate derivations are replaced by their description. The methods of functional analysis and the analytical theory of the solution to the Helmholtz equation are used very little.

The first edition of the book entitled "The approximability problem of electromagnetic field" was published in Russian (Nauka, Moscow 1996) [1]. The English edition is extended with new results (partly unpublished), in particular, with the new chapter "Long narrow beam of electromagnetic waves". The appendix "Antenna synthesis by amplitude radiation pattern and modified phase problem", written by N. N. Voitovich, is also included.

The author is deeply grateful to N. N. Voitovich for his careful translation of the manuscript into English and for his contribution in the scientific editing. Special thanks are due to Ulrike Werner and Melanie Rohn for editing the book thorougly and the permanent attention to the project. The author also thanks Oleg Kusyi for his technical assistance.

B.Z.Katsenelenbaum

Israel, July 5, 2003

Contents

1 Introduction

1.1 Introduction

1. In the following examples we briefly illustrate the problems considered in the book and the results obtained.

A. *A cylindrical mirror* is given; its director is an arc of a circle. The mirror is illuminated by a field with the polarization parallel to the cylinder's axis. The induced current has the same direction. For simplicity, we assume that both the field and the current do not depend on the coordinate in this direction. Can this current generate (at a suitable illumination) the pattern $1 + \cos\varphi$, or a pattern close to it?

The answer to this question depends on the value of the circle radius a. It is impossible to generate this pattern or even the one close to it if $J_0(ka) = 0$, where J_0 is the Bessel function and $k = \omega/c$ is the wave number (ω is the frequency). The closest pattern to the given one is $\cos\varphi$. This result does not depend on the mirror width.

The pattern can bear the stamp of the region in which the current is located. This stamp cannot be erased by changing the current distribution. If the condition $J_0(ka) = 0$ holds, then it is impossible for the pattern generated by the current, located on such a mirror, to be equal (even approximately) to a pattern, the Fourier series of which contains the constant term. If ka is not equal to a zero of the function J_0, then it is possible to approximate such a pattern or even realize it, but if ka is close to this zero, then the current generating the pattern must be very large.

B. *A metal screen* is a part of a sphere. It is illuminated by an electromagnetic impulse. Is it possible to verify whether the screen is really a part of a sphere, and to find its radius by the measured pattern of the scattered field? Both the spatial and time structures of the impulse as well as the shape of the mirror contour are unknown.

The answer is positive. To do it, the vector pattern generated (at any illumination) by the current on the screen should be multiplied by a certain weight vector function, and the product should be integrated over the solid angle 4π and Fourier-transformed over time. If the screen is a part of a sphere, then at some frequencies the Fourier transform will be equal to zero. The sphere radius can be calculated by values of these frequencies.

The above examples concerning the parts of cylindrical or spherical surfaces are not exotic. There are many such surfaces, moreover, in their neighborhood infinite numbers of other surfaces of this kind exist. This fact makes the study of the problem reasonable.

The shape of the surface, where the currents are located, can be decisive for the approximation not only of any pattern by the patterns of these currents, but it is also relevant to the fields in the near zone.

C. *A plane is illuminated by a beam* incoming from the antenna located in another plane parallel to the given one. Can some given field be created on it? The answer is positive only if the field fulfils the condition that some expression, containing this field, equals zero. The problem is a two-dimensional generalization of the problem on the existence condition for the field, located on the finite straight-line segment, which creates a given pattern. In both cases the problem is reduced to a first-kind integral equation. The above condition means that the *pseudo-solution* to this equation solves it, that is, the solution exists. In the case of the one-dimensional problem on the pattern of a linear current, the above condition means that the given pattern belongs to the class of functions defined by the Paley–Wiener theorem.

2. Using simple reasoning, one can easily explain the impossibility of approximating some class of patterns from the first example. Let us give such an explanation, emphasizing that it is not universal – in general, the physical explanation is more complicated.

As in most of the book, we will consider here the two-dimensional scalar formulation, because it is shorter and more demonstrative. In essence, the three-dimensional vector problems are not more complicated, but they are much more cumbersome. These problems will be considered in Chapter 5.

First, we prove that if $J_0(ka) = 0$, then no current on the whole circle $r = a$ generates a pattern having a constant term in the Fourier series. At this frequency there exists the solution to the homogeneous Helmholtz equation for the electric field, which equals zero on the circle and has no singularity inside it. This solution is $\widehat{u}(r, \varphi) = J_0(kr)$. In other words, the circle is a *resonant* one.

Consider an auxiliary interior problem on the field $\widehat{u}(r, \varphi)$ in a hollow metal cylinder of given radius $r = a$. At the resonant frequency the eigenoscillation $\widehat{u}(r, \varphi) = J_0(kr)$ can exist in such a volume. The current on the walls is proportional to $\partial u/\partial r$ and does not depend on the angle φ. This current generates a field equal to zero outside the cylinder. Below we will use only the fact that the current, independent of the angle φ, is not radiating at this frequency.

We return to the problem of the field generated by an arbitrary current on the circle arc. Starting with the case of the whole circle, expand the current in the Fourier series. Every term of the series is proportional to $\cos n\varphi$ or $\sin n\varphi$ $(n = 0, 1, ...)$ and generates the field with the same angular dependence. But the zero-order term does not generate a field outside the circle, therefore for $r > a$ such a term is absent in the field expansion of any current as well as in the pattern.

This result is also valid for any arc of the circle, in spite of the fact that the arc is not a closed line and, therefore, nonradiating current cannot be induced on it. Assume that some current on the arc generates a pattern with zero-order term in the Fourier series. Then we can supply the arc to the whole circle and set the current to be zero on the supplementary arc. In this way we have constructed the current on the whole circle, generating the pattern with nonzero constant term. But, as it follows from the above, it is impossible.

The direct proof of the above statement is elementary for this example. If C is a circle arc of radius a and $j(\theta)$ is a current, located on C, then the pattern generated by $j(s)$ is (with accuracy to a nonessential factor)

$$f(\varphi) = \int_C e^{ika\cos(\varphi-\theta)}\, j(\theta)d\theta. \tag{1.1}$$

The constant term in the Fourier series for the pattern is (with the same accuracy)

$$\int_0^{2\pi} f(\varphi)d\varphi = 2\pi J_0(ka) \int_C j(\vartheta)d\vartheta. \tag{1.2}$$

If $J_0(ka) = 0$, then the zeroth term in the series for $f(\varphi)$ is absent. This property does not depend on both the current and the length of arc C. If C is the whole circle, then this proves once more that, at the resonance, any current on the walls creates a field which does not have the zeroth Fourier term. However, the multiplier $J_0(ka)$ is factored out for the circle arc as well.

3. Let us state the problem in more specific terms, but still without aspiring to an exact formulation. Consider the current $j(s)$ located on the given line C (s is the coordinate on C). It generates the pattern

$$f(\varphi) = \int_C \mathcal{K}(s,\varphi)\, j(s)ds. \tag{1.3}$$

The form of a smooth kernel $\mathcal{K}$ in (1.3) is not important here. (In the three-dimensional vector case, the line C should be replaced by a surface, $\mathcal{K}$ – by a functional matrix, and so on.)

In the antenna synthesis theory, equation (1.3) is considered as the integral equation on the current $j(s)$. We are interested in the problem on the existence of a solution to this equation or to an equation in which $f(\varphi)$ is replaced by another function close to $f(\varphi)$ in the quadratic metric. The current $j(s)$ should have a finite norm. First of all, we will investigate how the *existence of a solution to this equation depends on the line* C.

For the validity of most of the results obtained below, it is not necessary for the norm of $j(s)$ to be finite. Moreover, it is acceptable that the current may have singularities and $|j(s)|^2$ may not be integrable. It is only significant that the current should be integrable itself, so that the integral on the right side of expression (2.6) exists (see below). This condition is fulfilled also for the current at the border of a semi-plane (for both polarizations), and for $j(s) \sim \delta(s - s_0)$, that is, for the approximation, usually used in the antenna array theory. However, for simplicity (particularly in Chapter 4), we will require $|j(s)|^2$ to be integrable.

There exist such lines, for which equation (1.3) has no solution, and (what is important) the equation, in which $f(\varphi)$ is replaced by a function close to it, has no solution, either. To obtain the solvable equation, we should change $f(\varphi)$ by a finite value. In terms of functional analysis this means that the complete set of currents $j(s)$ generates a noncomplete set of patterns $f(\varphi)$. It turns out, that line C possesses this property, if there exists a solution to the homogenous Helmholtz equation, equal to zero on C. In the above example “A” this solution is $J_0(kr)$. The investigation of such a solution turns out to be a very efficient method for solving the problem of approximability and the related ones.

The most important result is that there are “many” such lines and surfaces, and “many” patterns $f(\varphi)$, for which equation (1.3) has no solution even if $f(\varphi)$ is replaced by any function “close” to it. Therefore, the effect of nonapproximability and its consequences deserves a detailed analysis. We will explain below, what “many” and “close” mean .

1.2 Subject and Method of Investigation

1. In the first five chapters of the book, some properties of electromagnetic fields, it is three-dimensional vector problems are investigated. Almost all of the questions we are interested in are solved for these fields in much the same way as for the scalar ones. However, the equation itself as well as the material explanation and result formulations are, of course, much simpler in the scalar case. Therefore, the exposition will mainly be given for the scalar fields and, as a rule, for the two-dimensional ones. Transferring the methods and results to the three-dimensional case is almost always trivial; see Chapter 5.

Only one condition is nontrivial while transferring the results to the three-dimensional vector fields. It is connected with the fact that surfaces orthogonal to a given vector field $\mathbf{A}(x, y, z)$ do not always exist. Such surfaces exist only in the case, when

$$\mathbf{A}rot\mathbf{A} = 0. \tag{1.4}$$

There is no similar condition for existence of zero surfaces or zero lines in scalar fields. The methods developed in the book contain the construction procedure of such zero lines for some cases. While transferring the methods from $u(x, y)$ to $\mathbf{A}(x, y, z)$ (1.4) should be taken into account.

Thus, almost all the material in the book is related to properties of a scalar function of two variables. It will be denoted by $u(x, y)$ or $u(r, \varphi)$ depending on the coordinate system, in which the problem is investigated.

The monochromatic fields with time dependence $e^{i\omega t}$ are considered. Instead of the frequency ω, the wave number $k = \omega/c$ is involved in the formulas, and this number is called the frequency (c is the light velocity). We will consider the fields generated by the currents distributed on the surfaces.

This condition does not relate to the material of Chapter 6 where quasi-plane fields are investigated. They do not satisfy the condition (1.18) (see below). They are "generated" by some other fields (not by currents). We return to this question at the end of the subsection.

For the two-dimensional case, the line current (the current distributed along a line) is an analog of the surface one. It is more convenient to introduce the current, not as the right-hand side of the Helmholtz equation, but as the discontinuity of the function u or its derivative on the line.

The field $u(x, y)$ satisfies the homogenous Helmholtz equation

$$\Delta u + k^2 u = 0 \tag{1.5}$$

and one of two following conditions on a line C :

$$[u] \;=\; 0, \left[\frac{\partial u}{\partial N}\right] = j; \tag{1.6a}$$

$$[u] \;=\; j, \left[\frac{\partial u}{\partial N}\right] = 0, \tag{1.6b}$$

where Δ is the two-dimensional Laplace operator:

$$\Delta = \frac{\partial^2}{\partial x^2} + \frac{\partial^2}{\partial y^2} = \frac{\partial^2}{\partial r^2} + \frac{1}{r}\frac{\partial}{\partial r} + \frac{1}{r^2}\frac{\partial^2}{\partial \varphi^2}, \tag{1.7}$$

$[u]$ is the difference between the values of u on the sides of C, from and to which the normal N is directed.

The function $u(r, \varphi)$ satisfies the Sommerfeld radiation condition at infinity (i.e. at $r \to \infty$)

$$u(r, \varphi) = f(\varphi)\frac{e^{-ikr}}{\sqrt{kr}} + O\left[(kr)^{-\frac{3}{2}}\right]. \tag{1.8}$$

Here $f(\varphi)$ is a radiation pattern. Field $u(r, \varphi)$ is completely defined by equation (1.5), the given current $j(s)$ of the form (1.6a) or (1.6b) and condition (1.8) at infinity. The function $f(\varphi)$ in (1.8) is not a given one, it should be determined after the field $u(r, \varphi)$ is found. The field $u(r, \varphi)$ is complex, because any real field cannot have asymptotics as in (1.8). In the general case the function $f(\varphi)$ is also complex.

2. Let us explain the physical sense of the currents in (1.6). The two-dimensional scalar problem of the electrodynamics appears, when either the electrical or magnetic field has only one component (more exactly, when $E_x = E_y = H_z \equiv 0$ or $H_x = H_y = E_z \equiv 0$), and the fields do not depend on the z-coordinate: $\partial/\partial z \equiv 0$.

In this book, the first of these cases is considered, namely, the case, when $E_x = E_y = H_z \equiv 0$ and the fields do not depend on z. *The scalar function* $u(x, y)$ *is* E_z, two other components of the electric field equal zero identically. The derivative $\partial u/\partial N|_C$ is the component of the magnetic field, tangential to the line C. The expression $[\partial u/\partial N]$ is the jump of the magnetic field, or, what is the same, the z-component of the electric current. The electric current located on the line C is denoted in (1.6a) by j. The jump $[u]$ of the function u on C is the tangent to C component of the magnetic current $j^{(m)}$. One can write (1.6) in the form

$$\left[\frac{\partial u}{\partial N}\right] = j; \quad [u] = j^{(m)}. \tag{1.9}$$

While referring to $j_z^{(e)}$ and $j_z^{(m)}$ we omit the lower and upper indices z and (e), respectively. Nonessential factors are omitted in (1.9), too: in fact, the currents j and $j^{(m)}$ are not equal, but only proportional to the jumps of u and $\partial u/\partial N$, correspondingly.

There is another way to introduce the scalar function. One can consider a three-dimensional vector field with $E_z \equiv 0$ and $\partial/\partial z \equiv 0$. Then H_z depends only on x and y and it can be used as the above scalar function u. We will not use this approach in the book.

3. No special or very complicated mathematical methods are used for investigation of the problems, considered. Properties of solutions to equation (1.5) are investigated by the usual methods applied in typical diffraction problems. No diffraction problems themselves are considered in the book.

The functions considered below may have singularities – discontinuities and poles. Currents are given in the form of discontinuity of the functions or their normal derivatives. The existence or absence of singularities in the whole domain or in any part of it is very important in all the constructions below. This importance is connected with the wide use of the Green formula

$$\int\limits_{\mathcal{L}} \left(u_1 \frac{\partial u_2}{\partial N} - u_2 \frac{\partial u_1}{\partial N} \right) ds = 0 \tag{1.10}$$

for two solutions $u_1(x, y)$ and $u_2(x, y)$ to equation (1.5). This formula is valid only if both functions have no singularity inside the closed line (contour) $\mathcal{L}$.

All the lines mentioned below (either closed or not) are smooth or have only a finite number of angular points. We also assume that all the contours satisfy the conditions, necessary for the existence of a solution to the interior Dirichlet (u is zero on the contour) or Neumann ($\partial u/\partial N$ is zero on the contour) problems. At some frequencies the homogeneous problem for equation (1.5) with the corresponding boundary condition on C is supposed to have nonzero solutions.

The variational technique is also widely applied. Two variational methods are used most often. The first one is the Lagrangian multipliers method. It reduces the problem on a conditional extremum to the problem on an unconditional one. The second one is the Ritz method reducing the problem on the extremum of the two bilinear functionals ratio to an algebraic equation, involving equating some determinant to zero.

In some sections, the terms of functional analysis, such as: operators, completeness of the function set and so on, are used. However, only two theorems are referred to in the book. The first is the theorem on completeness of eigenfunctions of the self-adjoint integral operators; the second is on the tendency of eigenvalues of such operators to zero as the order number increases. In the problems on fields, generated in the free space by currents on surfaces, the nonself-adjoint operators are paramount. Some simple properties of these operators are briefly formulated as they are used in the book. The theory of analytical properties of solutions to the Helmholtz equation (or the Maxwell system) and functional analysis, are used simultaneously. This makes it possible to obtain some results in a simple and clear way.

The functional analysis formalism is used a little (mainly for shortness) in Chapter 4. The pseudo-solutions to the first kind equations, that is, the functions which minimize the mean square difference between the given function and the calculated one, are investigated. The solution to such an equation exists if and only if this minimal difference is zero.

All the material below is within the capacity of the layman in mathematics. In view of this the intermediate derivations are replaced by their descriptions where possible. As a rule, the technique used is trivial for mathematicians. However, the fact that an interesting branch of mathematical physics with unsolved problems, simple by formulation but still complicated by nature exists, may be of interest to them.

4. There are not many references to publications in the text. The general statement of the problem, denoted in the book title, and some essential results, were given in the papers published in 1988–2003. There are not references to all these papers in the book. The formulas and methods of diffraction theory used below are described, for instance, in [2]. Some formulations of functional analysis are in [3]. The formulas for special functions can be found in the reference book [4].

1.3 Realizability, Approximability, Amplitude Approximability

1. In the book some properties of a function $f(\varphi)$ are investigated. This function is determined by the current $j(s)$ and line C using (1.3). It is the radiation pattern of the field generated by this current. For brevity we will often use a term "the pattern generated by the line C" for the function $f(\varphi)$. Current $j(s)$ is an arbitrary function of the coordinate s on C. As a rule, we only assume that it has a finite norm N defined by

$$N^2 = \int_C |j(s)|^2 \, ds. \tag{1.11}$$

The typical formulation of our problem is the following. Given a function $F(\varphi)$ (in general, complex) with finite norm; for definiteness, we will usually put

$$\int_0^{2\pi} |F(\varphi)|^2 \, d\varphi = 1. \tag{1.12}$$

Can $F(\varphi)$ be a radiation pattern generated by the line C? If not, does a function, close to $F(\varphi)$, exist, which can be generated by C? Otherwise, does a current $j(s)$ on C exist, connected with the function $F(\varphi)$ or with that, close to it, by formula (1.3)?

The following four cases can occur.

1. *Realizability:* equality (1.3) (replacing $f(\varphi)$ by $F(\varphi)$), treated as an integral equation on $j(s)$, has a solution with finite norm. It means there exists a current $j(s)$ generating the pattern $f(\varphi)$ equal to $F(\varphi)$; more specifically,

$$\int_0^{2\pi} |F(\varphi) - f(\varphi)|^2 \, d\varphi = 0. \tag{1.13}$$

2. *Approximability:* for any given $\delta > 0$, a current $j(s)$ exists, such that the pattern $f(\varphi)$, generated by it, fulfils the condition :

$$\int_0^{2\pi} |F(\varphi) - f(\varphi)|^2 \, d\varphi \le \delta^2. \tag{1.14}$$

The realizability can be considered as a special case of approximability. It occurs when the above condition is valid for $\delta = 0$, too. Any function realizable by a line C is also approximable by it. If the function is not approximable, then it is not realizable, either.

3. *Amplitude approximability:* for any given $\delta > 0$, a current $j(s)$ and a real function (a phase) $\psi(\varphi)$ exist, such that

$$\int_0^{2\pi} \left| F(\varphi) e^{-i\psi(\varphi)} - f(\varphi) \right|^2 \, d\varphi \le \delta^2. \tag{1.15}$$

The approximability is a special case of the amplitude approximability, when (1.15) is valid for $\psi(\varphi) \equiv 0$.

4. *Amplitude nonapproximability:* the amplitude of any pattern generated by the line C is not close to $|F(\varphi)|$.

The nonapproximability of a given function $F(\varphi)$ means that the distance between any pattern generated by the line C and the given function $F(\varphi)$ (see below (1.32)) is finite. The amplitude nonapproximability means that the finite distance is between any pattern generated by the line C and an arbitrary function with amplitude $|F(\varphi)|$.

If the function $F(\varphi)$ is realizable, then the integral equation

$$\int_C \mathcal{K}(\varphi, s) j(s) ds = F(\varphi) \tag{1.16}$$

has a solution with a finite norm. We do not mention the last statement. If the function $F(\varphi)$ is approximable, then for any $\delta > 0$ there exists the function $\widetilde{F}(\varphi)$ satisfying the condition

$$\int_0^{2\pi} \left| F(\varphi) - \widetilde{F}(\varphi) \right|^2 d\varphi \le \delta^2, \tag{1.17}$$

so that the equation

$$\int_C \mathcal{K}(\varphi, s) j(s) = \widetilde{F}(\varphi) \tag{1.18}$$

has a solution. If the function $F(\varphi)$ is amplitude approximable, then for any $\delta > 0$ there exist the phase $\psi(\varphi)$ and the function $\widetilde{F}(\varphi)$ satisfying the condition

$$\int_0^{2\pi} \left| F(\varphi) e^{-i\psi(\varphi)} - \widetilde{F}(\varphi) \right|^2 d\varphi \le \delta^2, \tag{1.19}$$

so that equation (1.18) has a solution.

Below we will use the shortened expressions: "the function is realizable by the line C", "...is approximable by C", "...is amplitude approximable by C", omitting the words: "...by the patterns of currents with finite norm, located on along the line C".

2. The problem of realizability is not the subject of this book. Only a few results of this topic will be needed below. We will use the technique [5], based on the analysis of the convergence rate (with respect to n) of the Fourier series

$$F(\varphi) = \sum_{n=0}^{\infty} A_n \cos(n\varphi) \tag{1.20}$$

of the function $F(\varphi)$, the realizability of which is investigated, (see also Section 3.3, Subsection 7). In a similar way to many other cases, the terms with $\sin(n\varphi)$ are omitted in (1.20) for simplicity. The realizability depends on the existence and value of the limit

$$a_0 = \frac{2}{ek} \lim_{n\to\infty} \left(n \sqrt[n]{|A_n|}\right), \quad e = 2.718... \tag{1.21}$$

Strictly speaking, the term $\overline{\lim}$ should be used in (1.21) instead of lim, but this fact is not essential here. The value of a_0 does not depend on the norm of function $F(\varphi)$, because $\lim_{n\to\infty} \sqrt[n]{|c|} = 1$ for any given $c \neq 0$.

If the limit (1.21) does not exist (i.e., $a_0 = \infty$), then the function $F(\varphi)$ is not realizable by patterns of any currents, wherever they are located. The example is the Π-function, its Fourier coefficients A_n diminish as slowly as $1/n$ and it is not realizable by any line.

If ka_0 is a finite number, then the realizability of $F(\varphi)$ by the line C depends on the mutual location of the line C and the circle of radius a_0 with its centre at the coordinate origin. The function $F(\varphi)$ is not realizable by any line lying wholly inside this circle, but it is realizable by any closed nonresonant (with respect to the Dirichlet boundary condition) contour, encircling the circle. If $a_0 = 0$, then $F(\varphi)$ is realizable by any closed nonresonant contour containing the coordinate origin, and it is not realizable by any unclosed line. For the pattern of any current located on a nonclosed line, $a_0 \neq 0$ (see [5]). We do not analyze all the possible mutual locations of the line C and the circle here. Notice only that the closed contours realize a much wider class of patterns, than do the unclosed ones. As it will be shown later in the book, this is valid for the approximability problem, too.

As an example, consider the function (not normalized by (1.12))

$$F(\varphi) = e^{-A\sin^2(\varphi/2)}. \tag{1.22}$$

Its Fourier coefficients decrease as $I_n(A/2)$, where I_n is the modified Bessel function. It is asymptotically (at $n \to \infty$) equal to $(eA/2)^n n^{-n}$, so that (1.21) yields: $a_0 = A/k$. Any circle of the radius greater than A/k, or any contour, encircling this circle can realize the function (1.22). Any line, located inside the circle does not realize it. The narrower a pattern, the larger must be the contour which can realize it. The width of the pattern (1.22) is about $A^{-1/2}$.

It is significant for comparison of the notions of realizability and approximability, that the first one is defined by the behavior of A_n at large n according to (1.21). Therefore, the number ka_0 may be altered in an arbitrary way, by altering the function $F(\varphi)$ as little as desired. For example, if we truncate the series (1.20) at large n, we almost do not affect the function $F(\varphi)$, but the value ka_0 is zero. Otherwise, if we add the function $\sum_{n=N}^{\infty} (1/n^2)\cos(n\varphi)$ to $F(\varphi)$, then, at a large N, it will be altered by an arbitrary small value, but the Fourier series will converge so slowly that the limit (1.21) will not exist ($a_0 = \infty$), that means the new function will not be realizable by any line.

The realizability is a "gentle" property, the approximability is a more "rough" one. The nonapproximable function should be finitely altered to become an approximable one and vice versa. This statement will be justified below. It will be shown that if this alteration is small (although it is finite according to the approximability definition), then the difference between the presence and absence of the approximability will also be small. This corresponds to physical intuition.

3. Below, we often deal with the situation when all the patterns $f(\varphi)$, realizable by the line C, possess the property:

$$I = 0, \tag{1.23}$$

where

$$I = \int_{0}^{2\pi} f(\varphi)\hat{F}^{*}(\varphi)d\varphi. \tag{1.24}$$

Here $\hat{F}(\varphi)$ is a function, *determined by the line C* only. It has a finite norm and, by default, is normalized by the condition

$$\int_{0}^{2\pi} \left|\hat{F}(\varphi)\right|^{2} d\varphi = 1. \tag{1.25}$$

The function $\hat{F}(\varphi)$ and its connection with the line C are very significant for the problems considered in the book.

If the condition (1.23) is fulfilled, then the functions $F(\varphi)$ exist, which cannot be approximated by the line C. We often use this relation between the property (1.23) of the line C and the nonapproximability.

Let us give an elementary proof of this known fact. Suppose that for some function $F(\varphi)$, the integral

$$b = \int_{0}^{2\pi} F(\varphi)\hat{F}^{*}(\varphi)d\varphi \tag{1.26}$$

does not equal zero. Consider the integral

$$\int_{0}^{2\pi} \left[F(\varphi) - f(\varphi)\right] \hat{F}^{*}(\varphi)d\varphi. \tag{1.27}$$

According to (1.26) and (1.23), the first summand in (1.27) equals b, and the second one equals zero, so that the integral equals b. Applying the Cauchy inequality

$$\int_{a}^{b} |\alpha(x)|^{2}dx \int_{a}^{b} |\beta(x)|^{2}dx \geq \left|\int_{a}^{b} \alpha(x)\beta(x)dx\right|^{2}, \tag{1.28}$$

which is valid for any square integrable functions $\alpha(x)$, $\beta(x)$, to (1.27) gives

$$\int_{0}^{2\pi} |F(\varphi) - f(\varphi)|^{2} d\varphi \cdot \int_{0}^{2\pi} \left|\hat{F}(\varphi)\right|^{2} d\varphi \geq |b|^{2}. \tag{1.29}$$

The second factor in (1.29) equals one by (1.25), and for any function $f(\varphi)$ generated by C we have

$$\int_{0}^{2\pi} |F(\varphi) - f(\varphi)|^{2} d\varphi \geq |b|^{2}. \tag{1.30}$$

That is the condition of nonapproximability of the function $F(\varphi)$ by the line C.

The inverse statement is also valid. If the inequality (1.30) holds for any function $F(\varphi)$ and any pattern $f(\varphi)$, then it is also valid for $F(\varphi) = f(\varphi)$, where $f(\varphi)$ is a function generated by C, and (1.30) yields $b = 0$, that is, the condition (1.23).

4. Sometimes for the sake of brevity we will use the terms of functional analysis. However, we really do not need its technique. In these terms, condition (1.23) appears as the following: the set of functions $f(\varphi)$ is not complete, although it is generated by the integral operator (1.3) acting over the complete set of currents. There exist functions of the orthogonal complement. The above function $\hat{F}(\varphi)$ is an element of the space of these functions. It is orthogonal to all the functions $f(\varphi)$ of the form (1.3) . The orthogonality means that the integral (1.23) (the *inner product* of functions $f(\varphi)$ and $\hat{F}(\varphi)$) equals zero. The inner product of two functions $A(\varphi)$ and $B(\varphi)$ is the integral:

$$(A, B) = \int_0^{2\pi} A(\varphi)B^*(\varphi)d\varphi. \tag{1.31}$$

In particular, $(A, A) = ||A||^2$ is the squared norm of A. The distance between $A(\varphi)$ and $B(\varphi)$ is introduced as

$$\left\{ \int_0^{2\pi} |A(\varphi) - B(\varphi)|^2 \, d\varphi \right\}^{1/2} . \tag{1.32}$$

According to (1.23) and (1.30), the function is nonapproximable, if it is not orthogonal to $\hat{F}(\varphi)$. For instance, the function $\hat{F}(\varphi)$ itself is nonapproximable as well. The orthogonality of some function $F(\varphi)$ to $\hat{F}(\varphi)$ is the necessary condition for $F(\varphi)$ to be approximable. It is clear that this condition is not fulfilled for an arbitrary function. If there is only one function of the orthogonal complement for the given line C, then the above condition is also sufficient.

Let us give an obvious geometric illustration modelling the space of functions by the space of three-dimensional vectors. Then (1.31) is the usual scalar product of two vectors $\vec{A}$ and $\vec{B}$, and (1.32) is the length of the straight-line segment connecting the end points of these vectors if their origins coincide.

The condition (1.23) means that all the vectors $f(\varphi)$ lie in the plane orthogonal to $\hat{F}(\varphi)$. Only vectors lying in this plane are approximable. If the vector $F(\varphi)$ does not lie in this plane, then from comparison of (1.32) with (1.30) it follows: the distance from the end point of the vector to the plane, that is, the distance between $F(\varphi)$ and the vector $f(\varphi)$, closest to it, is equal to $|b|$. The statement "if the condition (1.23) is fulfilled, then almost all functions are nonapproximable" in terms of this illustration only, means that the dimension of the plane is less than the dimension of the whole space. The function $f(\varphi)$, closest to $F(\varphi)$, lying in the plane, orthogonal to $\hat{F}(\varphi)$, is the projection of $F(\varphi)$ onto the plane and equals $F(\varphi) - b\hat{F}(\varphi)$. In Chapter 4 we will obtain this result without referring to the three-dimensional illustration.

5. The orthogonal complement of functions $f(\varphi)$ may contain not only one function $\hat{F}(\varphi)$, but also a set of linear independent functions $\hat{F}_p(\varphi)$. In this case the three-dimensional illustration cannot be used. As we will see in Section 2.2, the number of functions $\hat{F}_p(\varphi)$ can even be infinite.

Give an obvious generalization of the formula (1.30) for this case. Let all the patterns $f(\varphi)$ satisfy the conditions

$$I_p = 0, \quad p = 1, 2, ..., P, \tag{1.33}$$

where

$$I_p = (f, \hat{F}_p). \tag{1.34}$$

Orthogonalize the linear independent functions $\hat{F}_p(\varphi)$ and normalize them to unity, so that

$$(\hat{F}_p, \hat{F}_q) = \delta_{pq}, \tag{1.35}$$

where δ_{pq} is the Kronecker symbol ($\delta_{pq} = 0$ for $p \neq q$, $\delta_{pp} = 1$).

Combine the expression, analogous to (1.27)

$$\left(F - f, \sum_{p=1}^{P} b_p \hat{F}_p \right), \tag{1.36}$$

where

$$b_p = (F, \hat{F}_p). \tag{1.37}$$

Applying the Cauchy inequality (1.28) to (1.36) with $\alpha = F - f$, $\beta = \sum\limits_{p=1}^{P} b_p \hat{F}_p$ and using (1.35), we have

$$\left| \left(F - f, \sum_{p=1}^{P} b_p \hat{F}_p \right) \right|^2 \leq \|F - f\|^2 \sum_{p=1}^{P} |b_p|^2 . \tag{1.38}$$

On the other hand, using (1.33) and (1.37) yields

$$\left(F - f, \sum_{p=1}^{P} b_p \hat{F}_p \right) = \sum_{p=1}^{P} |b_p|^2 , \tag{1.39}$$

which together with (1.38) gives

$$\int_0^{2\pi} |F(\varphi) - f(\varphi)|^2 \, d\varphi \geq \sum_{p=1}^{P} |b_p|^2 . \tag{1.40}$$

This is a generalization of the formula (1.30).

If there are only P linearly independent functions of the orthogonal complement, that is if there are not other functions to which all the patterns, generated by the line C, are orthogonal, then the equalities

$$b_p = 0, \quad p = 1, 2, ..., P \tag{1.41}$$

are not only the necessary conditions for the function $F(\varphi)$ to be approximable, but they are also the sufficient ones.

1.4 Outline of the Book

1. One of the purposes of the book is to investigate some properties of the *fields generated by surface currents*. The simplest two-dimensional model is mainly considered, which is based on the scalar Helmholtz equation with currents, located on a line. The property consists in the following: there are "many" lines such that the fields generated by them cannot approximate "many" functions themselves, as well as all functions close to them. This property is widely investigated throughout the book.

Such *specific* lines are zero lines of a *real* solution to the Helmholtz equation. As a matter of fact, this property is elementary, but it allows us to find and effectively investigate the above lines (surfaces). Even if a line is close to a specific one, very large currents are required for the approximation of almost all fields. The antenna surface should not be close to a specific one (see, however, Subsection 2.2.6).

The second purpose is to investigate the *two-dimensional Fourier transformation*: the main instrument for consideration of the fields in the form of a long narrow beam.

2. In Chapter 2 the approximability of the patterns (i.e., the fields at infinity) is analyzed. Different real fields, their zero lines, as well as prohibitions to the approximation connected with them, are under analysis. For some lines the prohibitions are so strict that many functions with a given amplitude and free phase cannot even be approximated by the fields of currents on these lines.

The existence of prohibitions to the approximation can be used for obtaining some information about the shape of a metallic body, if the scattered pattern is known but the incident field is unknown. This information can have the form of a probable statement. It is assumed that the incident field is unknown. In particular, if the measured pattern cannot be approximated by the patterns of currents, located on a specific surface, then the surface of the scatterer cannot be close to it.

3. The prohibitions to approximation exist not only for the patterns; they relate also to fields in the near zone (*near fields*). The questions connected with this problem are investigated in Chapter 3. In general, the prohibitions to the near field are stronger than to the field at infinity. They grow slowly while going away from the current. Similarly to the previous case, the problem is reduced to construction of a real field and determination of its zero lines.

If the Helmholtz equation has a solution, which equals zero on the given line, then the prohibition relates only to the part of the plane where the solution has no singularities. If the solution is analytical in the whole plane, then the prohibition relates to both the near field and the pattern. If singularities exist, then prohibitions relate only to the near field; any pattern can be approximated by currents on the line. When singularities exist at infinity only, then the prohibitions relate to the near field.

For this reason the problem of the *analytical extension* of the eigenoscillation field in the resonant domain outwards from its boundary is very important. The field should be real and continuous together with its normal derivative on the boundary. One should find the conditions for such an extension to be realizable in the case when singularities are absent, as well as to develop the method to locate these singularities, when they exist. This problem is not solved with the practically acceptable efficiency even in the two-dimensional scalar case. In the chapter several simple examples are given, when the singularities appear in a finite part of the plane, at infinity, or when they do not appear at all.

4. The concept of approximability is a generalization of the concept of realizability. The *optimal current synthesis* is a generalization of approximability. Even in the case, when a given pattern is approximable, it can require a very large current. It is often possible to alter the pattern *finitely but a little*, so that the new pattern can be realized by an essentially smaller current. The problem of finding the patterns, close to the given one but generated by currents with small norm, is considered in Chapter 4.

This problem is solved by using the known technique of *generalized functions of the double orthogonality*. This technique deals with two orthogonal sets of functions, mutually connected: the set of currents and the set of patterns. A current described by a function of the first set, generates the pattern described by an appropriate function of the second one.

Some closed line can have the two following properties. The first is to be the specific line. That means, there exists a pattern, which cannot be generated by a current located on the line. The second property consists in the existence of a current on the line (closed), which does not generate a field outside it. The usage of the function sets mentioned allows us to understand the connection between these properties. It results that they are independent, but mutually symmetrical in some sense.

In the last section of the chapter we give an expression for a lower bound of the norm of a current approximating a given pattern. To determine this estimation, it is not necessary to calculate preliminary the current. This bound can be calculated exactly as a result of some passing to the limit.

Each specific line has an influence on the norm of currents located on the nonspecific ones close to it. The mentioned estimation gives a value of this influence. The current norm is inversely proportional to a value having a meaning of the distance between the given line and the specific one near to it. To approximate almost any pattern, the current located on the line nearby a specific one should have a large norm. This statement is still more related to the approximation of fields in the near zone. It is in accordance with our physical intuition.

5. In Chapter 5 the results of the previous chapters are transferred to the case of three-dimensional vector fields satisfying the Maxwell equations. The transference is trivial almost everywhere. The first and second electrodynamic problems are formulated instead of the Dirichlet and Neumann ones; the Lorentz lemma is used instead of the Green formula, and so on. Section 5.1, devoted to this transference, is very short. The detailed exposition of results and intermediate derivations in the vector form are cumbersome but in fact do not provide anything new in the understanding of the nonapproximability phenomenon and the related prohibitions.

Section 5.2 is devoted to the essential difference between the three-dimensional vector problems and the related scalar ones (two- or three-dimensional). The auxiliary real field connected with a specific surface in the vector case is restricted by some conditions, which have no analogy in the scalar case. Some results, which are obvious or easy to prove in the scalar case, are formulated in the vector one only as probable hypothesis. That is the first effect caused by this difference. The second one is the limitation of the practical use of results obtained, in the problems, in which the vector nature of the field is essential. However, the general behavior of the nonapproximability phenomenon and the related problem of the optimal current synthesis differs a little in the scalar and vector cases.

6. In Chapter 6 the theory of *long narrow beams* of electromagnetic waves is considered. Such beams are supposed to be applied for transmitting the power from the orbiting solar power stations to the Earth. The beams are created by large plane antennas; they preserve a quasi-plane structure at a long distance, that is, on the receiving antenna (rectenna). With an accuracy to simple phase factors, the field in the rectenna plane is described by two-dimensional Fourier transformation of the field on the antenna. The field distributions on the antenna are determined, which provide either maximum of the power transmission coefficient or the maximal closeness of the field in the rectenna plane to an "ideal" one. The first demand reduces the problem to a homogeneous equation for the field on the antenna; the second one is fulfilled if the field on the antenna is the pseudo-solution to some first-kind integral equation.

In this problem the pseudo-solution differs from the identical zero. Note that the pseudo-solution technique is not expedient to be applied in Chapters 2-5, because in the most interesting case, when the surface is specific, that is, when the orthogonal complement functions exist, the pseudo-solution to the appropriate integral equation can be equal to zero ("If an operator gives rise to a noncomplete set of functions, then the kernel of the adjoint operator is nonempty").

The question of the antenna *shape* is of special interest in the beam problem. A hypothesis that the optimal shape is circular or elliptical is well founded (but not proved!).

7. The diffraction (scattering) problems on the nontransparent bodies (metallic ones or with impedance boundaries) as well as on the partially transparent obstacles or walls have been investigated for a long time. The key works in the mathematical theory of the patterns (*far fields*) completeness are [6], [7], [8]. This theory is explained in detail in the book [9]. In particular, it is shown there that some properties of the diffracted (scattered) field depend on the interior domain even for bodies with nontransparent boundaries, i.e. when the field does not penetrate into this domain. At its eigenfrequencies the space of the radiation patterns can be noncomplete, namely, at arbitrary incident field the pattern does not contain some elements of a complete set of the angle functions. This takes place in the case, when there exists an auxiliary field (the so-called *Herglotz wave function*) analytical in the whole space, which satisfies the Helmholtz equation (the Maxwell ones in the vector case) and equals zero on the scatterer boundary.

The material given in the book is closely related to the above problem. However, our aim is more to explain what physical sense these specific properties have and in which way they can show themselves in the practical problems, than to obtain some new mathematical properties of the scattered fields.

No diffraction problem is considered here. Common to the problems is the fact that the induced current is allocated on the surface and the volume currents do not arise. The *scattered field is a field of* the induced (*surface*) *currents*. The properties of the surface currents are investigated in the book without connection with the way in which these currents arise. This problem statement is based on the fact that the noncompleteness of the pattern set is not connected with any property of the surface currents.

The above noncompleteness is not obligatorily connected with the existence of the resonance of the interior domain. This property is also inherent in the fields scattered on nonclosed shells (screens), because these fields are also created by the surface currents at the diffraction on the screens; the auxiliary field mentioned above exists at all frequencies "almost always". Just the *existence of* such an *auxiliary field is the necessary and sufficient condition* for the *noncompleteness* of the *scattered patterns* set (see [9, Theorem 6.32]). Of course, for the nontransparent bodies or closed shells the auxiliary field exists at resonant frequencies only. Great attention is paid in the book to the *structure of the orthogonal complement function set.* The connection of these functions with the auxiliary field is stated. The situation is investigated in detail, when there are many such functions, for instance, when the set of them is countable.

The analogous properties of the patterns created by the magnetic currents are stated. The currents of both types together create the complete set of patterns.

All the results related to the patterns are valid (even in a reinforced form) for the near fields. The noncompleteness of the set of scattered patterns can be used for the reconstruction of the scatterer shape and position.

As a consequence of the nonapproximability phenomenon, the problem of the optimal current synthesis arises. It is solved using the eigenfunctions of iterated operators (see[10], [11]). This technique allows to state the symmetry between two independent properties of the surface: the noncompleteness of the fields of surface currents and the existence of the nonradiating surface currents (i.e., those connected with the eigenoscillations of the interior volume).

The book is based on the material partially published by the author and his colleagues in a series of papers [5], [12]-[29].

8. In the Appendix some questions related to the optimal current synthesis at the given amplitude pattern are considered. In general, the pattern is supposed to be nonapproximable, and the problem is formulated in the variational form. The appropriate Lagrange-Euler equation is nonlinear, it has many solutions describing all stationary points of the functional. The solutions bifurcate as the antenna size (or the frequency) increases. In the particular case of the linear antenna, with an appropriate metric in the functional space of the patterns, expression (1.3) becomes the Fourier transformation. The problem is close to the so-called phase problem and can be considered as a modification of it.

In this case the solutions to the mentioned nonlinear equation can be expressed in an explicit form, with a finite number of unknown complex parameters. The parameters are determined from a system of transcendental equations. This allows to investigate the equation completely including the calculation of all its branching points. The case of the discrete Fourier transformation describing the equidistant antenna array, is also considered.

All theoretical results are reinforced by numerical ones.

The nonlinear equations considered in the Appendix are also interesting from the mathematical point of view, not only owing to their connection with the Fourier transformation theory, but also as representatives of an unstudied class of nonlinear integral equations of the Hammerstein type.

2 Nonapproximability of Patterns

2.1 Nonapproximability and Zero Lines of the Real Wave Field

1. The function $f(\varphi)$ describes the radiation pattern generated by the current $j(s)$ located on the line C. This function is determined as the coefficient at the factor $\exp(-ikr)/\sqrt{kr}$ in the asymptotics (1.8) of the field $u(r,\varphi)$ at $r \to \infty$. The field satisfies the homogeneous Helmholtz equation (1.5), the mentioned radiation condition (1.8) at infinity, and a certain condition on the line C. The last one is introduced by (1.6a) as a jump of the normal derivative on C (the field itself is continuous on C):

$$\left[\frac{\partial u}{\partial N}\right] = j(s), \quad [u] = 0. \tag{2.1}$$

Introduce an *auxiliary field* $\hat{u}(r,\varphi)$. This field plays a very important role in all the constructions below. By definition, it satisfies the same homogeneous equation

$$\Delta\hat{u} + k^2\hat{u} = 0 \tag{2.2}$$

and has no singularities in the whole plane. At infinity the field $\hat{u}(r,\varphi)$ should certainly contain not only the term (1.8), but also the term proportional to $\exp(ikr)/\sqrt{kr}$ (otherwise it should be equal to zero identically). The asymptotics of $\hat{u}(r,\varphi)$ at infinity is

$$\hat{u}(r,\varphi) \simeq \hat{F}^*(\varphi)\frac{\exp(ikr)}{\sqrt{kr}} + \Phi(\varphi)\frac{\exp(-ikr)}{\sqrt{kr}}. \tag{2.3}$$

The terms of the order $(kr)^{-3/2}$ are omitted in (2.3). At $r \to \infty$ the field $\hat{u}(r,\varphi)$ contains the "nonsommerfeld" term (the first one in (2.3)). The field is generated by a cylindrical wave incoming from infinity. Its magnitude is the function $\hat{F}^*(\varphi)$. If we suppose that the field $\hat{u}(r,\varphi)$ is real, then $\Phi(\varphi) = \hat{F}(\varphi)$. In any case, this factor will not appear in further constructions. The function $\hat{F}(\varphi)$ has a finite norm. Normalize it by the condition

$$\int_0^{2\pi} \left|\hat{F}(\varphi)\right|^2 d\varphi = 1. \tag{2.4}$$

Thereby we have also normalized the function $\hat{u}(r,\varphi)$.

The real field $\hat{u}(r,\varphi)$ is completely determined by equation (2.2), the asymptotic condition (2.3) and the condition that it be analytical in the whole plane including the line C.

2. Apply the second Green formula (1.10) to the field $u(r,\varphi)$ generated by the current $j(s)$ and the auxiliary field $\hat{u}(r,\varphi)$ generated by the incoming cylindrical wave. Both functions satisfy the same Helmholtz equation (1.5) or (2.2). The field $u(r,\varphi)$ is nonanalytical on the line C and therefore this line should be removed from the integration domain. We should encircle the line C by an additional contour. If C is a closed line, then the additional contour consists of two closed lines lying inside and outside C, respectively. In our case the line $\mathcal{L}$ consists of this additional contour and the infinitely remote circle. Since the functions u and $\partial\hat{u}/\partial N$ are continuous on C, the first term in (1.10) will be canceled out while integrating over the additional contour, and the second term will take the form $[\partial u/\partial N]\,\hat{u}$. This yields

$$\lim_{r\to\infty}\int_0^{2\pi}\left(\hat{u}\frac{\partial u}{\partial r}-u\frac{\partial\hat{u}}{\partial r}\right)rd\varphi=-\int_C\left[\frac{\partial u}{\partial N}\right]\hat{u}ds. \tag{2.5}$$

After using the asymptotic conditions (1.8), (2.3) in the left-hand side of (2.5) only the crossed term remains under the integral. Using the condition (2.1) for the right-hand side, we finally obtain

$$\int_0^{2\pi} f(\varphi)\hat{F}^*(\varphi)d\varphi=\frac{i}{2}\int_C j(s)\hat{u}(s)ds. \tag{2.6}$$

This is the main formula for many of the constructions below.

In fact, (2.6) represents the *reciprocity theorem* (two-dimensional analog of the Lorentz lemma) for two fields $u(r,\varphi)$ and $\hat{u}(r,\varphi)$, generated by the current $j(s)$ on C and a "current" at infinity, respectively. The left-hand side of (2.6) contains the product of "current" $\hat{F}^*(\varphi)$ generating the field $\hat{u}(r,\varphi)$, and the field $u(r,\varphi)$. On the right-hand side we have such a product of the current $j(s)$ generating the field $u(r,\varphi)$, and the field $\hat{u}(r,\varphi)$. The integration is performed over the domains in which the currents are located: over the infinitely remote circle and the contour C, respectively.

Introduce a notion of "*the specific line*". The line C is a specific one (we denote such lines by $\hat{C}$), if there exists a function $\hat{u}(r,\varphi)$, analytical in the whole plane, which satisfies equation (2.2), the zero condition on the line $\hat{C}$

$$\hat{u}(r,\varphi)|_{\hat{C}}=0, \tag{2.7}$$

and the asymptotic condition (2.3) with a function $\hat{F}(\varphi)$ of a finite norm. Both the field $\hat{u}(r,\varphi)$ and its normal derivative are continuous on $\hat{C}$.

It follows from (2.6), (2.7), that the pattern of any current located on $\hat{C}$ fulfils the condition (1.23). In these conditions the function $\hat{F}(\varphi)$ is connected with the asymptotic condition (2.3) for the field $\hat{u}(r,\varphi)$. The function $\hat{F}^*(\varphi)$ is the factor in the nonsommerfeld term of the asymptotics of the real field, equal to zero on $\hat{C}$.

The existence of the field $\hat{u}(r,\varphi)$ fulfilling (2.7) is the sufficient condition for the pattern of any current located on $\hat{C}$ to satisfy (1.23) with the function $\hat{F}(\varphi)$ connected with $\hat{u}(r,\varphi)$ by (2.3). This condition is also necessary: if the patterns $f(\varphi)$ of all the currents located on some line fulfil the conditions (1.23) with some function $\hat{F}(\varphi)$, then there exists a field $\hat{u}(r,\varphi)$, which satisfies the Helmholtz equation, has no singularities, is equal to zero on this line and is connected with $\hat{F}(\varphi)$ by the condition (2.3). To prove the necessity, we construct the field $\hat{u}(r,\varphi)$ by its asymptotics (2.3), which is analytical in the whole plane and satisfies the Helmholtz equation. It is sufficient to prove that this field fulfils the condition (2.7). Let $\{j_n\}$ be a complete set of currents on $\hat{C}$ (an example of such a set is given in Section 4.2). According to (1.23), all the patterns, generated by the currents of this set, are orthogonal to $\hat{F}(\varphi)$. It follows from (2.6) that all the integrals $\int_C j_n(s)\hat{u}(s)ds$ are equal to zero, too. Since the set of $\{j_n\}$ is complete, this means that the condition (2.7) is valid.

As it was noticed in Section 1.3, the nonapproximability of any function $F(\varphi)$ nonorthogonal to $\hat{F}(\varphi)$ (i.e. for which the integral (1.26) is not equal to zero) by the line $\hat{C}$, follows from (1.23). This property is the most important in practice. However, direct checking of the condition (1.23) for any current on $\hat{C}$ is, as a rule, impossible. On the other hand, the existence of the field $\hat{u}(r,\varphi)$ with the property (2.7) can often be easily proved; sometimes this can be done in the simplest way by the direct construction of this field. That is the main significance of the equivalence between (1.23) and (2.7).

Note that the construction of the line $\hat{C}$ by the given function $\hat{F}(\varphi)$ can be considered as a determination of some parameter in the operator (1.1) by its orthogonal complement function.

3. Consider in detail the connection between the field $\hat{u}(r,\varphi)$ and the function $\hat{F}(\varphi)$, given by the formula (2.3). The general form of the field $\hat{u}(r,\varphi)$ satisfying the Helmholtz equation and having no singularities is

$$\hat{u}(r,\varphi) = \sum_{n=0}^{\infty} C_n J_n(kr)\cos(n\varphi) \tag{2.8}$$

(for simplicity, the terms with $\sin(n\varphi)$ are omitted here; of course, the consideration is valid also for the fields not being even). The coefficients C_n are supposed to be such that the series (2.8) converges everywhere. To have zero lines, the field $\hat{u}(r,\varphi)$ and, consequently, coefficients C_n should be real (with an accuracy to a constant factor $\exp(i\alpha)$; this nonessential generalization will be considered below).

According to the asymptotic behavior of the Bessel functions, the asymptotics of the field $\hat{u}(r,\varphi)$ contains both the term with $\exp(ikr)/\sqrt{kr}$ and $\exp(-ikr)/\sqrt{kr}$. Equating the first of them with the first term in (2.3) yields

$$\hat{F}^*(\varphi) = \frac{1}{\sqrt{2\pi i}} \sum_{n=0}^{\infty} C_n(-i)^n \cos(n\varphi). \tag{2.9}$$

The normalization condition (2.4) gives

$$\sum_{n=0}^{\infty}(1+\delta_{0n})|C_n|^2 = \frac{1}{\pi}. \tag{2.10}$$

Formulas (2.8), (2.9) establish the one-to-one correspondence between the functions $\hat{F}(\varphi)$ and $\hat{u}(r,\varphi)$: if one of them is given, then the second one is known.

The field satisfying the Helmholtz equation and having no singularities can be expressed not only in the cylindrical coordinate system, but also in the Cartesian one as an integral of the plane waves:

$$\hat{u}(x,y) = \int_0^{2\pi} D(\beta)\exp(ikx\cos\beta + iky\sin\beta)d\beta. \tag{2.11}$$

The asymptotics of (2.11) at $x^2 + y^2 \to \infty$ contains only values of $D(\beta)$ for $0 \le \beta \le 2\pi$ and therefore we have included in the consideration only the homogeneous plane waves, for which $|\cos\beta| \le 1$, $|\sin\beta| \le 1$.

Substituting $x = r\cos\varphi$, $y = r\sin\varphi$ into (2.11) we obtain

$$\hat{u}(r,\varphi) = \int_0^{2\pi} D(\beta)\exp(ikr\cos(\beta-\varphi))d\beta. \tag{2.12}$$

Note that the fields expressed in such a form with arbitrary $D(\beta) \in L_2(0, 2\pi)$ are called the *Herglotz wave functions* [9]. The asymptotics of the integral (2.12) at $r \to \infty$ can be evaluated by the stationary phase method, which gives

$$\hat{u}(r,\varphi) \simeq \sqrt{2\pi i}D(\varphi)\frac{\exp(ikr)}{\sqrt{kr}} + \sqrt{-2\pi i}D(\varphi+\pi)\frac{\exp(-ikr)}{\sqrt{kr}}. \tag{2.13}$$

Comparing (2.13) with (2.3) yields

$$\hat{F}^*(\varphi) = \sqrt{2\pi i}D(\varphi). \tag{2.14}$$

Both (2.9) and (2.14) give two simple ways of constructing the field $\hat{u}(r,\varphi)$ by $\hat{F}(\varphi)$: the coefficients C_n are determined from (2.9) and $\hat{u}(r,\varphi)$ is calculated by the series (2.8); the function $D(\varphi)$ is determined from (2.14) and $\hat{u}(r,\varphi)$ is calculated by the integral (2.11).

Formulas (2.11), (2.14) and (2.8), (2.9) are equivalent. Indeed, expanding $D(\varphi)$ into the Fourier series and using a known integral

$$\int_0^{2\pi} \exp(ikr\cos\varphi)\cos(n\varphi)d\varphi = 2\pi i^n J_n(kr), \tag{2.15}$$

we obtain from (2.12) together with (2.14), the same series (2.8).

4. The requirement for $\hat{u}(r,\varphi)$ to be real (in-phase) imposes some condition on the function $\hat{F}(\varphi)$. Substituting $\varphi + \pi$ for φ into (2.9) we obtain

$$\hat{F}(\varphi+\pi) = \frac{1}{\sqrt{-2\pi i}}\sum_{n=0}^{\infty} C_n^*(-i)^n\cos(n\varphi). \tag{2.16}$$

If $\hat{u}(r,\varphi)$ is real, then C_n are real as well, and comparing (2.16) with (2.9) gives

$$\hat{F}(\varphi+\pi) = -i\hat{F}^*(\varphi). \tag{2.17}$$

If $\hat{u}(r,\varphi)$ is in-phase, that is, $\hat{u}(r,\varphi) = |\hat{u}(r,\varphi)|\exp(i\alpha)$, $\alpha = const$, then $C_n = |C_n|\exp(i\alpha)$ and

$$\hat{F}(\varphi+\pi) = -i\hat{F}^*(\varphi)\exp(-2i\alpha). \tag{2.18}$$

Since α is an arbitrary real number, $\hat{F}(\varphi+\pi)$ differs from $\hat{F}^*(\varphi)$ only in a phase multiplier, independent of φ.

This result can be obtained straightforwardly from (2.2), (2.3), without using the asymptotics of the Bessel function. Suppose $\hat{u}(r,\varphi)$ is real. Let $\hat{u}(r,\varphi)$ have no singularities and be represented in the form (2.8). Apply the Green formula (1.10) to the field $\hat{u}(r,\varphi)$ and the plane wave $\exp(ikr\cos(\varphi-\varphi'))$ incoming from the direction $\varphi=\varphi'$. The integration domain in (1.10) is the infinitely remote circle. Calculate this integral by the stationary phase method. The integrand consists of two terms containing the multipliers $\exp\{ikr\left[1+\cos(\varphi-\varphi')\right]\}$ and $\exp\{ikr\left[-1+\cos(\varphi-\varphi')\right]\}$, respectively. The stationary phases in these terms are $\varphi=\varphi'+\pi$ and $\varphi=\varphi'$, respectively. As a result, we obtain equality (2.17).

We proved that the condition (2.18) is necessary for the function $\hat{F}(\varphi)$ to appear in the asymptotic condition (2.3) of a field $\hat{u}(r,\varphi)$. The inverse statement is also valid: if the factor $\hat{F}(\varphi)$ in (2.3) fulfills the condition (2.17), then, according to (2.9) and (2.16), the coefficients C_n are real, and the field $\hat{u}(r,\varphi)$ is real, too. If the condition (2.18) is fulfilled, then $C_n=|C_n|\exp(i\alpha)$ and $\hat{u}(r,\varphi)=|\hat{u}(r,\varphi)|\exp(i\alpha)$.

The above conditions are also fulfilled by the function $\hat{F}(\varphi)$, in the case when $\hat{u}(r,\varphi)$ is not an even function of φ, that is, when the series (2.8), (2.9) contain terms with $\sin(n\varphi)$.

It is easy to show that any field $\hat{u}(r,\varphi)$ with asymptotics (2.3) has zero lines (at any rate, for large r). Let $\hat{u}(r,\varphi)$ be real ($\alpha=0$). Then

$$\hat{u}(r,\varphi)=\left|\hat{F}(\varphi)\right|\frac{2}{\sqrt{kr}}\cos\left[kr-\beta(\varphi)\right]+O(kr)^{-3/2}, \tag{2.19}$$

where $\beta(\varphi)$ is the phase of the function $\hat{F}(\varphi)$ (see (2.20) below). It follows from (2.19), that at large r the function $\hat{u}(r,\varphi)$ has zero lines close to the contours $r=\left[(n+1/2)\pi+\beta(\varphi)\right]/k$, that is, to the circles $r=\pi n/k$.

According to (2.12), (2.14), the immediate (not using the Fourier coefficients) expression of the field $\hat{u}(r,\varphi)$ by the function $\hat{F}(\varphi)$ is

$$\hat{u}(r,\varphi)=\frac{1}{\sqrt{2\pi i}}\int_0^{2\pi}\hat{F}^*(\beta)e^{ikr\cos(\beta-\varphi)}d\beta. \tag{2.20}$$

This field satisfies the Helmholtz equation, because the kernel $e^{ikr\cos(\beta-\varphi)}$ satisfies it at all β. If we require $\hat{u}(r,\varphi)$ to be real, that is $\alpha=0$ in (2.7), then the function $\hat{F}(\varphi)$ should fulfil the following condition:

$$\hat{F}^*(\varphi)=i\hat{F}(\varphi+\pi). \tag{2.21}$$

This condition is not a necessary, but also a sufficient one for the field (2.20) to satisfy the asymptotic condition (2.3) replacing $\Phi(\varphi)$ by $F(\varphi)$:

$$\hat{u}(r,\varphi)\simeq\hat{F}^*(\varphi)\frac{\exp(ikr)}{\sqrt{kr}}+F(\varphi)\frac{\exp(-ikr)}{\sqrt{kr}}. \tag{2.22}$$

Indeed, if one substitutes (2.18) into (2.13) and replaces $\hat{F}^*(\varphi)$ according to (2.21), then (2.13) takes the form (2.22). The integral in (2.20) can be divided into two terms over $(0,\pi)$ and $(\pi,2\pi)$, respectively. Changing the variable of integration β in the second integral by $\gamma=\beta-\pi$ and using (2.21), one obtains an expression for $\hat{u}(r,\varphi)$, equivalent to (2.20):

$$\hat{u}(r,\varphi)=\sqrt{\frac{2}{\pi}}\operatorname{Re}\int_0^{\pi}\hat{F}(\varphi)e^{i\pi/4-ikr\cos(\beta-\varphi)}d\beta. \tag{2.23}$$

Hence, in order that some function $\hat{F}(\varphi)$ appears in the asymptotics (2.22) at $r \to \infty$ for a real field $\hat{u}(r,\varphi)$, which satisfies the Helmholtz equation (2.2) and is connected with $\hat{F}(\varphi)$ by (2.20), it is necessary and sufficient that $\hat{F}(\varphi)$ satisfies the condition (2.21) (or some more general condition (2.18)). As it was shown above, if $\hat{F}(\varphi)$ is connected with $\hat{u}(r,\varphi)$ by the condition (2.22), then it is a function of the orthogonal complement for the set of patterns generated by the currents on the zero lines of this field, hence it is orthogonal to all these patterns. If $\hat{F}(\varphi)$ is normalizable by (2.4), then the patterns cannot approximate any function, nonorthogonal to $\hat{F}(\varphi)$.

The main formulas (2.21), (2.20) have a simple physical meaning. According to (2.21), the magnitude $\hat{F}(\varphi+\pi)$ of the cylindrical wave $\exp(-ikr)/\sqrt{kr}$ outgoing into direction $(\varphi+\pi)$ is proportional to the magnitude $\hat{F}^*(\varphi)$ of the cylindrical wave $\exp(ikr)/\sqrt{kr}$ incoming from the direction φ (see (2.22)). The multiplier $-i$ describes the "loss of the quarter wave", usual in the diffraction theory, which arises while passing through the geometrical focus (the coordinate origin in our analogy). According to (2.20), the field at each point is the sum of fields of the plane waves incoming from all the directions $\varphi = \beta$ $(0 \le \beta \le 2\pi)$ and having the magnitudes $\hat{F}^*(\beta)$.

5. The above results are obtained under the assumption that there are no bodies in the field, that is, the field $\hat{u}(r,\varphi)$ exists in the whole plane and has no singularities: only then one can express them in the form (2.8) or (2.11). In this subsection we will give a generalization of the formulas (2.21), (2.20), for the case when a body is present in the field [27]. We confine ourselves to the case when the body is a metallic one with the boundary S, such that the field $\hat{u}(r,\varphi)$ exists only outside this boundary and equals zero on it:

$$\hat{u}(r,\varphi)|_S = 0. \tag{2.24}$$

The formula (2.24) differs in meaning from the similar one (2.7), it describes the boundary condition on the given contour S. The field is singular: its normal derivative is discontinuous on S. In (2.7) the line $\hat{C}$ is not given, it is determined by this condition (however, see Sections 3.2, 3.3 below), (2.7) is not the condition that the field $\hat{u}(r,\varphi)$ should fulfil. The field and its normal derivative are continuous on $\hat{C}$.

Let us express $\hat{u}(r,\varphi)$ by the function $\hat{F}(\varphi)$ which appeared in its asymptotics (2.22). The condition on $\hat{F}(\varphi)$ will be also found, at which this field is a real solution to the Helmholtz equation everywhere outside S, satisfying the condition (2.24) on S and having the asymptotics (2.22).

The sought expression will naturally contain the functions describing the solution to the auxiliary problem on diffraction of the plane wave on the metallic body with boundary S. Let $g(r,\varphi,\beta)$ denote the diffraction field (not the whole one!) generated by the plane wave of the unit magnitude, incoming from direction $\varphi = \beta$ onto the body. This field exists everywhere outside S, it satisfies the homogeneous Helmholtz equation in this domain, the radiation condition

$$g(r,\varphi,\beta)|_{r\to\infty} \simeq \Phi_{sc}(\varphi,\beta)\frac{\exp(-ikr)}{\sqrt{kr}} \tag{2.25}$$

at infinity and the boundary condition

$$g(r,\varphi,\beta) + e^{ik(x\cos\beta + y\sin\beta)}\Big|_S = 0 \tag{2.26}$$

on S, providing the whole field $\hat{u}(r,\varphi)$ fulfils the condition (2.24). The function $\Phi_{sc}(\varphi,\beta)$ in (2.25) is the scattered pattern of the plane wave $\exp\left[ik(x\cos\beta + y\sin\beta)\right]$ on the body S. This function is not defined by the formula (2.25), it is determined after the field $g(r,\varphi,\beta)$ has been found.

The diffraction field $g(r,\varphi,\beta)$ is generated by the currents induced on S. The whole field is the sum of the incident and diffracted fields:

$$\hat{u}(r,\varphi) = \int_0^{2\pi} D(\beta)e^{ikr\cos(\beta-\varphi)}d\beta + \int_0^{2\pi} D(\beta)g(r,\varphi,\beta)d\beta, \tag{2.27}$$

where $D(\beta)$ is the magnitude of the plane wave incoming from the direction $\varphi - \beta$. Substituting the expression for $D(\beta)$ by $\hat{F}^*(\varphi)$ under (2.14) into (2.27), we obtain the two-term formula generalizing (2.20):

$$\hat{u}(r,\varphi) = \frac{1}{\sqrt{2\pi i}}\int_0^{2\pi} \hat{F}^*(\beta)e^{ikr\cos(\beta-\varphi)}d\beta + \frac{1}{\sqrt{2\pi i}}\int_0^{2\pi} \hat{F}^*(\beta)g(r,\varphi,\beta)d\beta. \tag{2.28}$$

Let $r \to \infty$ in this formula. According to (2.13), (2.14), the first summand gives

$$\hat{F}^*(\varphi)\frac{\exp(ikr)}{\sqrt{kr}} - i\hat{F}^*(\varphi+\pi)\frac{\exp(-ikr)}{\sqrt{kr}}, \tag{2.29}$$

and, on account of (2.25), the second one is transformed into

$$\frac{1}{\sqrt{2\pi i}}\int_0^{2\pi} \hat{F}^*(\beta)\Phi_{sc}(\varphi,\beta)d\beta\frac{\exp(-ikr)}{\sqrt{kr}}. \tag{2.30}$$

Equating the coefficients at $\exp(ikr)/\sqrt{kr}$ and $\exp(-ikr)/\sqrt{kr}$ to corresponding ones in (2.22) we obtain the sought condition for $\hat{F}(\varphi)$:

$$\hat{F}(\varphi) = i\hat{F}^*(\varphi+\pi) + \frac{1}{\sqrt{2\pi i}}\int_0^{2\pi} \hat{F}^*(\beta)\Phi_{sc}(\varphi,\beta)d\beta. \tag{2.31}$$

Formulas (2.28), (2.31) generalize (2.20), (2.21). In one half of the interval $0 < \varphi \le 2\pi$ the function $\hat{F}(\varphi)$ can be chosen at will, and in the second one it is determined by equation (2.31).

6. The field $u(r,\varphi)$ considered above is generated by the "electrical current" $j = \left[\partial u/\partial N\right]|_C$ located on a line C. Now we will consider fields generated by "magnetic currents". Denote such a field by $u^{(m)}(r,\varphi)$, the current generating it by $j^{(m)}$, and the line where the current is located by $C^{(m)}$, so that $j^{(m)} = \left[u^{(m)}\right]\big|_{C^{(m)}}$. The current $j^{(m)}$ is tangential to $C^{(m)}$.

The field $u^{(m)}(r,\varphi)$ satisfies the Helmholtz equation (1.5), the asymptotic condition analogous to (1.8)

$$u^{(m)}(r,\varphi) = f^{(m)}(\varphi)\frac{e^{-ikr}}{\sqrt{kr}} + O\left[(kr)^{-\frac{3}{2}}\right], \tag{2.32}$$

and the following conditions on the line $C^{(m)}$:

$$\left[u^{(m)}\right]\Big|_{C^{(m)}} = j^{(m)}, \quad \left[\frac{\partial u^{(m)}}{\partial N}\right]\Bigg|_{C^{(m)}} = 0, \tag{2.33}$$

where $j^{(m)}$ is given.

Similarly to $\hat{u}(r,\varphi)$, introduce the auxiliary field $\hat{u}^{(m)}(r,\varphi)$ in the whole plane, which has no singularities and satisfies equation (1.5) and the condition

$$\left.\frac{\partial \hat{u}^{(m)}}{\partial N}\right|_{C^{(m)}} = 0. \tag{2.34}$$

The field $\hat{u}^{(m)}(r,\varphi)$ has the same asymptotic behavior (2.3) after replacing the function $\hat{F}(\varphi)$ by $\hat{F}^{(m)}(\varphi)$. This function is normalized by the condition of the type (2.4). Any line $\hat{C}^{(m)}$, for which such a field $\hat{u}(r,\varphi)$ exists, is called the "specific line of the magnetic type".

Apply the Green formula (1.10) to $u^{(m)}(r,\varphi)$ and $\hat{u}^{(m)}(r,\varphi)$, where the integration is made over the infinitely remote circle and an auxiliary contour encircling the line $\hat{C}^{(m)}$ (on $\hat{C}^{(m)}$ the function $u^{(m)}(r,\varphi)$ is not analytical). The integral over this contour equals

$$-\int_{\hat{C}^{(m)}} \left[u^{(m)}\right] \frac{\partial \hat{u}^{(m)}}{\partial N} ds. \tag{2.35}$$

Using the conditions (2.33), (2.34) on $\hat{C}^{(m)}$ and the asymptotic conditions at infinity for the functions $u^{(m)}(r,\varphi)$ and $\hat{u}^{(m)}(r,\varphi)$, we obtain

$$\int_0^{2\pi} f^{(m)}(\varphi)\hat{F}^{(m)*}(\varphi)d\varphi = -\int_{\hat{C}^{(m)}} j^{(m)}(s)\frac{\partial \hat{u}^{(m)}}{\partial N} ds. \tag{2.36}$$

According to (2.34), the right-hand side of (2.36) is equal to zero, and we have a formula analogous to (1.23)

$$\int_0^{2\pi} f^{(m)}(\varphi)\hat{F}^{(m)*}(\varphi)d\varphi = 0. \tag{2.37}$$

All the above results obtained for $f(\varphi)$ and $\hat{C}$ are transferred to $f^{(m)}(\varphi)$ and $\hat{C}^{(m)}$ in the obvious way.

7. As will be shown in Section 3.2, there exist nonclosed lines which simultaneously belong to both the $\hat{C}$-type and $\hat{C}^{(m)}$-type. It means, that there exist both a field $\hat{u}(r,\varphi)$ fulfilling the condition (2.7) and a field $\hat{u}^{(m)}(r,\varphi)$ fulfilling the condition (2.37). The patterns generated by currents $j(s)$ and $j^{(m)}(s)$ cannot separately approximate some functions $F(\varphi)$. But, except for a specific case, when the functions $\hat{F}(\varphi)$ and $\hat{F}^{(m)}(\varphi)$ are linearly dependent, the currents of both types together generate a complete set of patterns. We give an elementary proof of this fact.

It is sufficient to prove, that any function $F(\varphi)$ can be represented in the form

$$F(\varphi) = f(\varphi) + f^{(m)}(\varphi), \tag{2.38}$$

where $f(\varphi)$ and $f^{(m)}(\varphi)$ are functions generated by some currents $j(s)$ and $j^{(m)}(s)$ on the mentioned line, and they satisfy the conditions (1.23), (2.37), respectively. In order for (2.38) to be fulfilled, one can express $f(\varphi)$ and $f^{(m)}(\varphi)$ in the form

$$f(\varphi) = \frac{1}{2}F(\varphi) + \alpha\hat{F}(\varphi) + \beta\hat{F}^{(m)}(\varphi), \tag{2.39a}$$

$$f^{(m)}(\varphi) = \frac{1}{2}F(\varphi) - \alpha\hat{F}(\varphi) - \beta\hat{F}^{(m)}(\varphi) \tag{2.39b}$$

with arbitrary coefficients α, β. If their values are

$$\alpha = -\frac{I + I^{(m)}J}{2(J^2 - 1)}, \qquad \beta = \frac{I^{(m)} + IJ}{2(J^2 - 1)}, \tag{2.40}$$

where

$$I = (F, \hat{F}), \quad I^{(m)} = (F, \hat{F}^{(m)}), \quad J = (\hat{F}, \hat{F}^{(m)}), \tag{2.41}$$

then the conditions (1.23), (2.37) are also fulfilled. Since the functions $\hat{F}, \hat{F}^{(m)}$ are linearly independent and normalized by (2.4), $J^2 - 1 \neq 0$.

It follows from the statement we have proved, that an arbitrary line (a nonspecific one, as well as a specific one of both electric and magnetic types separately or simultaneously) can approximate any function $F(\varphi)$ when using the currents $j(s)$ and $j^{(m)}(s)$ together.

The formula (2.38) has a simple interpretation in a three-dimensional model of the functional space. Namely, in such a model every vector $F(\varphi)$ is the sum of two vectors lying in the planes, orthogonal to the non-complanar vectors $\hat{F}$ and $\hat{F}^{(m)}$, respectively.

2.2 Examples of Specific Lines. "Prohibited" Antenna Shapes

In this section we give some simple examples of the specific lines $\hat{C}$ and related functions $\hat{F}(\varphi)$ of the orthogonal complement. The lines and functions are connected by the conditions (2.3), (2.7) imposed on the auxiliary field $\hat{u}(r, \varphi)$. The examples are constructed in an elementary way: first of all the field $\hat{u}(r, \varphi)$ is chosen as a solution to the Helmholtz equation with separated variables in an appropriate coordinate system, and then zero lines of this field are investigated.

1. Consider the function

$$\hat{u}(r, \varphi) = J_n(kr)\cos(n\varphi), \quad n = 0, 1, 2, \ldots \tag{2.42}$$

The index n is taken as integer numbers so that $\hat{u}(r, \varphi)$ is continuous at $\varphi = 2\pi$. For a fixed n, the line $\hat{C}$, on which $\hat{u}(r, \varphi)$ equals zero, consists of a countable set of the circles of radii

$$r = \mu_{nm}/k \tag{2.43}$$

(where μ_{nm} is the mth zero of the Bessel function J_n, $J_n(\mu_{nm}) = 0$) and $2n$ (at $n > 0$) rays

$$\varphi = \frac{(m + 1/2)\pi}{n}, \quad m = 0, 1, \ldots, 2n - 1. \tag{2.44}$$

The function

$$\hat{F}(\varphi) = \cos(n\varphi) \tag{2.45}$$

is the orthogonal complement function which relates both to the whole line $\hat{C}$ and to any part of $\hat{C}$ consisting of a combination of the circle arcs and the ray intercepts; the normalization factor (see (2.4)) is omitted in (2.45). For example, a cylindrical mirror with director in the form of an arc of an above circle cannot approximate the patterns, the Fourier series of which contain the term with $\cos(n\varphi)$. The pattern $f(\varphi)$, closest to a function $F(\varphi)$, generated by currents on such a mirror, differs from $F(\varphi)$ only in the term $\cos(n\varphi)$ which is absent in the Fourier series of the pattern. This fact is in accord with the general formula $f(\varphi) = F(\varphi) - b\hat{F}(\varphi)$ obtained in Subsection 1.3.4. If $n = 0$, then $\hat{F}(\varphi) = const$ and, for instance, the nearest pattern to $F(\varphi) = 1 + \cos(\varphi)$ is $f(\varphi) = \cos(\varphi)$.

As it will be shown in Section 4.4, if the mirror shape is close to that given by (2.43), then a pattern, non-orthogonal to $\hat{F}(\varphi)$, is approximable, but it can be generated only by a very large current.

If azimuthal currents (instead of axis ones) are located on the cylindrical mirror, then the function $u(r, \varphi)$ itself (instead of its normal derivative) is discontinuous on the contour. In our terms this means that the pattern is generated by the magnetic current. In this case, on account of (2.34), the specific line contains the circles $r = \nu_{nm}/k$, $m = 1, 2, \ldots$ (where ν_{nm} are the roots of equation $J_n'(\nu_{nm}) = 0$) and the rays $\varphi = m\pi/n$, $m = 0, 1, 2, \ldots, 2n - 1$, on which $\sin(n\varphi) = 0$.

The impossibility of approximating the functions $\cos(n\varphi)$ at $n > 0$ by patterns generated by electrical currents on the rays (2.44) is a consequence of the following obvious fact: as follows from (1.1), in the case when C is a straight line, all the patterns generated by the currents located on any part of such a line are symmetrical with respect to the line. Hence, only such patterns can be approximated by this line or any part of it; all the asymmetrical functions are the functions of the orthogonal complement. The formula (2.45) is only a statement of this result. Indeed, the function (2.45) can be written in the form $\hat{F}(\varphi) = (-1)^{n-1}\sin(n\psi)$, where $\psi = \varphi - (m + 1/2)\pi/n$, it means it is asymmetrical with respect to the rays (2.44).

The connection between a specific line and the orthogonal complement functions can be considered as a generalization of the connection between a part of the straight line and the functions, odd with respect to the polar angle, measured from this line. The last connection is based on the trivial idea of symmetry. The only nontrivial fact is that there exist many specific lines.

2. Consider the case, when the contour consists of two intersecting straight lines. The angle between two lines of the set (2.44), corresponding to $m = m_1$ and $m = m_2$, is

$$\alpha = \frac{t}{n}\pi, \tag{2.46}$$

where $t = m_1 - m_2$. We can confine ourselves to the values $\alpha < \pi$, which correspond to $t = 1, 2, \ldots n - 1$. These two lines combine a specific line $\hat{C}$. The function of the orthogonal complement for this specific line is $\sin(n\psi)$, where ψ is the angle measured from one of the lines. This function can also be written in the form $\sin(n\bar{\psi})$, where $\bar{\psi}$ is the angle measured from the second line. Since $\psi = \bar{\psi} + \alpha$, these two functions may differ only in sign.

According to (2.46), any two straight lines, intersecting at the angle which is a rational part of π, make up a specific line. If the angle between two intersecting lines is not of the form (2.46) (that is, its value is not a rational part of π), then these lines are not specific. A complete set of currents, located on both these lines, generates a complete set of patterns. This is the principal difference between the lines intersecting at the angle which is or is not a rational part of π. The contrast appears non-physical, because any value α/π can be approximated by a rational fraction.

The paradox is apparent. The larger n in (2.45), the smaller (in general) the product $b = (F, \tilde{F})$, and, therefore, the smaller the mean-square distance $|b|$ (see (1.30)) between any given function $F(\varphi)$ and the approximable one $f(\varphi)$ closest to it. At large n in (2.45) only very indented functions (having large terms of the large numbers in the Fourier series) will be practically nonapproximable by two intersecting lines with angles (2.46). The angles with an irrational ratio α/π and those with a rational one but at $n >> 1$ are practically indistinguishable in the approximability problem. In the first case (the line is a non-specific one), any pattern is approximable; in the second case (a specific line), the measure of nonapproximability $|b|$ is very small. If α/π is an irrational value, but close to a rational one, then the approximation demands a very large current.

Some *restrictions* on the *horn antennas* with n not large ($n = 1, 2, 3, 4$) follow from the above. These restrictions have the same nature as those in the case of antennas in the form of a part of a circular cylinder. Suppose, that the antenna shape is a specific line associated with the orthogonal complement function (2.45) at not large n. Then many patterns cannot be approximated by the currents on the antenna surface. The divergence angles equal, or close, to

$$\alpha = 90^0 \ (n = 2); \ \alpha = 60^0, 120^0 \ (n = 3); \ \alpha = 45^0, 135^0 \ (n = 4), \tag{2.47}$$

should be avoided.

The two intersecting straight lines with angles (2.45) at given integers n and t are an example of the specific line associated with many functions of the orthogonal complement. Return to the variable ψ – the angle measured from one of the lines. Not only the function $\hat{F}(\varphi) = \pi^{-1/2}\sin(n\psi)$ (this is a normalized form of $\hat{F}(\psi)$), but also all the functions of the infinite sequence

$$\hat{F}(\psi) = \pi^{-1/2}\sin(pn\psi), \quad p = 1, 2, \ldots \tag{2.48}$$

are the orthogonal complement functions associated with this specific line. This follows from the obvious fact that at any p the functions (2.48) are odd with respect to both ψ and $\bar{\psi} = \psi - \alpha$.

This result can also be obtained by the general method. All the auxiliary functions

$$\hat{u}_p(r, \psi) = J_{pn}(kr)\sin(pn\psi) \tag{2.49}$$

equal zero on the lines $\psi = 0$ and $\psi = \pi t/n$. These two lines make up a specific line $\hat{C}$. The associated orthogonal complement functions, that is, the functions determining the ψ-dependence of $\lim\limits_{r\to\infty} \hat{u}_p(r, \psi)$ are the functions (2.48).

The existence of infinitely many functions of the orthogonal complement will be used in the next section.

3. In the elliptic coordinate system (ξ, η):

$$x = c\cosh\xi\cos\eta, \qquad y = c\sinh\xi\sin\eta \tag{2.50}$$

the lines $\xi = const$ are the ellipses with the half-axes $c\cosh\xi$ and $c\sinh\xi$ ($2c$ is the distance between the foci), and eccentricity $e = 1/\cosh\xi$. The lines $\eta = const$ are the hyperboles having the same foci and the angle $\alpha = 2\eta$ between the asymptotes. At $(x^2 + y^2) \to \infty$ the coordinate η is transformed into the polar angle φ, and $c\cosh\xi \approx c\sinh\xi$ is transformed into the radial coordinate r.

Similarly to (2.42), choose the function $\hat{u}(\xi, \eta)$ in the simplest form

$$\hat{u}(\xi, \eta) = \Psi_n(\xi)\Phi_n(\eta), \quad n = 0, 1, \ldots, \tag{2.51}$$

where $\Phi_n(\eta)$ are the "azimuthal" Mathieu functions (analogs of the functions $\sin(n\varphi)$ and $\cos(n\varphi)$) and $\Psi_n(\xi)$ are the "radial" Mathieu functions (analogs of the Bessel functions $J_n(kr)$).

If $\Psi_n(\xi_0) = 0$ at some integer n, then the ellipse $\xi = \xi_0$ is a specific line $\hat{C}$; the orthogonal complement function, associated with this line, is $\hat{F}(\varphi) = \Phi_n(\varphi)$. The hyperbola $\eta = \eta_0$ is a specific line, when $\Phi_n(\eta_0) = 0$, and the associated orthogonal complement function is the same function $\Phi_n(\varphi)$.

Both Mathieu functions contain the parameter $q = (kc/2)^2$ including the frequency k. The Bessel functions $J_n(kr)$, analogous to $\Psi_n(\xi)$, also contain the frequency, whereas the trigonometric functions, analogous to $\Phi_n(\eta)$, do not depend on k. The hyperbolas can be specific lines only at the discrete values of the frequency (similarly to ellipses and circles), although they are nonclosed lines. This follows from the fact that the equation for $\Phi_n(\eta)$contains the dimensional parameter c. That is a principal distinction between the intersecting straight lines and the hyperbola. The properties of two straight lines depend only on the angle of crossing, hence these lines have no dimensional parameter.

The "undesirable" parameters for the elliptic or hyperbolic antennas (i.e. cylinders with the director in the form of an ellipse or hyperbola) depend on the frequency. In other words, there exist frequencies, for which any given antenna of this type is undesirable. At each of these frequencies the antenna surface is a specific one.

We give some numerical examples. Let the parameter q be equal to 3.75. Then the first zero of the radial Mathieu function $\Psi_0(\xi)$ is $\xi = 0.46$ [4]. This value corresponds to the eccentricity $e = 0.91$. Then $kc = 3.86$, and the wavelength $\lambda = 2\pi/k = 1.64c$ (i.e. 0.82 of the foci distance $2c$). At this frequency the elliptic mirror with $e = 0.91$ is a specific surface, and the orthogonal complement function is the azimuthal Mathieu function $\Phi_0(\eta)$ with the same value of q. At $q = 0.75$, the first zero of $\Psi_0(\xi)$ is $\xi = 1.04$, which corresponds to $e = 0.63$. This gives $kc = 1.73$, and the undesirable wavelength is $\lambda = 3.6c$ (1.8 of the foci distance).

Consider an antenna in the form of a hyperbolic horn. It can be characterized by the angle α between the asymptotes. The equation of the hyperbola is $\eta = \alpha/2$. The hyperbola is a specific line when $\alpha/2$ is a zero of the azimuthal Mathieu function $\Phi_n(\eta)$. This undesirable value of α depends on the frequency by the parameter q. For instance, at $q = 1$ $(kc = 2)$ and $q = 10$ $(kc = 6.3)$ these values are

$$\begin{aligned} &\alpha = 106^0; \quad 66^0; \quad 124^0; \quad 94^0 \qquad (q = 1), \\ &\alpha = 144^0; \quad 116^0; \quad 146^0; \quad 118^0 \qquad (q = 10). \end{aligned} \tag{2.52}$$

These four values α for each q are connected with four different functions $\hat{F}_n(\varphi)$, coinciding with the azimuthal Mathieu functions at the given q, with some first values of n and both types of evenness. In usual notation [4], these functions are: $ce_2(q,\varphi)$, $ce_3(q,\varphi)$, $se_3(q,\varphi)$, $se_4(q,\varphi)$.

4. In the Cartesian coordinates the simplest auxiliary field $\hat{u}(x,y)$ is

$$\hat{u}(x,y) = \cos(kx\cos\beta)\cos(ky\sin\beta). \tag{2.53}$$

Here β is a given parameter; at each β the function (2.53) satisfies the Helmholtz equation.

The zero lines of $\hat{u}(x,y)$ complete the rectangular net

$$x = \frac{\left(n+\frac{1}{2}\right)\pi}{k\cos\beta}, \quad y = \frac{\left(m+\frac{1}{2}\right)\pi}{k\sin\beta}, \quad n,m = 0,\pm1,\pm2,\cdots \tag{2.54}$$

Each cell of the net is a rectangle with sides

$$a = \frac{\pi}{k\cos\beta}, \quad b = \frac{\pi}{k\sin\beta}. \tag{2.55}$$

In order to determine the function $\hat{F}(\varphi)$, one has to calculate the asymptotics (2.3) of $\hat{u}(x,y)$ at $r\to\infty$. The field $\hat{u}(x,y)$ is the sum of four plane waves $\exp[ik(\pm x\cos\beta \pm y\sin\beta)]$. In polar coordinates these waves have the form $\exp[ikr\cos(\varphi-\varphi_n)]$, $n=1,2,3,4$, where $\varphi_{1,2} = \pm\beta$, $\varphi_{3,4} = \pi\pm\beta$. It is known that the radiation pattern of any plane wave contains the δ-function. To explain this fact, consider the limit of the integral $\int_0^{2\pi} F(\varphi)\exp[ikr\cos(\varphi-\varphi_n)]d\varphi$ at $r\to\infty$. This integral can be calculated by the stationary phase method, we used while investigating the expression (2.12). This method yields

$$\lim_{kr\to\infty}\int_0^{2\pi} F(\varphi)\exp[ikr\cos(\varphi-\varphi_n)]d\varphi = \sqrt{2\pi i}F(\varphi_n)\frac{\exp(ikr)}{\sqrt{kr}} + \sqrt{-2\pi i}F(\varphi_n+\pi)\frac{\exp(-ikr)}{\sqrt{kr}}. \tag{2.56}$$

The same result can be obtained after calculation of the following integral

$$\int_0^{2\pi} F(\varphi)\left[\sqrt{2\pi i}\delta(\varphi-\varphi_n)\frac{\exp(ikr)}{\sqrt{kr}} + \sqrt{-2\pi i}\delta(\varphi-(\varphi_n+\pi))\frac{\exp(-ikr)}{\sqrt{kr}}\right]d\varphi. \tag{2.57}$$

The symbolic form of this equality is

$$\lim_{kr\to\infty}\exp[ikr\cos(\varphi-\varphi_n)] = \sqrt{2\pi i}\delta(\varphi-\varphi_n)\frac{\exp(ikr)}{\sqrt{kr}} + \sqrt{-2\pi i}\delta(\varphi-(\varphi_n+\pi))\frac{\exp(-ikr)}{\sqrt{kr}}. \tag{2.58}$$

This formula makes sense when it is used under the integral as (2.57). Since in the book $\hat{F}(\varphi)$ is used only in such a role, namely as a multiplier in the scalar product, we can use (2.58) to express this function in the cases when the field $\hat{u}(x,y)$ itself is not a generalized function at $r\to\infty$.

triangles are not $\hat{C}$ or $\hat{C}^{(m)}$ lines, either. In Chapter 4 the near field will be investigated. The concept of the specific lines is generalized for this field. "Almost all" resonant contours are specific lines in this new conception. And "almost all" unclosed lines are specific lines, not only of the $\hat{C}$ type, but also of the $\hat{C}^{(m)}$ type, and not only at a single frequency, but also in a frequency band. The term "almost all" will be explained below.

6. As it is emphasized in the title of this section, the antenna surface should not coincide with the specific one. Otherwise, "almost every" pattern cannot be approximated by the field of currents located on this surface (such a situation will be considered in detail in Section 4.4). Therefore it seems that the surface close to the specific one cannot be the antenna surface.

However, there exists the class of antennas, namely, the so-called *resonant antennas,* where this condition is violated. The resonant antenna is a volume resonator with walls in which the radiating slots are cut [30]. The feeder located inside the resonator excites a large field within it, and large currents flow along the walls. The electric field arises on the slots cutting off these currents; this field creates the radiation.

The resonator with semitransparent walls [32], where the radiating electric field is created on the whole surface, is a variant of such an antenna. If there are not many slots and they are small, then the antenna is a high-quality open resonator.

Compare this resonator with a closed one, having approximately the same shape and the eigenfrequency coinciding with the working frequency of the antenna. The surface of the closed resonator is a specific one. The higher the quality factor of the antenna, the closer this specific line to the antenna surface (see the detailed analysis in [33]). Just the closeness of the antenna surface to that of the closed resonator, provides the efficient action of the resonant antennas.

Above we have followed the usual concept of the antenna as an assembly of currents radiating the exterior field and, in particular, the radiation pattern. According to (1.1), each element of the current radiates as in vacuum, independently of the other currents. The existence of any surfaces (e.g. metallic) does not show itself immediately (i.e., by some boundary conditions), but in the fact that the currents are induced on them and the fields of these currents contribute to the total field (and, of course, provide the fulfilment of the mentioned boundary conditions).

In these terms the action of the high-quality resonant antenna can be explained as follows: the currents on the feeder induce those on the surface, and the fields of both types of currents together, create the exterior field and its pattern. Since the working frequency coincides with (or is close to) the eigenfrequency of the high-quality resonator, a large field arises in the volume and the currents, large in comparison with the feeding one, flow along the walls. However, these currents are located on the surface close to the specific one, and they do not create large fields, but only finite ones (because they cancel themselves almost completely). The current distribution on the surface is like that of the eigenoscillation currents in the close resonator; the currents are disturbed only owing to the existence of the slots or the partial transparency of the walls, respectively.

In such an exposition the theory of resonant antennas is very complicated. This is connected with the fact that the notions (e.g. "the current creates the field") used here are inadequate to the processes occurring in the antenna. This theory is essentially simpler if the *electric fields on the antenna surface* (instead of the currents) are considered to be the radiating elements. These fields are easy to find: they are the fields on the slots or on the semitransparent walls, excited by the currents of approximately the same distribution as in the eigenoscillation of the close resonator. Therefore, the direct problem of the resonant antenna theory consists in finding the field outside the closed surface by the given tangential components of the electric field on it.

Construct the formal solution of this problem using the standard method based on the Green functions [29]. Let $u(r, \varphi)$ be the field satisfying equation (1.5) outside the contour C and the condition (1.8) at $r \to \infty$. The value $u|_C$ is known. Calculate the pattern $f(\varphi)$ involved in (1.8). To this end, introduce the function $G(r, \varphi, \varphi_0)$ as a field which has arisen by the incidence of the plane wave of the unit magnitude incoming from the direction $\varphi = \varphi_0$ onto the metallic body with the same contour C. The field of this wave is $\exp[ikr\cos(\varphi - \varphi_0)]$. The function $G(r, \varphi, \varphi_0)$ satisfies the Helmholtz equation and equals zero on C; the "diffracted field" $G(r, \varphi, \varphi_0) - \exp[ikr\cos(\varphi - \varphi_0)]$ satisfies the radiation condition. The existence and uniqueness of the function $G(r, \varphi, \varphi_0)$ are proved in the diffraction theory; they are obvious from the physical point of view.

Apply the formula (1.10) to the functions $u(r, \varphi)$ and $G(r, \varphi, \varphi_0)$ in the domain bounded by the contour C and the circle of the large radius $r = R$. The integral over C contains only one term $\int_C u(s)\partial g/\partial N ds$, where N is the outward normal to C. The integral (1.10) over the circle $r = R$ is proportional to $f(\varphi_0)$. Indeed, the waves of the same direction (outgoing) are canceled while constructing the integrand in (1.10), and the contraflow ones give an expression proportional to

$$[1 + \cos(\varphi - \varphi_0)]e^{ikr[\cos(\varphi - \varphi_0) - 1]}. \tag{2.66}$$

Calculating this integral by the stationary phase method yields

$$f(\varphi_0) = A \int_C u(s) \frac{\partial G(r, \varphi, \varphi_0)}{\partial N} ds, \quad A = \frac{1}{4\sqrt{2\pi i}}. \tag{2.67}$$

Thus, to find the radiation pattern, one should solve in advance the auxiliary problem of the diffraction of the plane wave on the "metallized" antenna and calculate the integral (2.67).

Similarly to (1.1), the operator in (2.67) is an integral one, but it has an essentially complicated kernel. However, in contrast to (1.1), this operator generates the complete set of functions $f(\varphi_0)$ after acting onto any complete set of the functions $u(s)$ on the arbitrary contour C; any orthogonal complement function does not correspond to this operator. This is connected with the analytical properties of solutions to the Helmholtz equation: if a function $u(r, \varphi)$, analytical between a close contour C and any close contour containing C, is different from zero on C identically, then it is such on the mentioned exterior contour, too. No analogue to the "nonradiating current", that is, to the nonradiating discontinuity of the field exists for the field itself.

The calculation of $f(\varphi)$ by the given $u|_C$ can be made not only by formula (2.67). The theory of the resonant antennas and, in particular, exposition of different ways of calculating $f(\varphi)$, is not a subject covered in this book. There also exist different kinds of optimization problems concerning the resonant antennas (see, e.g. [34]).

2.3 Amplitude Nonapproximability

In practice, the phase of the pattern is usually not very essential, its amplitude being of the main significance. For this reason, one can give only the amplitude of the function $F(\varphi)$ to be approximated, and the phase can *be chosen at will.* The question to be investigated in this section is: in which cases is the function, given by its amplitude, approximable by the currents on the line? As one would expect, the answer depends on the structure of the set of orthogonal complement functions associated with this line. The more linearly independent functions are contained in this set, the more difficult it is to approximate a function with given amplitude. As an example, the case of the amplitude approximation by the currents located on two perpendicular lines is investigated in detail. In this case the mentioned set of functions is countable.

1. Assume, that the amplitude of function $F(\varphi)$ is given, and its phase is free. The amplitude approximability problem is to state the possibility of approximation of this function by patterns of currents located on a specific line $\hat{C}$. The number of orthogonal complement functions associated with this line is principal to this possibility.

Emphasize the principal difference between the notions of realizability of the pattern (existence of contour and current on it, creating the pattern) and its approximability by the given line (existence of current on the line, creating a pattern infinitely close to the given one). The latter is determined by the analytical properties of function $F(\varphi)$, for instance, by the convergence rate of its Fourier series (see (1.20)). Using these properties, one can reconstruct the phase of the pattern by its amplitude (see Appendix). Remember that this procedure is ill-posed, that is, a small variation of the amplitude can cause a large variation of the phase.

A pattern can be approximable irrespectively of its analytical properties. For instance, this is valid for the Π-form amplitude pattern. The restriction on the approximability by a specific line is of a "rough" form. Even in the case when the given amplitude pattern is sufficiently smooth, it can become nonapproximable by a given specific line, after it is supplied with the phase, which is chosen to be realizable (but not by this line). In this case another technique of choosing the phase can lead to a pattern, approximable (but not realizable) by this line.

It was shown above (see (1.41)) that the function $F(\varphi)$ is approximable by the specific line $\hat{C}$ if it is orthogonal to all the orthogonal complement functions $\hat{F}_p(\varphi)$, associated with this line. Let the function $F(\varphi)$ not satisfy this condition, that is, not all the values

$$b_p = \int_0^{2\pi} F(\varphi)\hat{F}_p^*(\varphi)d\varphi, \quad p = 1, 2, \ldots, P \tag{2.68}$$

are equal to zero. It is necessary to find a phase $\Psi(\varphi)$ so that the replacement $F(\varphi)$ in (2.68) by $F(\varphi)\exp[-i\Psi(\varphi)]$ yields all zero values b_p. This problem can be reduced to an algebraic one if there are only a finite number of functions $\hat{F}_p(\varphi)$.

At first, consider the case when there is only one orthogonal complement function $\hat{F}_1(\varphi)$, so that only one value b_1 should be equal to zero:

$$b_1 = \int_0^{2\pi} F(\varphi)e^{-i\Psi(\varphi)}\hat{F}_1^*(\varphi)d\varphi = 0. \tag{2.69}$$

Choose the sought phase $\Psi(\varphi)$ in the form

$$\Psi(\varphi) = \begin{cases} 0, & 0 \le \varphi < \varphi_1, \\ -\alpha, & \varphi_1 \le \varphi < 2\pi, \end{cases} \tag{2.70}$$

where $\alpha = const$, and the angle φ_1 is chosen from the condition

$$\left|b_1^{(1)}\right| - \left|b_1^{(2)}\right| = 0, \tag{2.71}$$

where

$$b_1^{(1)} = \int_0^{\varphi_1} F(\varphi)\hat{F}_1^*(\varphi)d\varphi, \quad b_1^{(2)} = \int_{\varphi_1}^{2\pi} F(\varphi)\hat{F}_1^*(\varphi)d\varphi. \tag{2.72}$$

Such an angle exists, because, according to (2.69), the left-hand side of (2.71), as a function of φ_1, equals $-\left|b_1^{(2)}\right|$ (and is negative) at $\varphi_1 = 0$, and it equals $\left|b_1^{(1)}\right|$ (and is positive) at $\varphi_1 = 2\pi$.

Denote the phases of values $b_1^{(1)}, b_1^{(2)}$ by $\beta_1^{(1)}, \beta_1^{(2)}$, so that

$$b_1^{(1)} = \left|b_1^{(1)}\right| e^{-i\beta_1^{(1)}}, \quad b_1^{(2)} = \left|b_1^{(2)}\right| e^{-i\beta_1^{(2)}}. \tag{2.73}$$

Then

$$b_1 = \left|b_1^{(1)}\right| e^{-i\beta_1^{(1)}} \left[1 + e^{i(\beta_1^{(1)} - \beta_1^{(2)} + \alpha)}\right]. \tag{2.74}$$

If α is chosen as

$$\alpha = \beta_1^{(2)} - \beta_1^{(1)} + \pi, \tag{2.75}$$

then $b_1 = 0$. Consequently, the function $F(\varphi)$ with such a discontinuous phase $\Psi(\varphi)$ is approximable by the line $\hat{C}$.

This phase is not unique: there exist also non-jump functions $\Psi(\varphi)$, which give $b_1 = 0$.

The above technique can be generalized to the case $P > 1$. Then the following P equalities should be fulfilled:

$$b_p = \int_0^{2\pi} F(\varphi)e^{-i\Psi(\varphi)}\hat{F}_p^*(\varphi)d\varphi = 0, \quad p = 1, 2, \ldots, P. \tag{2.76}$$

Introduce the angles φ_q, $q = 0, 1, \ldots, Q$, $0 = \varphi_0 < \varphi_1 < \ldots < \varphi_{Q-1} < \varphi_Q = 2\pi$, and denote

$$b_p^{(q)} = \int_{\varphi_{q-1}}^{\varphi_q} F(\varphi)\hat{F}_p^*(\varphi)d\varphi = \left|b_p^{(q)}\right| e^{-i\beta_p^{(q)}}, \quad p = 1, 2, \ldots, P; \quad q = 1, 2, \ldots, Q. \tag{2.77}$$

The angles φ_q can be chosen arbitrarily but they (together with the number Q) should satisfy some additional conditions, stated below.

Choose the phase $\Psi(\varphi)$ in the piecewise-constant form

$$\Psi(\varphi) = -\alpha_q, \quad \varphi_{q-1} \le \varphi < \varphi_q, \quad q = 1, 2, \dots, Q. \tag{2.78}$$

One can put $\alpha_1 = 0$.

Substituting (2.78) into (2.76), we obtain P complex equalities

$$\sum_{q=1}^{Q} \left| b_p^{(q)} \right| e^{-i(\beta_p^{(q)} - \alpha_q)} = 0, \quad p = 1, 2, \dots, P, \tag{2.79}$$

which are equivalent to $2P$ real ones

$$\sum_{q=1}^{Q} \left| b_p^{(q)} \right| \cos(\beta_p^{(q)} - \alpha_q) = 0, \quad \sum_{q=1}^{Q} \left| b_p^{(q)} \right| \sin(\beta_p^{(q)} - \alpha_q) = 0, \quad p = 1, 2, \dots, P. \tag{2.80}$$

When φ_q are chosen, then (2.80) is the set of $2P$ nonlinear equations in Q real unknown α_q. For this set to have a solution, two necessary conditions must be fulfilled. The first one is obvious: the number of the unknowns must be not less than the number of the equations: $Q \ge 2P$. The second one follows from a polynomial inequality (a generalization of the triangle one): each side of the polygon cannot be larger than the sum of all the others. Since equation (2.79) describes a closed polygon in the complex plane, compounded from the vectors with lengths $\left| b_p^{(q)} \right|$, $q = 1, 2, \dots, Q$, the above inequality means that at fixed p the modulus of each value $b_p^{(q)}$ cannot be larger than the sum of the moduli of the rest of $b_p^{(j)}$, $j \ne q$. Of course, it is sufficient to check this condition for the largest $\left| b_p^{(j)} \right|$, $j = 1, 2, \dots, Q$. Assuming that the largest $\left| b_p^{(j)} \right|$ is $\left| b_p^{(1)} \right|$, one can write the second necessary condition for (2.80) to have a solution, in the form

$$\left| b_p^{(1)} \right| \le \sum_{q=2}^{Q} \left| b_p^{(q)} \right|. \tag{2.81}$$

One can also give an algebraic proof of (2.81). Write equations (2.80) in the form

$$\left| b_p^{(1)} \right| \cos(\beta_p^{(1)} - \alpha_1) = \sum_{q=2}^{Q} \left| b_p^{(q)} \right| \cos(\beta_p^{(q)} - \alpha_q), \tag{2.82a}$$

$$\left| b_p^{(1)} \right| \sin(\beta_p^{(1)} - \alpha_1) = \sum_{q=2}^{Q} \left| b_p^{(q)} \right| \sin(\beta_p^{(q)} - \alpha_q). \tag{2.82b}$$

Squaring both sides of equations (2.82b) and adding the results, we have

$$\begin{aligned} \left| b_p^{(1)} \right|^2 &= \sum_{q=2}^{Q} \sum_{q'=2}^{Q} \left| b_p^{(q)} \right| \left| b_p^{(q')} \right| \cos(\beta_p^{(q)} - \beta_p^{(q')} - \alpha_q + \alpha_{q'}) \\ &\le \sum_{q=2}^{Q} \sum_{q'=2}^{Q} \left| b_p^{(q)} \right| \left| b_p^{(q')} \right| = \left[\sum_{q=2}^{Q} \left| b_p^{(q)} \right| \right]^2. \end{aligned} \tag{2.83}$$

Inequality (2.83) is the condition for choosing the angles φ_q: if this condition is not fulfilled for some $b_p^{(q)}$, then the sector $(\varphi_{q-1}, \varphi_q)$ should be dissected into two parts. Introduction of additional angles φ_q (that is, increasing the number of the unknowns) is also necessary in the case when the equation set (2.80) has no solution.

Similarly to the case $P = 1$, the phase $\Psi(\varphi)$, calculated in such a way, is not unique: there also exist non-jump functions $\Psi(\varphi)$, which give $b_p = 0, p = 1, 2, \ldots, P$.

If the number P of the orthogonal functions increases, then the phase $\Psi(\varphi)$ (and, consequently, the function $F(\varphi)\exp[-i\Psi(\varphi)]$) becomes more indented. The Fourier series of the function converges more slowly, and its approximation demands a larger current. However, formally it remains approximable: if the number P of the orthogonal complement functions is finite, then each function is amplitude approximable by the line $\hat{C}$.

Two different situations can occur in the problem of amplitude approximability, if the number P is infinite: the phase $\Psi(\varphi)$ can either exist or not exist depending on the given function $F(\varphi)$. Both these situations will be considered in the next subsection. In the following simple example the second case is illustrated.

Let $\hat{C}$ be a segment of the straight line $\varphi = 0$. In this case all realizable functions are even functions of φ. All the odd functions make up the orthogonal complement. Choose the linearly independent set of them in the form $\hat{F}_p(\varphi) = \sin(p\varphi)/\sqrt{\pi}$, $p = 1, 2, \ldots$. Then b_p are the sin-coefficients in the Fourier series of the function $F(\varphi)\exp[-i\Psi(\varphi)]$. Let $F(\varphi)$ be a positive function. If it is not even, then all these coefficients cannot be equal to zero: there is no phase $\Psi(\varphi)$, for which $F(\varphi)\exp[-i\Psi(\varphi)]$ is even. In this case the amplitude approximability is impossible: any even function cannot be infinitely close to an uneven one.

2. The amplitude approximability problem is more meaningful in the case when the specific line $\hat{C}$ consists of two straight lines intercrossed at an angle α which is a rational part of π: $\alpha/\pi = t/n$, $n > 1$ (see (2.46)). In this case the amplitude approximability depends on an additional condition which should be fulfilled by the amplitude $\Phi(\varphi)$ of the pattern

$$F(\varphi) = \Phi(\varphi)e^{-i\Psi(\varphi)}, \quad \Phi(\varphi) \geq 0. \tag{2.84}$$

This condition will be formulated in the next subsection.

Let the intercrossed lines have the equations $\varphi = 0$ and $\varphi = \alpha$ in the polar coordinate system. (In the previous section the angle coordinate in such a system was denoted by ψ.) The complete set of the orthogonal complement functions is (see (2.48))

$$\hat{F}_p(\varphi) = \pi^{-1/2}\sin(pn\varphi), \quad p = 1, 2, \ldots \tag{2.85}$$

Similarly to the set $\hat{F}_p(\varphi) = \pi^{-1/2}\sin(p\varphi)$ for the single line $\varphi = 0$, the set (2.85) is countable but these functions are located "less densely". This means that between every two functions of the set (2.85) for the two intercrossed lines, there are $n - 1$ functions of the set $\hat{F}_p(\varphi)$ for the single line. Therefore, the condition for amplitude approximability of the pattern $\Phi(\varphi)$ by two intercrossed lines is milder than in the case of one line. The larger the number n, the milder this condition.

The amplitude approximability condition consists in satisfying the infinite system of equalities

$$\int_0^{2\pi} F(\varphi')\sin(pn\varphi')d\varphi' = 0, \quad p = 1, 2, \ldots \tag{2.86}$$

One can show that this system is equivalent to a functional condition on $F(\varphi)$ (in the next subsection we will reduce it to the condition on the function $\Phi(\varphi)$). To this end, multiply each equality (2.86) by $\sin(pn\varphi)$ and sum over p. Interchanging summation with integration we obtain

$$\int_0^{2\pi} F(\varphi')\sum_{p=1}^{\infty}\sin(pn\varphi')\sin(pn\varphi)d\varphi' = 0. \tag{2.87}$$

Use the formula

$$\sum_{p=1}^{\infty}\sin(pn\varphi')\sin(pn\varphi) = \frac{\pi}{2n}\sum_{s=0}^{n-1}\left[\delta(\varphi-\varphi'+2s\alpha)-\delta(\varphi+\varphi'-2s\alpha)\right], \tag{2.88}$$

where α and n are connected by (2.46), so that $2\alpha n$ is a multiple of 2π; δ-function is assumed to be applied to the 2π-periodical functions. This formula will be proved at the end of the subsection. Substituting (2.88) into (2.87) yields the sought condition

$$\sum_{s=0}^{n-1}\left[F(\varphi-2s\alpha)-F(2s\alpha-\varphi)\right] = 0. \tag{2.89}$$

It is easy to prove that the condition (2.89) is also sufficient for (2.86) to be satisfied. To show this, one should multiply (2.89) by $\sin(pn\varphi)$ and integrate over $[0, 2\pi]$. Introducing new variables $\mu = \varphi - 2s\alpha$, $\nu = 2s\alpha - \varphi$ and taking into account the periodicity of integrands we obtain

$$\sum_{s=0}^{n-1}\left[\int_{-2s\alpha}^{-2s\alpha+2\pi} F(\mu)\sin(pn\mu)d\mu + \int_{2s\alpha}^{2s\alpha+2\pi} F(\nu)\sin(pn\nu)d\nu\right]$$
$$= 2\sum_{s=0}^{n-1}\int_0^{2\pi} F(\varphi)\sin(pn\varphi)d\varphi = 0. \tag{2.90}$$

The summands in the last sum are equal to each other and do not depend on s; hence they are zero. This means that (2.86) is valid for all p.

The system (2.86) is the necessary and sufficient condition for the function $F(\varphi)$ to be approximable by the line $\hat{C}$. The condition (2.89) is equivalent to (2.86). Every pattern $f(\varphi)$ generated by the line $\hat{C}$ fulfils (2.86) (see(1.33)). Therefore, each pattern $f(\varphi)$ fulfils the condition (2.89), that is,

$$\sum_{s=0}^{n-1}\left[f(\varphi-2s\alpha)-f(2s\alpha-\varphi)\right] = 0. \tag{2.91}$$

It is easy to obtain (2.91) directly, applying the method used in the proof of (2.60). Indeed, (2.91) is a generalization of the symmetry property of the patterns generated by a straight line. Denote the patterns generated by the lines $\varphi = 0, \pi$ and $\varphi = \alpha,\ \alpha+\pi$, by $f_1(\varphi)$ and $f_2(\varphi)$, respectively. They are symmetrical with respect to these lines, that is

$$f_1(\varphi) - f_1(-\varphi) = 0, \quad f_2(\varphi) - f_2(2\alpha - \varphi) = 0. \tag{2.92}$$

Replace φ by $\varphi - 2s\alpha,\ s = 0, 1, \dots, n-1$ in (2.92) and sum the $2n$ obtained equations. The sum differs from $f(\varphi) = f_1(\varphi) + f_2(\varphi)$ in the absence of term $f_2(-\varphi)$ and the presence of a superfluous term $f_2(2n\alpha - \varphi)$. The arguments of these terms differ in the number multiple to 2π and hence these terms coincide. Thus, we obtain (2.91) as a consequence of (2.92).

In the general case of arbitrary line $\hat{C}$, the necessary and sufficient condition for approximability of the function $F(\varphi)$

$$\int_0^{2\pi} F(\varphi')\hat{F}_p^*(\varphi')d\varphi' = 0, \quad p = 1, 2, \dots \tag{2.93}$$

(see (1.41)) can be written in the form, analogous to (2.87):

$$\int_0^{2\pi} F(\varphi')\mathcal{P}(\varphi, \varphi')d\varphi' \equiv 0, \quad 0 \le \varphi < 2\pi, \tag{2.94}$$

where

$$\mathcal{P}(\varphi, \varphi') = \sum_p F_p(\varphi)\hat{F}_p^*(\varphi'). \tag{2.95}$$

To obtain (2.94) from (2.93), one should sum all equalities (2.93) multiplied by $F_p^*(\varphi)$, respectively, and interchange the summation with integration.

The identity (2.94) is not only the necessary, but also the sufficient condition for (2.93) to be fulfilled. To check this, one can interchange the integration and summation in (2.94) :

$$\sum_p F_p(\varphi) \int_0^{2\pi} F(\varphi')\hat{F}_p^*(\varphi')d\varphi' = 0, \quad 0 \le \varphi < 2\pi, \tag{2.96}$$

and use the linear independence of the orthogonal complement functions $F_p(\varphi)$. Of course, each pattern $f(\varphi)$ of the current located on $\hat{C}$ satisfies both (2.93) and (2.94), as a particular case of the approximable functions $F(\varphi)$.

The necessary and sufficient condition of approximability can be written in two equivalent forms (2.93) and (2.94). However, only in the case $P = \infty$ can the kernel $\mathcal{P}(\varphi, \varphi')$ be reduced to a simple implicit form, analogous to the right-hand side of (2.88).

It remains to prove the identity (2.88) used above while investigating the line $\hat{C}$ in the form of two intercrossing straight lines. Both sides of (2.88) are odd 2π-periodical functions of φ. It is sufficient to prove that these functions have the same Fourier coefficients at $\sin(l\varphi)$, $l = 1, 2, \dots$ These coefficients are functions of φ'. On the left-hand side these functions are $\pi\sin(l\varphi')$, if $l = pn$ at an integer p, and zero in all other cases. Multiplying the right-hand side of (2.88) by $\sin(l\varphi)$, integrating over $0 \le \varphi < 2\pi$ and using properties of the δ-function, one can see that these coefficients are $\pi/n \sum_{s=0}^{n-1} \sin\left[l(\varphi' - 2s\alpha)\right]$. If $l = pn$, and therefore $2pn\alpha$ is multiple to 2π, then the summands do not depend on s and the coefficients are $\pi\sin(l\varphi')$. In the opposite case, when $l \ne pn$, the coefficients can be reduced to the form (see [35])

$$\sin\left[l\varphi' - (n-1)l\alpha\right] \sin(nl\alpha) \,/\, \sin(l\alpha). \tag{2.97}$$

These values are zeros, because $nl\alpha$ is a multiple of 2π, and therefore the Fourier coefficients on both sides coincide.

Let us explain the term "reflection" which will sometimes be used below. Compare the problem considered here, in which the field is created by the currents located on two rays: $\varphi = 0$ and $\varphi = \alpha$, with the following auxiliary problem. Let two mirrors be located on the mentioned lines and the field be created by the source located at the point with the angular coordinate φ_0, $(0 < \varphi_0 < \alpha)$. As it is known, the field in the whole space in this situation equals the sum of this source and the fields of a series of "imaginary" sources obtained by the reflection of the real source in these mirrors. This is a generalization of the known property of the field in the presence of one plane mirror. If α is of the form (2.46) so that $2n\alpha$ is a multiple of 2π, then the number of these imaginary mirrors is finite, it equals $2n - 1$, and their angular coordinates are $\pm(\varphi_0 - 2s\alpha)$, $s = 0, 1, ...n - 1$. In this sense the terms $F(-\varphi)$, $F(\pi - \varphi)$ in our problem are the results of the singlefold reflection, $F(-2\alpha + \varphi)$ is obtained by the twofold reflection and so on. Of course, we deal here only with the virtual reflections, there are no mirrors in our problem.

3. The general approximability condition in the form (2.89) allows to formulate the amplitude approximability condition, that is, the condition on the function $\Phi(\varphi)$. At each fixed φ, $0 \leq \varphi < 2\pi$, the condition (2.89) means that the sum of $2n$ complex values $F(\varphi - 2s\alpha)$, $F(2s\alpha - \varphi)$, $s = 0, 1, \ldots, n - 1$ is zero. Associate these values with the vectors in the complex plane. The lengths of these vectors are

$$\Phi(\varphi - 2s\alpha), \quad \Phi(2s\alpha - \varphi), \quad s = 0, 1, \ldots, n - 1, \tag{2.98}$$

and the directions are not fixed, they can be chosen at will. The condition (2.89) means that these vectors should compose a *closed polygon*. Repeat the considerations of the previous subsection. The necessary and sufficient condition of the amplitude approximability by the specific line $\hat{C}$ is: the largest number of the set (2.98) should not be larger than the sum of the others.

This hinge polygon is not a unique one, there are many degrees of freedom, their quantity grows as n increases. The fulfilment of the above condition is easier, if n is large.

Let us estimate the probability of the following event: the maximal of $2n$ real nonnegative numbers is not smaller than the sum of the rest of them (amplitude nonapproximability). The estimation is highly relative, because it depends on the chosen model. The general conclusion following from this estimation is: the probability of this event decreases very quickly as n grows.

Consider the *model* in which $2n$ random real nonnegative numbers are given. Without loss of generality, assume that the last of them is maximal. Denote the ratios of these numbers to the last one by x_j, $(j = 1, 2, \ldots, N;\ N = 2n - 1)$, $0 \leq x_j \leq 1$. Let the density of the probability distribution of x_j be even. Calculate the probability of inequality

$$\sum_{j=1}^{N} x_j \leq 1 \tag{2.99}$$

(in our case it will be the probability of amplitude nonapproximability). Imagine the N-dimensional cube with sides equal to one; its volume is one. The probability of inequality (2.99) is the volume of the cube part containing all points $\{x_j\}$, for which (2.99) is valid. The volume is the N-dimensional integral

$$V_N = \int \ldots \int dx_1 \ldots dx_N; \tag{2.100}$$

the lower limit in the integral (2.100) is the planes $x_j = 0, j = 1, 2, \ldots, N$, the upper one is the $N - 1$-dimensional plane

$$\sum_{j=1}^{N} x_j = 1. \tag{2.101}$$

Write the integral (2.100) in the form

$$V_N = \int_0^1 dx_N \int \ldots \int dx_1 \ldots dx_{N-1}, \tag{2.102}$$

where the inner $(N-1)$-dimensional integral is calculated from $x_j = 0, j = 1, 2, \ldots, N-1$, to the N-dimensional plane

$$\sum_{j=1}^{N-1} x_j = 1 - x_N. \tag{2.103}$$

Introduce new variables $y_j = x_j/(1 - x_N)$, $j = 1, 2, \ldots, N-1$ in the inner integral of (2.103); the volume element is $dx_1 \ldots dx_{N-1} = (1 - x_N)^{N-1} dy_1 \ldots dy_{N-1}$. The upper limit in the integral is the $(N-1)$-dimensional plane

$$\sum_{j=1}^{N-1} y_j = 1. \tag{2.104}$$

This results in $V_N = V_{N-1} \int_0^1 (1 - x_N)^{N-1} dx_N = V_{N-1}/N$. Repeating this procedure $N - 1$ times gives

$$V_N = \frac{1}{N!} \tag{2.105}$$

Therefore, the probability of the amplitude approximability is

$$1 - \frac{1}{(2n-1)!}. \tag{2.106}$$

Of course, this quantitative result is valid only for the chosen model. However, it shows that the probability of the amplitude approximability is large enough even at small n. For instance, at $n = 2$ and $n = 3$, (2.106) gives the values 0.83 and 0.92, respectively. The difference between the angles (2.46) which are or are not the rational parts of π, disappears with growing n even more quickly in the amplitude approximation problem than in the general approximation one.

4. Let us give an example of the function which is amplitude approximable by two straight lines intercrossing at the angle $\alpha = \pi/2$; this example corresponds to $n = 2$ in (2.46). Let the given amplitude pattern be

$$\Phi(\varphi) = e^{-A\sin^2(\varphi/2-\pi/8)}. \tag{2.107}$$

This is a bell-shaped function; it has a maximum in the direction $\varphi = \pi/4$, that is, in the direction of the bisector of the angle between two lines $\varphi = 0$ and $\varphi = \pi/2$, of which the specific line $\hat{C}$ consists. The function (2.107) is symmetric with respect to the line $\varphi = \pi/4$. Its maximal value (at $\varphi = \pi/4$) equals one. The larger the parameter A, the narrower is the pattern. Its half-width δ, determined from the equation $\Phi^2(\pi/4 \pm \delta) = 0.5$, is connected with A by the formula $\sin^2(\delta/2) = 0.346/A$.

At $n = 2$ the set (2.98) contains the four following functions:

$$\Phi(\varphi), \quad \Phi(-\varphi), \quad \Phi(\pi-\varphi), \quad \Phi(\varphi-\pi). \tag{2.108}$$

The second function can be considered as the first one "reflected in the mirror $\varphi = 0$ ", the third one – as the first one "reflected in the mirror $\varphi = \pi/2$ ", and the fourth one – as the first one "double reflected in the mirrors $\varphi = \pi/2$ and $\varphi = \pi$". This is only a method of interpretation, there are no reflectors, in fact.

Which function of the set (2.108) has the largest value at fixed φ, depends on the quadrant φ lies in. For definiteness, one can assume that $0 < \varphi < \pi/2$. Then the first function has the largest value. Indeed, this is valid for $\varphi = \pi/4$. The first function falls down while $|\varphi - \pi/4|$ increases, the others grow. At the quadrant borders ($|\varphi - \pi/4| = \pi/4$) the two first functions are equal to each other, the two others are smaller than these.

Therefore, the inequality

$$\Phi(-\varphi) + \Phi(\pi-\varphi) + \Phi(\varphi-\pi) > \Phi(\varphi) \tag{2.109}$$

will hold at all φ, if it holds at $\varphi = \pi/4$. The necessary and sufficient condition for $\Phi(\varphi)$ to be amplitude approximable is

$$\Phi(-\pi/4) + \Phi(3\pi/4) + \Phi(-3\pi/4) > 1. \tag{2.110}$$

For instance, at $A = 1$ ($\delta = 72^0$) the summands in the left-hand side of (2.110) are 0.60, 0.60, 0.37, and their sum is larger than one. The pattern, having such a large width, can be amplitude approximated by the currents on two perpendicular lines. At $A = 2$ ($\delta = 49^0$) the left-hand side of (2.110) equals 0.87 and such a pattern cannot be amplitude approximated by these lines.

The critical value $A = A_0$ which separates these two cases, is determined from the equation

$$\Phi(-\pi/4) + \Phi(3\pi/4) + \Phi(-3\pi/4) = 1. \tag{2.111}$$

This equation is solved by $A_0 = 1.76$ which corresponds to the half-width $\delta_0 = 53^0$. The patterns of the type (2.107) with the wider form ($\delta > \delta_0$) are amplitude approximable, but those with the narrower form ($\delta < \delta_0$) are not amplitude approximable by currents on two perpendicular lines.

Similar results are valid for the Π-shaped pattern:

$$\Phi(\varphi) = \begin{cases} 1, & |\varphi - \pi/4| \leq \delta, \\ 0, & |\varphi - \pi/4| > \delta. \end{cases} \tag{2.112}$$

If $\delta < \pi/2$, then only the first function of the set (2.108) is not zero at $\varphi = \pi/4$. In this case inequality (2.110) does not hold and the pattern is not amplitude approximable. If $\delta > \pi/2$, then the first, third and fourth functions of (2.108) are equal to one simultaneously, and (2.110) holds. In this case $\Phi(\varphi)$ is amplitude approximable.

The two above examples are different with respect to the *realizability* problem. The pattern with the amplitude (2.107) is realizable at $\delta < \delta_0$ (if the phase is appropriately chosen), the function (2.112) cannot be made realizable by multiplying it by any phase function. However, there is a complete analogy between these cases in the problem of amplitude approximability: the critical angles δ_0 exist in both cases.

5. The amplitude approximability problem can be formulated in a more general form: the pattern $F(\varphi)$ should be amplitude approximated by the currents located on the line $\hat{C}$ in the *presence of additional currents*. These currents are characterized by the pattern $F_0(\varphi)$ generated in the absence of any other currents. As it will be shown below, in this case the approximability condition on the amplitude $\Phi(\varphi)$ is weaker than at $F_0(\varphi) \equiv 0$.

In the problem the function to be approximated by the line $\hat{C}$ is $F(\varphi) - F_0(\varphi)$. The necessary and sufficient condition for this is the fulfilment of the identity analogous to (2.89):

$$\sum_{s=0}^{n-1} [F(\varphi - 2s\alpha) - F(2s\alpha - \varphi)] + G_0(\varphi) = 0, \tag{2.113}$$

where

$$G_0(\varphi) = -\sum_{s=0}^{n-1} [F_0(\varphi - 2s\alpha) - F_0(2s\alpha - \varphi)]\,. \tag{2.114}$$

Repeat the consideration of Subsection 3. The equality (2.113) can be fulfilled by an appropriate choice of the phase $\Psi(\varphi)$ if and only if $2n + 1$ nonnegative functions

$$\Phi(\varphi - 2s\alpha), \quad \Phi(2s\alpha - \varphi), \quad (s = 0, 1, \ldots, n-1); \quad |G_0(\varphi)| \tag{2.115}$$

possess the property formulated earlier for $2n$ functions of the set (2.98). These functions are the same as the first $2n$ functions of the set (2.115). An additional condition arises for the functions (2.115): at each φ the sum of the first $2n$ functions should be greater than $|G_0(\varphi)|$. If this condition is not fulfilled, then the pattern with the amplitude $\Phi(\varphi)$ is not approximable. This means that in the presence of a large additional illumination the given pattern cannot be generated by the currents on $\hat{C}$. Note that the scale of this illumination is the function $G_0(\varphi)$ (but not $F_0(\varphi)$).

The problem can be solved if the amplitude pattern is given in the form $c\Phi(\varphi)$ with an indefinite coefficient c. This coefficient can always be chosen sufficiently large for the additional condition to be fulfilled. It is worthwhile to choose c as small as possible. Then the principal condition on $\Phi(\varphi)$, namely, that the sum of $2n$ first functions of set (2.115), excluding the largest of them, should be greater than $|G_0(\varphi)|$ subtracted from the largest function, will be the weakest. This condition is less strong then that in the absence of additional currents (i.e. at $F_0(\varphi) \equiv 0$).

6. Let us consider the case when the condition of the amplitude approximability is not fulfilled by the function $\Phi(\varphi)$. The problem is to find the approximable function $\tilde{F}(\varphi)$ which is *the closest* (among all the approximable functions), in the mean-square metric, to $F(\varphi)$ with the free phase $\Psi(\varphi)$. This problem belongs to the general problem on the optimal current synthesis, which will be considered in Sections 4.1, 4.3 below. However, it is reasonable to consider the problem here for a specific line associated with many complement functions $\hat{F}_p(\varphi)$ of the form (2.85).

According to (1.32), the distance between $F(\varphi)$ and $\tilde{F}(\varphi)$ is

$$\Delta = \left[\int_0^{2\pi} \left|F(\varphi) - \tilde{F}(\varphi)\right|^2 d\varphi\right]^{1/2}. \tag{2.116}$$

This distance should be minimized by choosing the approximable function $\tilde{F}(\varphi)$ and the phase $\Psi(\varphi)$ of $F(\varphi)$. The norm of $\tilde{F}(\varphi)$ is not fixed.

Begin with determination of $\tilde{F}(\varphi)$. To this end, express both functions in (2.116) in the Fourier series form. Owing to (2.93), the series of $\tilde{F}(\varphi)$ does not contain the functions $\hat{F}_p(\varphi) = \sin(pn\varphi)/\sqrt{\pi}$, $p = 1, 2, \ldots$ All other terms in this series should coincide with appropriate terms in the series of $F(\varphi)$. The difference between both these series contains only a part of the series of $F(\varphi)$, equal to $\int_0^{2\pi} F(\varphi')\mathcal{P}(\varphi,\varphi')d\varphi'$(see (2.95)). Hence (see (2.88))

$$\Delta^2 = \frac{1}{4n^2}\int_0^{2\pi} \left|\sum_{s=0}^{n-1} [F(\varphi - 2s\alpha) - F(2s\alpha - \varphi)]\right|^2 d\varphi. \tag{2.117}$$

The minimum was reached by choosing the function $\tilde{F}(\varphi)$ only. This minimal value should be further minimized by choosing the phase of $F(\varphi)$, such that the expression (2.117) for the residual will diminish.

The amplitude nonapproximability means that the amplitude $\Phi(\varphi)$ is such that the integrand in (2.117) cannot vanish at all φ. There exist sectors Ω_j in which the largest of $2n$ nonnegative values of (2.98) is larger than the sum of the other $2n - 1$ ones. In this sector the sum in (2.117) cannot be made zero by choosing the phase $\Psi(\varphi)$. Only this sector should be included in the integral limits in (2.117) for the minimal Δ.

At each φ from the intervals Ω_j the integrand of (2.117) are minimal if all complex numbers in the sum are made in-phase (e.g., real). The open $2n$-link polygon in the complex plane, composed by $2n$ vectors from (2.117) should be deflated into a straight line. Then the sum will equal the difference between the term with maximal modulus and the sum of moduli of the rest. The minimal value of Δ can easily be calculated at given $\Phi(\varphi)$.

Let us find the minimal Δ for the above examples. The specific line consists of two perpendicular lines, $n = 2$. The pattern to be approximated has the form (2.107), but we normalize it as (1.25) to obtain the relative value of Δ. This normalized function is

$$F(\varphi)=\frac{e^{A/2}}{\sqrt{2\pi I_0(A)}}e^{-A\sin^2(\varphi/2-\pi/8)}, \tag{2.118}$$

where I_0 is the modified Bessel function. The given value A is larger than the critical value $A_0 = 1.76$, the pattern is too narrow ($\delta < \delta_0$). The function (2.118) with such A is not amplitude approximable.

The minimal value of Δ^2 is the integral over Ω which consists of four sectors Ω_j, $j = 1, ..., 4$, located around directions $\varphi = \pm\pi/4$ and $\varphi = \pm 3\pi/4$.The integrals over separate sectors are equal to each other and have the form

$$\frac{1}{16}\int_{\Omega_1}\left[\Phi(\varphi)-\Phi(\varphi-\pi)-\Phi(-\varphi)-\Phi(\pi-\varphi)\right]^2 d\varphi, \tag{2.119}$$

Ω_1 is the sector in the first quadrant, where the expression in the brackets in (2.119) is positive. With accuracy to a positive factor, this expression is equal to

$$\sinh\left(\frac{A}{2}\cos(\varphi-\pi/4)\right)-\cosh\left(\frac{A}{2}\sin(\varphi-\pi/4)\right). \tag{2.120}$$

At $\delta = \delta_0$, the sector Ω_1 contains no point, $\Delta = 0$ and amplitude approximation is possible. At $\delta < \delta_0$ the sector Ω_1 is finite, and amplitude approximation is impossible. When $\delta \to 0$ (very narrow pattern), then $\Omega_1 \to [0, \pi/2]$.

The minimal value of Δ^2 is

$$\Delta^2=\frac{1}{\pi I_0(A)}\int_0^{\bar\gamma}\left[\sinh\left(\frac{A}{2}\cos\gamma\right)-\cosh\left(\frac{A}{2}\sin\gamma\right)\right]^2 d\gamma, \tag{2.121}$$

where $\bar\gamma$ is the first positive root of the equation

$$\sinh\left(\frac{A}{2}\cos\gamma\right)=\cosh\left(\frac{A}{2}\sin\gamma\right). \tag{2.122}$$

In Fig.2.1 the solid line describes the minimal value of the squared distance Δ^2 from the function (2.118) to the nearest amplitude approximable one in dependence on the ratio δ/δ_0, where δ is the half-width of the given amplitude pattern and δ_0 is its critical value. In fact, the function can be approximated (i.e. Δ can be made very small) if $\delta_0 - \delta$ is small. Then Δ^2 is not zero but it is very small in comparison with unity, which is the norm of $F(\varphi)$. The condition $\delta > \delta_0$ has no critical behavior, at least, for the bell-shaped functions. For very narrow functions, Δ^2 tends to finite value 0.25, and this value is approximately reached even at $\delta = \delta_0/4$.

Similar results are valid for the Π-shaped function normalized by (1.25). In this case Ω consists of the similar four sectors Ω_j , the critical half-width equals $\delta_0 = \pi/2$. The sectors Ω_j grow as $\pi - \delta/2$ when δ decreases from $\pi/2$ to $\pi/4$, and then they remain constant. The dependence of the squared distance Δ^2 on δ is of the form

$$\Delta^2=\begin{cases} 0, & \delta>\delta_0, \\ (\delta_0-\delta)/(4\delta), & \delta/2<\delta<\delta_0, \\ 0.25, & \delta<\delta_0/2. \end{cases} \tag{2.123}$$

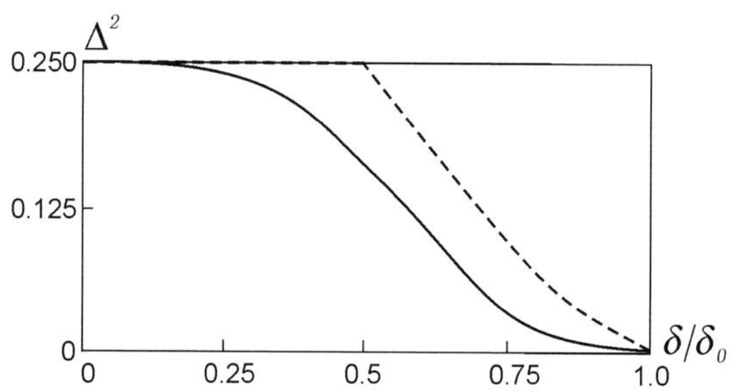

Figure 2.1:

(dashed line in Fig.2.1). The Π-shaped function is steeper than the bell-shaped one, and its narrowing results in the fastest increasing the minimal distance Δ.

2.4 Probabilistic Description of Body Shape by the Likeness Principle

In the next two sections, an inverse problem is considered. It is formulated in the following way: the pattern scattered by a metallic body is known (measured), the body shape is sought. This problem can be easily solved if the full illuminating field is known (including its phase). In this case the scattered field, given by its pattern, can be analytically continued up to its singularities similarly to [5]. The total field outside the scatterer is the sum of the illuminating field and the scattered one. The boundary of the metallic scatterer coincides with a zero line of this field. In the vector case, the scatterer surface is one of the surfaces normal to the electric field.

The method considered below is designed for the case when *the illuminating field is unknown*. In some such cases one can obtain a probabilistic information of the type "the body is like..." or "the body is not like..." This information is more reliable when more of the scattered patterns are measured at different illuminations and positions of the scatterer. In this approach we deal with a specific line $\hat{C}$ as the prospective contour of the scatterer.

1. Below we use the main property of the specific lines $\hat{C}$ established in Section 2.1: the pattern $f(\varphi)$ generated by any current located on $\hat{C}$, is orthogonal to the *orthogonal complement function* $\hat{F}(\varphi)$ associated with $\hat{C}$ (in this section we assume the existence of one such function only):

$$I \equiv (f, \hat{F}) = 0 \tag{2.124}$$

(see definition (1.31)). The function $\hat{F}(\varphi)$ satisfies the condition

$$\hat{F}^*(\varphi) = \hat{F}(\varphi + \pi) \tag{2.125}$$

following from (2.18) at $\alpha = -\pi/4$. This means that the coefficients C_n^c and C_n^s in the Fourier expansion

$$\hat{F}(\varphi) = \sum_n \left[C_n^c \cos(n\varphi) + C_n^s \sin(n\varphi)\right] (-i)^n \tag{2.126}$$

are real. In comparison with (2.9) the sin-terms are added and the normalizing factor is omitted in (2.126). The Fourier coefficients in (2.126) differ from those in (2.9) by a nonessential factor.

It is assumed that there are no other bodies in the field, therefore the orthogonal complement function satisfies the simple condition (2.125). In general case it should satisfy the more complicated one (2.31). An appropriate modification of the methods discussed in this section leads to complicated derivations but it does not require new ideas.

The specific line $\hat{C}$ (or a part of such a line) is a zero line of the field

$$\hat{u}(r,\varphi) = 2\pi i \sum_n \left[C_n^c \cos(n\varphi) + C_n^s \sin(n\varphi)\right] J_n(kr). \tag{2.127}$$

The coefficient $2\pi i$ is introduced for simplification of the calculations below.

For reasons of continuity it follows that if a line C does not coincide with the specific one associated with the given function $\hat{F}(\varphi)$, but such a line $\hat{C}$ is close to C, then the inner product of the pattern $f(\varphi)$ generated by the line C and the function $\hat{F}(\varphi)$ is small:

$$|(f,\hat{F})| << 1; \tag{2.128}$$

here both functions are assumed to be normalized according to (1.25):

$$(f,f) = 1, \tag{2.129a}$$
$$(\hat{F},\hat{F}) = 1. \tag{2.129b}$$

These properties of $\hat{F}(\varphi)$ and $\hat{u}(r,\varphi)$ are sufficient for the purposes stated in this section.

2. The idea used below is based on the fact that the scattered patterns are generated by currents located on the surface of the metallic bodies. If the condition (2.128) is violated, then the line on which the current is located (i.e. the boundary of the scatterer) is not close to the specific one associated with the function $\hat{F}$.

Let the *hypothesis:* "The sought body boundary C is close to a given contour $\hat{C}$" be verified. One should calculate the inner product of $\hat{F}(\varphi)$ and the measured scattered pattern $f(\varphi)$. Since the body can be shifted from $\hat{C}$ in the direction δ at some distance d and turned by an angle γ, the function $\hat{F}(\varphi-\gamma)\exp\left[ikd\cos(\varphi-\delta)\right]$ should be used in (2.128) instead of $\hat{F}(\varphi)$. The function of three variables

$$I(\gamma,\delta,d) = \left|\int_0^{2\pi} f(\varphi)\hat{F}(\varphi-\gamma)e^{-ikd\cos(\varphi-\delta)}d\varphi\right| \tag{2.130}$$

should be minimized with respect to γ, δ, d. The two following cases are possible:

A. The value $I_{\min}$ is not small, it does not satisfy the condition (2.128). Then the body shape is not close to the contour $\hat{C}$ at its any position. This statement is not probabilistic, it is a deterministic one.

B. The value $I_{\min}$ is small, $I_{\min} << 1$. This only means that the scatterer shape can be close to the contour $\hat{C}$ at some position. The statement has a probabilistic sense, because the condition (2.128) is only necessary, but not sufficient for C to be close to $\hat{C}$. The likelihood of this result grows if the above condition is fulfilled at some different illuminations of the same body. The contours C and $\hat{C}$ are not close to each other if $I_{\min}$ is not small even at one illumination.

The property (2.124) which is the basis of all considerations in this subsection, is also valid for nonclosed lines $\hat{C}$. Such a situation occurs when the unknown body has the form of a thin screen. Let, for instance, $\hat{C}$ be an ellipse; the function $\hat{F}(\varphi)$ is an azimuthal Mathieu function. The product I is small not only for the patterns scattered by the cylinder with the director close to $\hat{C}$, but also for the patterns scattered by a screen of the form of any part of such a cylinder.

The condition (2.124) is valid for the patterns generated by the line $\hat{C}$ even if the function $\hat{F}(\varphi)$ associated with $\hat{C}$ is nonnormalizable, and therefore it is not the function of the orthogonal complement. The *recognition* problems investigated in this section differ in this sense from the *approximability* problems. For instance, the above technique is also applicable to the rectangular contours. In this case the function $\hat{F}(\varphi)$ is the sum of four δ-functions (2.59). The product I consists of four terms like (2.60) rewritten in the shifted and turned cylindrical coordinate systems.

A closed contour can be the specific line only if it is the resonant one, that is, if the interior Dirichlet problem has nontrivial solution at the given frequency. The inverse statement is, in general, not valid: not every resonant contour is a specific line (see Section 4.2 below). But both these properties (to be resonant and specific lines) usually coincide for smooth contours such as the ellipse. The scattered pattern should be measured at the resonant frequency of the body with the boundary $\hat{C}$ when we want to verify the hypothesis that the scatterer contour C is close to $\hat{C}$.

The recognition problem can also be solved at illumination of the body by the non-monochromatic field. In this case the scattered pattern $f(\varphi, t)$ is the function of two variables: the angle φ and time t. Correspondingly, its Fourier time transformation

$$\tilde{f}(\varphi, k) = \int_{-\infty}^{\infty} f(\varphi, t)e^{-ikct}dt \tag{2.131}$$

is also the function of the frequency k. It keeps all the information contained in all scattered patterns, measured at the illumination by monochromatic fields in the frequency band where $\tilde{f}(\varphi, k)$ is not small. To determine the body shape, the product

$$I(k) = \int_{0}^{2\pi} \tilde{f}(\varphi, k)\hat{F}^{*}(\varphi)d\varphi \tag{2.132}$$

should be calculated at the resonant frequency $k = k_0$ of the body with boundary $\hat{C}$. This number characterizes the "closeness" of the contours C and $\hat{C}$. It is not necessary to know the time dependence and the structure of the field. Similarly to the monochromatic case, the value $I(k_0)$ should be minimized with respect to the position of $\hat{C}$.

3. In this subsection, a technique of finding the specific lines $\hat{C}$, the *most different* from the scatterer shape C is described. In contrast to the technique described above, this one does not allow to find the contour C. It has an auxiliary sense and allows to *estimate the probability* of the statement whether C is close to $\hat{C}$ or not.

Let the pattern $f(\varphi)$ be given (measured). Find the orthogonal complement function $\hat{F}(\varphi)$ for which the product $I = |(f, \hat{F})|$ is as large as possible. According to (2.129), $I \leq 1$. The function $\hat{F}(\varphi)$ should satisfy the condition (2.125). If $f(\varphi)$ also satisfies this condition, then the sought function $\hat{F}(\varphi)$ equals $f(\varphi)$. Otherwise, $\hat{F}(\varphi)$ should be found from the condition that it satisfies the condition (2.125) and *maximizes* the functional

$$L(F) = \frac{|(f, F)|^2}{(F, F)}. \tag{2.133}$$

Since the functional $L(F)$ is homogeneous with respect to functions $F(\varphi)$, they can be non-normalized.

Expand the function $f(\varphi)$ in the Fourier series

$$f(\varphi) = \sum_{n=0}^{\infty} f_n \cos(n\varphi); \tag{2.134}$$

the sin-terms are omitted here for simplicity. Let $f(\varphi)$ be normalized by (2.129a).

We will seek the function $\hat{F}(\varphi)$ in the form

$$\hat{F}(\varphi) = \sum_{n=0}^{\infty} C_n i^n \cos(n\varphi) \tag{2.135}$$

with real coefficients C_n. The coefficients C_n should be chosen from the condition for the functional $L(F)$ to be maximal.

After substituting (2.135) into (2.133) instead of F, the functional $L(F)$ becomes the ratio of two quadratic forms:

$$L(\{C_n\}) = \frac{\left|\sum_{n=0}^{\infty}(1+\delta_{01})C_n f_n\right|^2}{\sum_{n=0}^{\infty}(1+\delta_{01})\left|C_n\right|^2 \sum_{n=0}^{\infty}(1+\delta_{01})|f_n|^2}. \tag{2.136}$$

According to the Lagrange theorem the maximization of (2.133) is equivalent to the problem of finding the stationary points of the function of an infinite number of variables

$$\mathcal{L}(\{C_n\}) = \left|\sum_{n=0}^{\infty}(1+\delta_{01})C_n f_n\right|^2 - \Lambda \sum_{n=0}^{\infty}(1+\delta_{01})\left|C_n\right|^2 \sum_{n=0}^{\infty}(1+\delta_{01})|f_n|^2. \tag{2.137}$$

The value of Λ is chosen as the largest number at which the stationary point exists.

The problem is solved by equating to zero the derivatives of $\mathcal{L}$ with respect to C_n, $n = 0, 1, \ldots$ and calculating the largest eigenvalue Λ of the obtained algebraic equation system. It can be easily seen that $L_{\max} = \Lambda$.

If the Fourier series (2.134) contains only M terms, then $\hat{F}(\varphi)$ should also have M terms. The next terms do not change the numerator of the functional(2.133), they only increase its denominator, that is, decrease the value of the functional.

After the eigenvalue problem is solved, namely, when coefficients C_n are found as components of the respective eigenvector, then the real auxiliary field $\hat{u}(r, \varphi)$ should be constructed by (2.8), and its zero lines should be found. The scatterer contour C cannot coincide with $\hat{C}$, because $I \neq 0$, that is, $f(\varphi)$ is not orthogonal to $\hat{F}(\varphi)$. Only the "infinitely large" current induced on $\hat{C}$ can generate the pattern $f(\varphi)$ (normalized by (2.129a)), for which $|I|$ is not

the field $u^{\text{in}}(r(\varphi),\varphi)$ as well as the energy P and the induced current should be found. The value of P can be smaller than unity, because (f,f) is not the "energy of the scattered field", and equation (2.146) contains no limitations on P.

Expand the functions $H_n^{(2)}(kr(\varphi))\cos\varphi$ in the Fourier series:

$$H_n^{(2)}(kr(\varphi))\cos(n\varphi)=\sum_{m=0}^{\infty}\alpha_{nm}\cos(m\varphi),\quad n=0,1,... \tag{2.148}$$

where

$$\alpha_{nm}=\frac{1}{\pi(1+\delta_{0m})}\int_0^{2\pi}H_n^{(2)}(kr(\varphi))\cos(n\varphi)\cos(m\varphi)d\varphi. \tag{2.149}$$

Extract the real part from both parts of (2.148):

$$J_n(kr(\varphi))\cos(n\varphi)=\sum_{m=0}^{\infty}\operatorname{Re}\alpha_{nm}\cos(m\varphi),\quad n=0,1,... \tag{2.150}$$

Substituting the series (2.138), (2.143) into equality (2.147) and using the expansions (2.148), (2.150), give the equation system in the unknown coefficients C_n^{in}:

$$\sum_{n=0}^{\infty}\operatorname{Re}\alpha_{nm}C_n^{\text{in}}=-\sum_{n=0}^{\infty}\alpha_{nm}C_n^{\text{sc}},\quad n=0,1,... \tag{2.151}$$

The coefficients of this system depend on the contour C through the function $r(\varphi)$ in the argument of the Hankel functions in (2.149). The system is the sought result of our investigation.

Before giving an example, we make a following comment. From the above technique it follows that if C coincides with $\hat{C}$ and $(f,\hat{F})\neq 0$, then only an infinitely large incident field can create the scattered wave having the pattern normalized by (2.129a). To show this, express the pattern $f(\varphi)$ as the sum of two functions: $f_\perp(\varphi)$, orthogonal to $\hat{F}(\varphi)$, and $f_\parallel(\varphi)$ proportional to $\hat{F}(\varphi)$, so that

$$f_\perp(\varphi) = f(\varphi)-(f,\hat{F})\hat{F}(\varphi), \tag{2.152a}$$
$$f_\parallel(\varphi) = (f,\hat{F})\hat{F}(\varphi) \tag{2.152b}$$

(here $\hat{F}(\varphi)$ is assumed to be normalized by (2.129b)). One can also divide the fields $u^{\text{in}}(r,\varphi)$ and $u^{\text{sc}}(r,\varphi)$ into two corresponding summands and request that the condition (2.147) be fulfilled for the fields of both types:

$$u_\perp^{\text{in}}(r(\varphi),\varphi)+u_\perp^{\text{sc}}(r(\varphi),\varphi) = 0, \tag{2.153a}$$
$$u_\parallel^{\text{in}}(r(\varphi),\varphi)+u_\parallel^{\text{sc}}(r(\varphi),\varphi) = 0. \tag{2.153b}$$

The condition (2.153a) can be also fulfilled on the specific line, but (2.153b) cannot. Indeed, if C is a specific line, then the auxiliary field $\hat{u}(r,\varphi)$ corresponding to it fulfils the same conditions as $u^{\text{in}}(r(\varphi),\varphi)$ does. It has no singularities and creates the divergent wave with the pattern $\hat{F}(\varphi)$ (2.3), proportional to $f_\parallel(\varphi)$ (2.152b), in the presence of the scatterer. However, the field $\hat{u}(r,\varphi)$ equals zero on C, and it follows from (2.153b) that either $u_{\|}^{\rm sc}(r,\varphi)\Big|_C = 0$ (then $u_{\|}^{\rm sc}(r,\varphi)$ equals zero identically and the scattered pattern equals zero as well) or $u_{\|}^{\rm in}(r,\varphi)$ has an infinitely large magnitude.

Consider an elementary example. Let the scattered pattern be

$$f(\varphi) = \frac{1}{\sqrt{3\pi}}(1+\cos\varphi). \tag{2.154}$$

It does not fulfil the condition (2.125), hence the sought function $\hat{F}(\varphi)$ cannot coincide with $f(\varphi)$. Similarly to (2.154), the function $\hat{F}(\varphi)$ should not contain more than two summands (see (2.126)), that is, it should be of the form $\hat{F}(\varphi) = C_0 - iC_1\cos\varphi$ with real coefficients C_0, C_1. The value of (2.133) on this function is equal to

$$L(C_0, C_1) = \frac{1}{3}\frac{4C_0^2 + C_1^2}{2C_0^2 + C_1^2}. \tag{2.155}$$

It is obvious that it has a maximum (equal to $2/3$) at $C_1 = 0$, so that $\hat{F}(\varphi) = 1/\sqrt{2\pi}$. This orthogonal complement function corresponds to the auxiliary field $\hat{u}(r,\varphi)$ proportional to $J_0(kr)$.

The specific line closest to the origin is the circle of radius μ_{01}/k, where $J_0(\mu_{01}) = 0$, $\mu_{01} = 2.40$. This result is obvious for the pattern (2.154) without using the general theory. It is also obvious that the circle of the radius μ_{11}/k, where $J_1(\mu_1) = 0$, $\mu_{11} = 3.83$ is a specific line as well. The function $\hat{F}(\varphi)$ corresponding to it is $\pi^{-1/2}\cos\varphi$. The modulus of the product of $f(\varphi)$ and this function is $1/\sqrt{3\pi}$. It is smaller than that for the function $\hat{F}(\varphi) = 1/\sqrt{2\pi}$, where it equals $\sqrt{2/3}$.

Assume that the scatterer contour is a circle of the radius a lying in the range $2.40/k < a < 3.83/k$. Calculating the energy P as a function of ka, estimate the probability of different values of a in this interval.

If $r(\varphi) = a$, then, $\alpha_{nm} = \delta_{nm}H_n^{(2)}(ka)$ due to (2.149), and it follows from (2.154) that

$$C_n^{\rm in} = -\frac{H_n^{(2)}(ka)}{J_n(ka)}C_n^{\rm sc}, \quad n = 0, 1. \tag{2.156}$$

According to (2.145), (2.154), we have

$$\left|C_n^{\rm in}\right|^2 = \frac{1}{12\pi}\left|\frac{H_n^{(2)}(ka)}{J_n(ka)}\right|^2 \quad n = 0, 1. \tag{2.157}$$

Substituting this result into (2.142) gives

$$P = \frac{1}{4} + \frac{1}{12}\left\{2\left[\frac{N_0(ka)}{J_0(ka)}\right]^2 + \left[\frac{N_1(ka)}{J_1(ka)}\right]^2\right\}. \tag{2.158}$$

If ka is close to 2.40 or to 3.83, then P is very large, and it grows faster at $ka \to 2.40 + 0$, than at $ka \to 3.83 - 0$, owing to the difference between two products of $f(\varphi)$ and the

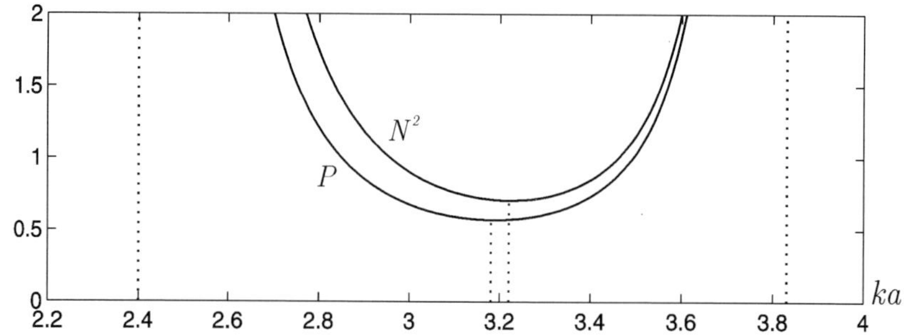

Figure 2.2:

corresponding functions $\hat{F}(\varphi)$. The minimum of P is reached at $ka = 3.18$ (see Fig.2.2) and this is the most probable value of ka for the circle-shaped scatterer.

This probability can also be estimated by the value of the norm of current (1.11), induced on the scatterer surface,

$$j(\varphi) = \frac{1}{k}\frac{\partial}{\partial n}\left[u^{\text{in}}(r,\varphi) + u^{\text{sc}}(r,\varphi)\right]\Big|_C , \tag{2.159}$$

where $\vec{n}$ is the normal to C. The calculations by the formulas obtained above give for our example

$$N^2 = \int_0^{2\pi} |j(\varphi)|^2 d\varphi = \frac{2}{3\pi}\frac{1}{(ka)^2}\left[\frac{2}{J_0^2(ka)} + \frac{1}{J_1^2(ka)}\right]. \tag{2.160}$$

As can be seen from Fig.(2.2), the dependence of $N^2(ka)$, in fact, repeats that of $P(ka)$; the minimal N^2 is reached at $ka = 3.22$.

5. Let us make two additional remarks.

a. In the approach described above all conclusions concerning the body shape are drawn not from the detailed behavior of the pattern $f(\varphi)$, but only from the value I which is a smoothing functional of f. Therefore small errors in the measured pattern should not lead to large ones in the result.

This fact makes possible the reconstruction of the pattern in the whole angle range $0 \leq \varphi < 2\pi$, by its measured values in the part of this range where most of the energy is scattered. The methods for such reconstruction are based on the analytical properties of the function $f(\varphi)$ and therefore are mathematically improperly posed. However, they can be applicable in our problem owing to the smoothing properties of the functional I. Besides, the property of the line to be a specific one is "rough", which shows itself, in particular, in the fact that practically identical results follow from the formally different statements $I = 0$ and $|I| << 1$. Some conclusions can be made also from the measured amplitude pattern only, but in this case, probably, the measurements should be carried out with extreme accuracy.

b. The scattered pattern should be generated only by the current located on the body surface. If the scattered field cannot be measured separately, then the field illuminating the receiver in the absence of the body should be measured and subtracted from the total field. In many cases the illumination is realized by the plane wave; then the scattered pattern cannot be measured in the directions the plane wave comes from and goes to. Remember that the measured pattern in these directions is the sum of this wave and the scattered pattern. It is also valid in the case when the pattern of the total field is small in such a direction, for instance, when the direction is the shaded one. Then the scattered pattern should be large in this direction to cancel the pattern of the plane wave.

2.5 Body Shape Reconstruction by its Scattered Patterns [1]

As in the previous section, an inverse problem is considered here. However, the problem formulation is different: we try to reconstruct the body shape (without "probabilistic" aspects) by a set of the measured patterns scattered at different illuminations. Any a priori information concerning these illuminations and the scatterer shape, is not assumed to be known. A variational statement, based on a generalization of the functional (2.133), is used. To determine the body eigenfrequency, at which its contour is a specific line (or is close to such a line), the method first requires to illuminate the body at different frequencies. The case of a uniformly moving body can be considered as well. The method is illustrated on the model examples, including those in which the body contour is not a specific line (the case of a rectangle is considered).

1. The subject of consideration in this section is the problem of finding the body shape C by a set of scattered patterns measured at different illuminations and frequencies. As above, the illuminating field is assumed to be unknown. The idea of the method consists in usage of the fact that any smooth contour is specific at some frequency (one of its eigenfrequencies). At this frequency all the scattered patterns satisfy the condition like (2.124) with unknown function $\hat{F}(\varphi)$ associated with the contour.

The problem is formulated as a variational one consisting in minimization of the functional

$$L(F) = \frac{\sum_{m=1}^{\infty} |(f_m, F)|^2}{(F, F)}, \tag{2.161}$$

where f_m are scattered patterns measured at some frequency k. The patterns $f_m(\varphi)$ should be generated by a complete linear independent set of currents on C. They are assumed to be normalized by (2.129a). Minimization is carried out with respect to the function $F(\varphi)$.

The frequency k should be chosen such that the minimal value of L is zero. This is possible only when k is an eigenfrequency of the body; then $F(\varphi)$ minimizing the functional is the orthogonal complement function $\hat{F}(\varphi)$ associated with the body boundary as a specific line.

Applying the Lagrange method to (2.161) leads to the functional

$$\mathcal{L}(F) = \sum_{m=1}^{\infty} |(f_m, F)|^2 - \Lambda(F, F). \tag{2.162}$$

The least Λ at which $\mathcal{L}(F)$ has a stationary value coincides with the minimal value of (2.161), and the function, on which this stationary value is reached, minimizes the functional $L(F)$. At

[1] This section is written by O. Kusyi and contains his results [36].

each frequency k, except for the eigenfrequencies of the body, the value $L_{\min}(k) = \min_F L(F)$ differs from zero (recall that $L(F)$ depends on k by means of the measured patterns f_m). The equation for determination of the eigenfrequencies is

$$L_{\min}(k) = 0. \tag{2.163}$$

In principle it is sufficient to find only one eigenfrequency of the body (say, the first one), after that we can use the patterns measured at this frequency for determining the body shape.

The stationary points of the functional (2.162) can be found in two ways. The first one consists in expanding unknown function into the series by a complete set of functions (in particular, in the Fourier series) and determining its coefficients from the necessary condition of stationarity: equaling to zero all the partial derivatives of $\mathcal{L}$ with respect to these coefficients. This approach leads to an infinite system of the homogeneous algebraic equations for unknown expansion coefficients. The coefficients making up eigenvectors are the stationary points of $\mathcal{L}$, the eigenvalues corresponding to these vectors being the stationary values of L; the minimal of them is $L_{\min}(k)$. If we have a finite number of the measured patterns, we can take only a finite number of terms in the series, and the infinite equation system is reduced to the finite one.

The second approach reduces the variational problem to the Lagrange–Euler homogeneous equation. Its eigenfunctions are the stationary points of (2.162), the corresponding eigenvalues are equal to the values of the functional (2.161) on these functions.

Describe these two approaches in detail.

2. Take a complete linear independent system of functions $\psi_n(\varphi)$ and express the sought function, minimizing L, in the form

$$F(\varphi) = \sum_{n=1}^{\infty} C_n \psi_n(\varphi). \tag{2.164}$$

Substituting (2.164) into (2.162) and differentiating with respect to $\operatorname{Re} C_n$, $\operatorname{Im} C_n$ we obtain an infinite homogeneous system of linear equations

$$\sum_{n=1}^{\infty} \alpha_{nq} C_n - \Lambda \sum_{n=1}^{\infty} \beta_{nq} C_q = 0, \quad q = 1, 2, \ldots, \tag{2.165}$$

where

$$\alpha_{nq} = \sum_{m=1}^{\infty} (f_m, \psi_q)(\psi_n, f_m); \quad \beta_{nq} = (\psi_n, \psi_q). \tag{2.166}$$

The matrices $A = \{\alpha_{nq}\}$, $B = \{\beta_{nq}\}$ are Hermitian, the eigenvalues Λ_n are real and nonnegative. The smallest of them Λ_1 is the minimum of L, and the eigenvector $\{C_n\}$ corresponding to it gives the coefficients of the expression (2.164).

While $\Lambda_1 \neq 0$, $F(\varphi)$ is not a function of orthogonal complement $\hat{F}(\varphi)$. To achieve this, one should find the eigenfrequency k_0 of the body, that is, solve equation (2.163) rewritten as

$$\Lambda_1(k) = 0. \tag{2.167}$$

After the function $\hat{F}(\varphi)$ is found, one should construct the field $\hat{u}(r,\varphi)$; to this end, the formula (2.12) (or its equivalent (2.11)) can be used. Then the contour $\hat{C}$ should be determined as a zero line of $\hat{u}(r,\varphi)$. This contour can be separated from other zero lines if we know which order number had its eigenfrequency k_0 calculated before. Otherwise, we should have some information about linear sizes of the body.

If the functions ψ_n satisfy the condition as (2.125), then the coefficients C_n can be chosen to be real and the system (2.165) transforms to

$$\sum_{n=1}^{\infty} \operatorname{Re}\alpha_{nq}C_n - \Lambda \sum_{n=1}^{\infty} \operatorname{Re}\beta_{nq}C_q = 0, \quad q = 1, 2, \ldots \tag{2.168}$$

In practice one has a limited number of measured (complex) patterns, $m \leq M$. Then the infinite equation systems (2.165), (2.168) have not more than M and $2M$ nonzero eigenvalues, respectively. One of the ways to obtain a finite system consists in limiting the number of coefficients in the series (2.164) by $n \leq M$ for the system (2.165) and by $n \leq 2M$ for the system (2.168). In this case, equation (2.167) can in general have no solution. To calculate approximately the eigenfrequency $k = k_0$, we should select the minimal eigenvalue at each k and then minimize it with respect to k.

A difficulty can arise in this problem, caused by the fact that besides Λ_1, the homogeneous linear system can have other small eigenvalues; the last ones correspond to the fast varying eigenfunctions $F(\varphi)$, which are not $\hat{F}(\varphi)$. To identify Λ_1 among small Λ_n, we should analyze the structure of eigenfunctions corresponding to these eigenvalues.

3. To reduce the functional (2.162) to the Lagrange–Euler equation in the second approach, one should take the function $F + \delta F$ instead of F in (2.162) and equate to zero the first variation of $\mathcal{L}(F)$, that is, the linear (with respect to δF) part of its increment. After some derivations we obtain

$$\sum_{m=0}^{\infty} (F, f_m) f_m = \Lambda F. \tag{2.169}$$

This is the sought Lagrange–Euler equation with respect to eigenvalues Λ_n and eigenfunctions F_n, equivalent to the problem of finding the stationary points of the functional (2.162).

Other equation can be obtained for the case if the solution is found in a class of functions satisfying the condition (2.125). Then such a condition can be imposed on the δF and the Lagrange–Euler equation becomes of the form

$$\frac{1}{2} \sum_{m=0}^{\infty} [\beta_m f_m(\varphi) + \beta_m^* f_m^*(\varphi + \pi)] = \Lambda F(\varphi), \tag{2.170}$$

where $\beta_m = (F, f_m)$.

One can use, for instance, the Galerkin method for solving these equations numerically. It consists in substituting (2.164) into (2.169) or (2.170) (including β_m) and satisfying the orthogonality condition of the residual to all the functions $\psi_n(\varphi)$. If the same basis functions $\psi_n(\varphi)$ are used in this method, then the above algebraic systems (2.165), (2.168) are obtained from (2.169), (2.170), respectively.

In the numerical examples demonstrated below we use the trigonometrical basis functions

$$\psi_n(\varphi) = \frac{(-i)^n}{\pi(1+\delta_{01})}\cos(n\varphi), \quad n = 0, 1, ... \tag{2.171}$$

satisfying the condition (2.125) (as above, the sin-terms are omitted for simplicity), that is, the unknown function $F(\varphi)$ is expressed in the form of the Fourier series

$$F(\varphi) = \sum_{n=0}^{\infty} \frac{(-i)^n C_n}{\pi(1+\delta_{01})}\cos(n\varphi). \tag{2.172}$$

In this case the system (2.168) (or, equivalent to it (2.170)) can be used, the coefficients C_n are real, and the field $\hat{u}(r,\varphi)$ obtained at the eigenfrequency by (2.12) with D connected with $\hat{F}$ by (2.14) has an explicit form

$$\hat{u}(r,\varphi) = \sum_{n=0}^{\infty} C_n J_n(kr)\cos(n\varphi). \tag{2.173}$$

This series converges in the whole plane if (2.172) converges. Indeed, at each fixed r the coefficients in the series (2.173) differ from those in (2.172) by the multipliers $J_n(kr)$ which decrease as n^{-n} with n increasing.

4. The first task to be solved before applying the above technique consists in choosing the coordinate origin, so that it would be located in the body "centre". This task can be solved (with a great probability) by using only one of the measured patterns. For this we can use the simple formula

$$f^{\text{sh}}(\varphi) = f(\varphi)e^{ikd\cos(\varphi-\alpha)} \tag{2.174}$$

connecting the pattern $f(\varphi)$ created by some current $j(s)$ on the contour C with the pattern $f(\varphi; d, \alpha)$ created by the same current $j(s)$ located on the contour C, shifted parallel to itself in the direction α at the value d. It is seen that the information on the unknown shift is contained in the phase of the pattern. One can expect that the position of the coordinate origin will be optimal for the method described above if the values α, d are chosen in such a way that the difference between the maximal and minimal values of the phase pattern $\arg f(\varphi)$ is minimal. Of course, the phase must be calculated as a continuous function without jumps in 2π. If $f_1(\varphi)$ is given, then one can first calculate the approximate values $\tilde{\alpha}$, such that

$$\left|\arg f^{\text{sh}}(\tilde{\alpha}) - \arg f^{\text{sh}}(\tilde{\alpha}+\pi)\right| = \max_{0\le\varphi\le\pi}\left|\arg f^{\text{sh}}(\varphi) - \arg f^{\text{sh}}(\varphi+\pi)\right|. \tag{2.175}$$

After that the value $\tilde{d}$ is calculated by the formula

$$2k\tilde{d} = \arg f^{\text{sh}}(\tilde{\alpha}) - \arg f^{\text{sh}}(\tilde{\alpha}+\pi). \tag{2.176}$$

If several patterns $f_m^{\text{sh}}(\varphi)$ measured at the same position of the body are given, then $\tilde{\alpha}$ can be calculated as the average value $\bar{\alpha}$ of $\tilde{\alpha}_m$ calculated by (2.175) for all m. Then the shift $\tilde{d}$ is calculated as the average value $\bar{d}$ of $\tilde{d}_m$ calculated by (2.176) for all m at $\tilde{\alpha} = \bar{\alpha}$. The pattern $f(\varphi)$ in "reconstructed" coordinates has the form

$$f^{\text{rec}}(\varphi) = f^{\text{sh}}(\varphi)e^{-ik\tilde{d}\cos(\varphi-\tilde{\alpha})} \tag{2.177}$$

In Fig. 2.3 the phase of the patterns $f(\varphi)$, $f^{\text{sh}}(\varphi)$ created by the same random current, distributed on the rectangle with the sides $2a$, $2b$, $ka = 1$, $kb = 0.3$ with the center at the origin and shifted in the direction $\alpha = \pi/6 = 0.523$ at $kd = 2.5$, are described by lines 1 and 2, respectively. The approximate values of α, d, calculated by (2.175), (2.176) are $\tilde{\alpha} = 0.471$, $k\tilde{d} = 2.702$. Line 3 describes the reconstructed pattern $f^{\text{rec}}(\varphi)$. It is seen that these parameters are calculated with the precision which allows the coordinate origin to be around the rectangle center. Such precision is sufficient to apply our technique for the contour C reconstruction.

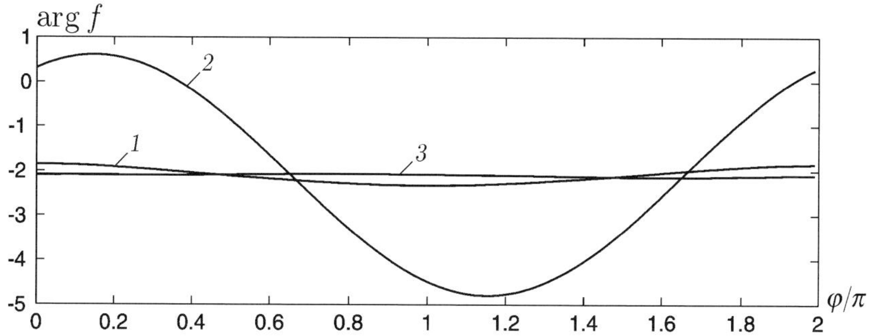

Figure 2.3:

Determining the resonant frequency k_0 is facilitated when the patterns f_m, measured at different k, correspond to illuminations continuously dependent on k. In this case the eigenvalues Λ_n are also continuous functions of k and the selection of Λ_1 is a noncomplicate problem. At this stage (determining the resonant frequency) one should not have many measured patterns at each frequency.

If illuminations are not continuous as k changes, then one can divide all the measured patterns at each frequency into two groups and solve the problem of the shape reconstruction for both groups independently. The frequency at which the calculated contours coincide (with some accuracy) is the resonant one, and the contours describe the sought shape of the body (with the same accuracy).

To improve the accuracy of the shape reconstruction, one should increase the number M of the pattern measured at the resonant frequency. However, only a few patterns measured at each frequency are sufficient for determination of the eigenfrequency in both approaches. Of course, while calculating the zero line of the field $\hat{u}(r,\varphi)$, constructed by $\hat{F}(\varphi)$, we should choose the one, which corresponds to the eigenfrequency we used in the linear equation systems (2.165), (2.168) or (2.170).

The method described above was applied to some model recognition problems. The results are shown in Figs.2.4–2.7. They concern bodies with three shapes: a) ellipse with the half-axes ratio $b/a = 0.3$; b) rectangle with the same the half-axes ratio $b/a = 0.3$; c) "kite" described by the parametric equations

$$x(t) = \cos t + a(1 - \cos(2t)), \quad y = b \sin t; \quad 0 \leq t \leq 2\pi \tag{2.178}$$

with $b/a = 1.5/0.65$. In the models the "measured" patterns were calculated by the formula (1.1), where the currents j_m were chosen in the form of the Fourier series with a limited number of random coefficients, and integration was made over the contours to be recognized.

Fig. 2.4 demonstrates the procedure of eigenfrequency determination by dependencies $\Lambda_1(k)$. Only three patterns $f_m(\varphi)$ were taken at each frequency k in the cases a) (curve Λ_1^e), b) (curve Λ_1^r), and four in the case c) (curve Λ_1^k). The currents generating these patterns were chosen randomly, they did not depend on k. In all the cases the minimal eigenvalue Λ_1 has clearly seen minima. The precision of eigenfrequency determination comes to about 0.01 percent.

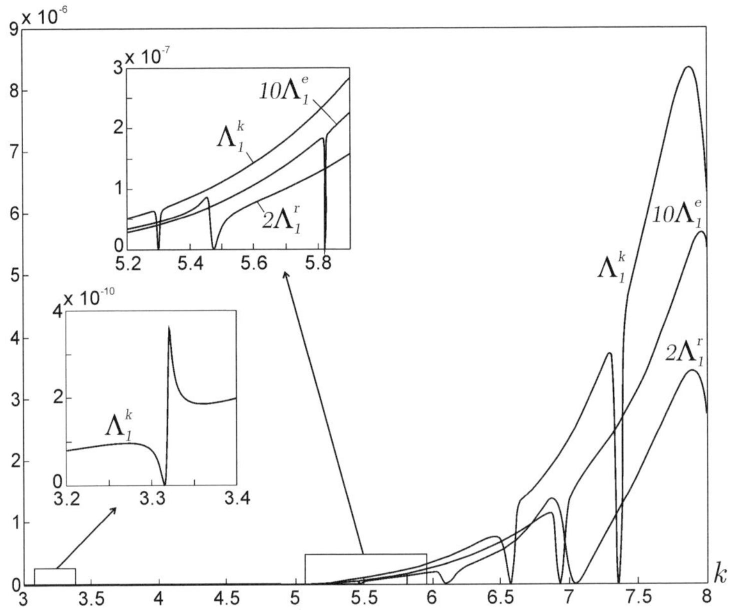

Figure 2.4:

Fig. 2.5 shows the typical behavior of zero lines of the field $\hat{u}(r, \varphi)$ for the case of a rectangle at its first eigenfrequency $k_1 = 5.446$ (a) and at $k = 5.500$ (b) shifted from k_1 by 1 percent. The exact shapes of the contours are shown by dotted lines. These results have shown that even a small shift of the frequency from the resonant one cardinally reconstructs the structure of zero lines. In particular, the closed line describing the body contour fails to exist. In this sense the procedure of finding the body shape is unstable, it is very sensitive to the precision of determining the eigenfrequency (more specifically, to the precision of determining the minimum of Λ_1). The algorithm can be modified a little to make it more stable. The modification consists in finding the line $r = r(\varphi)$ on which $|\hat{u}(r, \varphi)|$ has local minimum, instead of being zero.

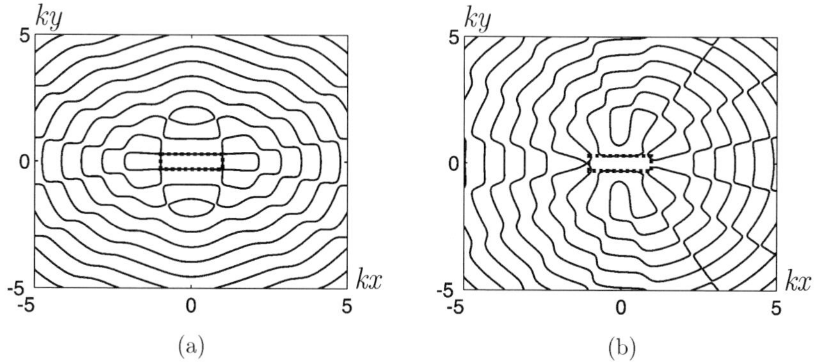

Figure 2.5:

Note that, according to Section 2.2, the resonant rectangle is not a specific line. The series (2.172) is divergent for it. But the usage of short segments of the series allows to reconstruct the contour with an acceptable accuracy. In fact, we reconstruct a specific line which is very close to the rectangle.

Similar results, related to the bodies of the form of ellipse and kite, are presented in Figs. 2.6, 2.7, respectively. The ellipse is reconstructed with a very high accuracy at its eigenfrequency. However, the frequency shift leads to the same effect as for the rectangle: the zero lines are qualitatively rebuilt.

To obtain a closed contour for the bodies with the concave segments, as in the kite-shaped case, we should take more measured patterns (and, probably, with more special illuminations), and, as a result, more terms of the Fourier series (2.172). This leads to some computation problems connected with very small values of the minimal eigenvalue Λ_1. However, the modified algorithm reconstructs this shape with a sufficient accuracy even at small values of the patterns. Such a shape is shown by the dashed line in Fig. 2.7(a).

A surprising result was obtained at the fourth eigenfrequency of the kite-shaped body (see Fig.2.7(c)): the shape was practically reconstructed by 4 measured patterns only. This fact suggests that more complicated shapes can be effectively reconstructed at their higher eigenfrequencies.

As we have already mentioned, the scattered patterns satisfy the same condition (2.124) also in the case, when the closed contour is substituted with any of its arcs. This occurs on the frequency, resonant for the interior volume, bounded by the whole contour. It follows from this, that after the shape of the closed contour is determined by the above method, one cannot know whether the scattering was made by the whole contour or by an unclosed screen coinciding with a part of its surface. To answer this question one should use the measurements made at the frequencies close to k_0.

5. The above method can be modified for reconstruction of the moving body. We explain this modification for the case when the body moves with a constant velocity and the measurements are made at constant time periods. In this case the measured patterns should be corrected by the phase factor $\exp(ikmd\cos(\varphi-\alpha))$, where d is the unknown shift of the body for one time period and α is the moving direction. The elementary shift d and direction

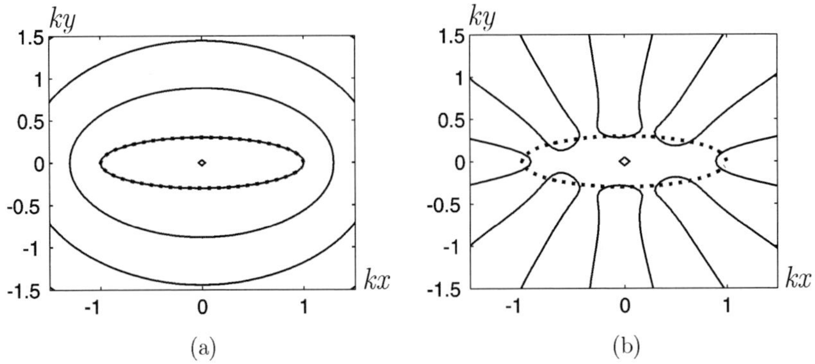

Figure 2.6:

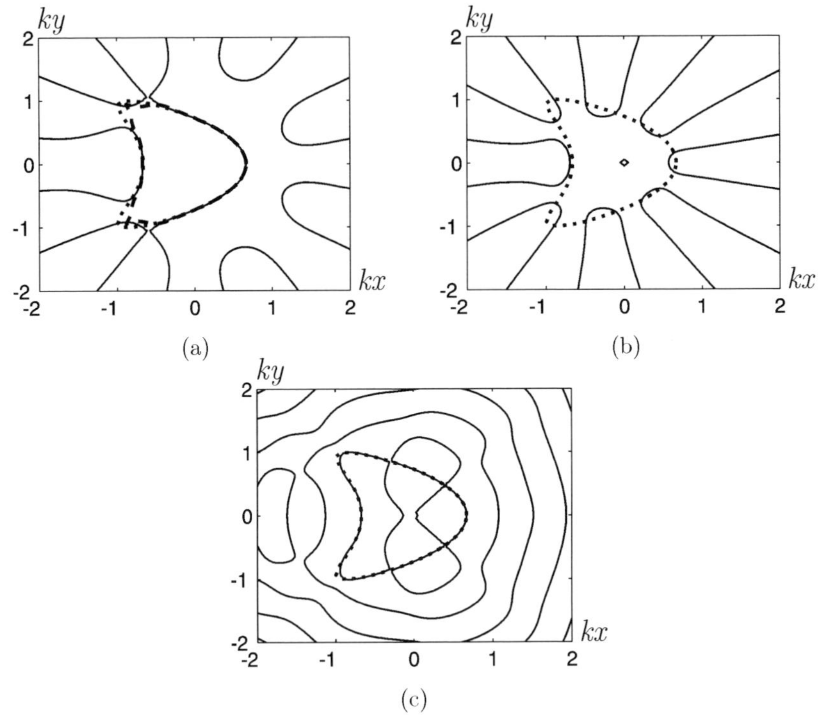

Figure 2.7:

α should be calculated in the first stage, during determination of the resonant frequency. The approximate values of these numbers are calculated making use of the technique described at the beginning of the previous subsection. To adjust these values, one should construct the linear equation system like (2.168) depending on α, d and minimize the eigenvalue Λ_1, with respect to these values.

2.6 Properties of Specific Lines

In this section some elementary properties of specific lines $\hat{C}$ are shown. Remember that, by the definition, the specific line is a zero line of an auxiliary real field $\hat{u}(r,\varphi)$ satisfying the homogeneous Helmholtz equation in the whole plane and having no singularity in it.

1. *Each closed specific line is a resonant contour*, that is, inside it there exists a nonidentical solution to the homogeneous Dirichlet problem with zero boundary condition. Indeed, the contour $\hat{C}$ is the zero line of some real field $\hat{u}(r,\varphi)$. If $\hat{u}(r,\varphi)$ were equal to zero identically inside $\hat{C}$, then its normal derivation would also be zero, and this would result in the field being zero in the exterior domain, too. This contradicts the definition of $\hat{C}$.

2. The specific line *cannot end in some point*, it should go to infinity or be closed.

This statement can be proved in a way analogous to the proof of the maximum principle for Laplace equation. Let $\hat{C}$ end at a point O. Construct a circle with the center O, inside which the field $\hat{u}(r,\varphi)$ does not change its sign. For instance, let the field be positive at $r < R$, where r is the radial coordinate in the polar set with the center at the point O. Introduce the Green function of the Laplace equation: $G = (2\pi)^{-1}\ln(r/R)$. It is negative at $r < R$, equal to zero at $r = R$, and $\Delta G = \delta(r)$. Applying the second Green formula to the functions G and $\hat{u}(r,\varphi)$ gives

$$\hat{u}(0) = -k^2 \int_{r<R} \hat{u} G dS + \int_{r=R} \left(\hat{u}\frac{\partial G}{\partial N} - G\frac{\partial \hat{u}}{\partial N} \right) ds, \tag{2.179}$$

where dS and ds are the domain and boundary elements, respectively. The integrand in the first integral is negative owing to the above properties of the functions $\hat{u}$ and G. In the second integral the first term is positive owing to positivity of both $\hat{u}$ and $\partial G/\partial N = 1/(2\pi R)$; the second term is zero because $G|_{r=R} = 0$. From this, the left-hand side is positive. However, $\hat{u}(0) = 0$ because the point O lies on the zero line of $\hat{u}$. Consequently, the supposition that $\hat{u}(r,\varphi)$ retains its sign in some neighborhood of a point located on $\hat{C}$ leads to contradiction.

3. There exists a specific line *passing through any M given points* in the plane.

The proof is elementary. Express a field $\hat{u}(r,\varphi)$ in the form

$$\hat{u}(r,\varphi) = \sum_{n=0}^{M} C_n J_n(kr)\cos(n\varphi) \tag{2.180}$$

with real coefficients C_n. Calculate the ratios C_n/C_0, $n = 1, 2, ..., M$ from the linear equation system

$$\sum_{n=1}^{M} C_n J_n(kr_m)\cos(n\varphi_m) = -C_0 J_0(kr_m), \quad m = 1, 2, ..., M, \tag{2.181}$$

where r_m, φ_m are the coordinates of the mth given point. If the matrix of (2.181) is degenerate, then an analogous system should be written for the ratios C_n/C_1, and so on. The function (2.180) satisfies all the required conditions: it is a real solution to the Helmholtz equation, has no singularity and M given points lie on its zero line. This line is a sought specific one. (Note that a specific line constructed in such a way can be multiply connected).

Of course, the above result does not mean that one can pass to infinity in (2.181) at $M \to \infty$. In general, when passing, the series can be divergent at some points of the plane; moreover the coefficients C_n can tend to no limits themselves. Indeed, if such a passage to the limit were admissible, we would choose the points on a nonresonant contour uniformly and have in the limit the whole of this contour, densely filled with the zero points. However, it is impossible, because nonresonant contours cannot be specific lines.

4. If we have *two specific lines* $\hat{C}_1$ *and* $\hat{C}_2$, being zero lines of two different fields $\hat{u}_1(r, \varphi)$, $\hat{u}_2(r, \varphi)$, then there exists a *sequence of specific lines depending on a parameter* α, which are zero lines of the fields depending on the same parameter. For instance, such a sequence can consist of zero lines C_α $(1 \le \alpha \le 2)$ of the field

$$\hat{u}_\alpha(r, \varphi) = (2 - \alpha)\hat{u}_1(r, \varphi) + (\alpha - 1)\hat{u}_2(r, \varphi). \tag{2.182}$$

These lines transform into $\hat{C}_1$ and $\hat{C}_2$ at $\alpha = 1$, $\alpha = 2$, respectively. One can construct an infinite number of such sequences, adding to $\hat{u}_\alpha(r, \varphi)$ the summand $(2 - \alpha)(\alpha - 1)\hat{v}(r, \varphi)$, where $\hat{v}(r, \varphi)$ is any real nonsingular solution to the Helmholtz equation with asymptotics (2.3). There exists a sequence of the orthogonal complement functions $\hat{F}_\alpha(\varphi)$, associated with the sequence of the specific lines.

5. *Any neighborhood* of a specific line *contains another specific line* with another orthogonal complement function. We do not consider the trivial way of obtaining a neighboring specific line by shifting or turning the field $\hat{u}(r, \varphi)$ as a whole. In this case the function is only multiplied by an nonessential phase factor or the coordinate φ is shifted to a constant angle.

The above statement follows from the fact that the series (2.128) with real coefficients C_n represents a field $\hat{u}(r, \varphi)$, when it converges in the whole plane. The value of the field at any point is a continuous function of all the coefficients C_n. Fix an angle $\varphi = \varphi_0$. The value $r(\varphi_0)$ at which the field is zero, is also the continuous function of C_n. Zero of $\hat{u}(r, \varphi)$ on the ray $\varphi = \varphi_0$ is displaced infinitely little when one of C_n receives an infinitesimal increment. At this the zero line turns into the one infinitely close to it.

The same proof is also valid for the statement that any neighborhood (in the L_2 metric) of function $\hat{F}(\varphi)$, contains an orthogonal complement function of another field. This follows from the fact that the series (2.9) with real coefficients C_n represents an orthogonal complement function $\hat{F}(\varphi)$ if it converges at all values of φ. The value of $\hat{F}(\varphi)$ at each point φ continuously depends on all C_n.

We often use the statement: “There are many specific lines”. It means that these lines have the properties stated in the last two subsections. In fact, they follow from the continuous dependence of the field on coefficients C_n only.

6. Let us mention a property related to the auxiliary field $\hat{u}(r, \varphi)$. For any line C one can construct an infinite sequence of the fields $\hat{u}_m(r, \varphi)$, $m = 1, 2, \ldots$, so that

$$\lim_{m \to \infty} \int_C \hat{u}_m^2 ds = 0. \tag{2.183}$$

Here all the fields $\hat{u}_m(r, \varphi)$ are the solutions to the Helmholtz equation, they have no singularities in the whole plane and are normalized by the condition (2.4) for its asymptotics $\hat{F}_m(\varphi)$. The proof of this statement will be given in Section 4.2. It does not follow from (2.183) that the function $\lim_{m\to\infty} \hat{u}_m(r,\varphi)$ exists, and if it does exist, that the line C (an arbitrary line!) is its zero line.

7. The properties of the magnetic specific lines $\hat{C}^{(m)}$ are analogous to those of the specific lines $\hat{C}$. Of course, the closed lines $\hat{C}^{(m)}$ must be resonant contours for the Neuman boundary problem (not for the Dirichlet one).

Let us point a property of the lines $\hat{C}^{(m)}$suitable for their practical construction. The definition of these line (2.34) means that the vector $grad\hat{u}^{(m)}(r,\varphi)$ is tangent to $\hat{C}^{(m)}$ at all its points. The condition

$$\frac{dy}{dx} = \frac{\partial \hat{u}^{(m)}(x,y)/\partial y}{\partial \hat{u}^{(m)}(x,y)/\partial x} \tag{2.184}$$

is valid, where $y = y(x)$ is the equation of the line $\hat{C}^{(m)}$. The lines $\hat{C}^{(m)}$ are the integral curves of equation (2.184). In particular, it follows from this that at given $\hat{u}^{(m)}(r,\varphi)$ the lines $\hat{C}^{(m)}$ can pass through almost each point of the plane.

3 Nonapproximability of Near Fields

3.1 Approximability Condition for Near Fields

1. We have stated above the existence of specific line $\hat{C}$ which possess the property that the pattern $f(\varphi)$ created by any current on them obeys the restriction (1.23). It turns out that not only is the restriction on the fields of these currents at infinity, but also an analogical restriction (even a stronger one) on the fields at *finite distances* from $\hat{C}$ is imposed. This restriction has the same form, as (1.23), but it is related to the field $u(r,\varphi)$ itself, instead of its asymptotics $f(\varphi)$.

Let us define this restriction. Construct a closed contour Σ encircling the specific line $\hat{C}$. The contour should be sufficiently smooth for the Dirichlet and Neumann exterior problems with radiation conditions to have a solution; in other aspects it can be arbitrary. We find the function on Σ which has the same meaning as $\hat{F}(\varphi)$ in (1.23).

Since line $\hat{C}$ is a specific one, there exists a real field $\hat{u}(r,\varphi)$ which equals zero on the line and has no singularity inside Σ. Here we use the absence of singularities only inside Σ, not in the whole plane; this circumstance will be important for the constructions in the next two sections. Denote the value of $\hat{u}(r,\varphi)$ on Σ by $\hat{u}(\sigma)$, where σ is the coordinate on the contour Σ $(0 \leq \sigma \leq 2\pi)$. Solve the following Dirichlet problem in the domain, exterior with respect to Σ: seek the field which solves the homogeneous Helmholtz equation, equals $\hat{u}(\sigma)$ on Σ, and satisfies the Sommerfeld radiation condition at infinity $(r \to \infty)$. This field is complex because it is complex at $r \to \infty$ owing to the radiation condition. This Dirichlet problem is always solvable because the frequency k is real.

Denote by $\hat{v}(r,\varphi)$ the auxiliary field which is equal to $\hat{u}(r,\varphi)$ inside contour Σ and to the solution of the above Dirichlet problem outside Σ. The field $\hat{v}(r,\varphi)$ satisfies the Helmholtz equation everywhere and the radiation condition at infinity; it is continuous and has the only singularity in the form of discontinuity of the normal derivative on Σ. Denote this discontinuity by $\hat{V}^*(\sigma) = [\partial\,\hat{v}/\partial N]$, where N is the outward normal to Σ. It is the "current" creating the field $\hat{v}(r,\varphi)$.

One can seek $\hat{V}(\sigma)$ without solving the above exterior Dirichlet problem. This function can be found immediately by the given value of field $\hat{V}$ on Σ without extending it into the exterior domain. In terms of the *potential theory* the function $\hat{V}(\sigma)$ is the density of the single layer, distributed along Σ (it is not a double layer potential, because the field $\hat{v}$ is continuous on Σ). The field created by the single layer potential of the density $\hat{V}(\sigma)$, satisfying the radiation condition, is

$$\widehat{v}(r,\varphi) = \frac{1}{4i}\int_{\Sigma} \hat{V}(\sigma) H_0^{(2)}(k|\vec{r} - \vec{r}_0|)d\sigma, \tag{3.1}$$

where $\vec{r}$ has coordinates (r, φ), and $\vec{r}_0$ is a point on Σ (see, e.g. [37, formulas (3,51), (3.53)]). When r also lies on Σ, the left-hand side of (3.1) is the known function $\hat{u}(\sigma)$ and (3.1) is the integral equation with respect to $\hat{V}(\sigma)$. This equation differs only in its left-hand side from the equation for the current, which has arisen on Σ in the diffraction on the metallic body with the same boundary Σ (see, e.g. [38, formulas (12.23)]). This equation is ill-posed, but a weak (logarithmic) singularity in its kernel permits to solve it with an acceptable accuracy by relatively simple methods and to define the "current" $\hat{V}(\sigma)$.

This "current" is involved in the conditions imposed on the field on Σ created by the current $j(s)$ on $\hat{C}$ (see (2.6a)). The proof is similar to that in Subsection 2.1.2. Apply the second Green formula to the functions $u(r, \varphi)$, $\widehat{v}(r, \varphi)$ in the whole plane. Both functions satisfy the same Helmholtz equation and conditions at $r \to \infty$. The field $u(r, \varphi)$ has the jump of the derivative on $\hat{C}$, which is equal to $j(s)$. The field $\widehat{v}(r, \varphi)$ has the jump of the derivative on Σ, which is equal to $V^*(\sigma)$. The second Green formula yields

$$\int_{\Sigma} u(\sigma)\hat{V}^*(\sigma)d\sigma = \int_{\hat{C}} \widehat{v}(s)j(s)ds, \tag{3.2}$$

where $u(\sigma)$ is the value of function $u(r, \varphi)$ on Σ, and $\widehat{v}(s)$ is the value of function $\widehat{v}(r, \varphi)$ on $\hat{C}$. The function $\widehat{v}(r, \varphi)$ coincides with $\hat{u}(r, \varphi)$ on $\hat{C}$, and, according to (5.7), equals zero there. Therefore the right-hand side of (3.2) is also equal to zero and hence

$$\int_{\Sigma} u(\sigma)\hat{V}^*(\sigma)d\sigma = 0. \tag{3.3}$$

Properties of the set of fields $u(\sigma)$, following from equation (3.3), are the same as properties of the set of patterns $f(\varphi)$, following from (3.15). The complete set of currents $j(s)$ on $\hat{C}$ creates *a noncomplete set of fields* $u(\sigma)$ on an arbitrary contour Σ encircling $\hat{C}$. The function $\hat{V}(\sigma)$ is the orthogonal complement to this set (see (3.1)). The functions $u(\sigma)$ can approximate only the functions $U(\sigma)$ given on Σ, which satisfy conditions

$$b_{\Sigma} \equiv \int_{\Sigma} U(\sigma)\hat{V}^*(\sigma)d\sigma = 0. \tag{3.4}$$

Repeating the reasons of Subsection 1.3.5, we note the following. If there are several functions $\hat{u}(r, \varphi)$, equal to zero on $\hat{C}$, then several functions of orthogonal complement to $\hat{V}(\sigma)$ on the same contour Σ, exist. In this case, some function $U(\sigma)$ can be approximated only if it is orthogonal to all these functions. The orthogonality of $U(\sigma)$ to all linear independent functions of orthogonal complement is the sufficient condition for its approximability.

Normalize the orthogonal complement functions by

$$\int_{\Sigma} \left|\hat{V}(\sigma)\right|^2 d\sigma = 1; \tag{3.5}$$

(this normalization differs from that following from (3.1) if the function $\hat{u}(r, \varphi)$ is normalized by (2.4)). Then the minimal residual, namely, the minimal mean-square difference between some given function $U(\sigma)$ and the field $u(\sigma)$ created by an arbitrary current on $\hat{C}$ (analogous to (1.32)) is equal to $|b_{\Sigma}|^2$ (see (3.20a)):

$$\int_{\Sigma} |U(\sigma) - u(\sigma)|^2 d\sigma \geq |b_{\Sigma}|^2 . \tag{3.6}$$

Consider the case when some current is located on a circle arc (i.e. it is the current induced on a cylindrical surface in the three-dimensional case), and for some value of n we have $J_n(ka) = 0$ (a is the circle radius). Then on an arbitrary larger circle concentric with the above one, the reflected field can be approximated with accuracy limited from below by the squared modulus of its nth Fourier coefficient.

2. If the contour Σ encircling the specific line $\hat{C}$ is a circle, then one can solve the above Dirichlet problem analytically and obtain the explicit expression for $\hat{V}(\sigma)$ in terms of Fourier coefficients of the field $\hat{u}(r,\varphi)$, analogous to the expression (2.9) for the function $\hat{F}(\varphi)$.

Denote the radius of circle Σ by R. Since the field $\hat{u}(r,\varphi)$ is given as the series (2.8) in the whole plane, it has the following form on Σ:

$$\hat{u}(\varphi) = \sum_{n=0}^{\infty} C_n J_n(kR)\cos(n\varphi), \tag{3.7}$$

where φ is the angle introduced instead of σ. At $r \geq R$ the solution to the Dirichlet problem, satisfying the radiation condition at infinity and continuity condition at $r = R$, is

$$\hat{v}(r,\varphi) = \sum_{n=0}^{\infty} C_n J_n(kR)\frac{H_n^{(2)}(kr)}{H_n^{(2)}(kR)}\cos(n\varphi). \tag{3.8}$$

Taking into account the expression for the Wronskian of cylindrical functions, seek the function $\hat{V}(\sigma)$, that is the difference between the normal derivatives of the fields $\hat{u}(r,\varphi)$ (2.8) at $r \leq R$ and $\hat{v}(r,\varphi)$ (3.8) at $r \geq R$, in the form

$$\hat{V}(\varphi) = -\frac{2i}{\pi R}\sum_{n=0}^{\infty} C_n \frac{1}{H_n^{(2)}(kR)}\cos(n\varphi). \tag{3.9}$$

This is a sought expression for $\hat{V}(\varphi)$.

According to (2.3), at $R \to \infty$, that is when the circle Σ is extended toward infinity, the field $\hat{u}(r,\varphi)$ and its normal derivative on Σ are

$$\hat{u}(\sigma) = \frac{1}{\sqrt{kR}}\left[\hat{F}^*(\varphi)e^{ikR} + \hat{F}(\varphi)e^{-ikR}\right], \tag{3.10}$$

$$\frac{\partial \hat{u}(\sigma)}{\partial r} = \frac{ik}{\sqrt{kR}}\left[\hat{F}^*(\varphi)e^{ikR} - \hat{F}(\varphi)e^{-ikR}\right]. \tag{3.11}$$

The field $\hat{v}(r,\varphi)$ equals $\hat{u}(\sigma)$ at $r = R$. At $r \geq R$ it is

$$\hat{v}(r,\varphi) = \hat{u}(\sigma)\sqrt{\frac{R}{r}}e^{-ik(r-R)}. \tag{3.12}$$

The normal derivative of this field at $r = R$ equals

$$\frac{\partial \hat{v}(\sigma)}{\partial r} = -ik\hat{u}(\sigma). \tag{3.13}$$

Subtracting (3.13) from (3.11), we obtain

$$\hat{V}(\varphi) = 2ik\frac{e^{ikR}}{\sqrt{kR}}\hat{F}^*(\varphi). \tag{3.14}$$

Of course, the Sommerfeld part of the field $\hat{u}(r,\varphi)$ is not contained in $\hat{V}(\varphi)$.

At any finite R, the multiplier $1/H_n^{(2)}(kR)$ decreases with increasing n. Therefore, the Fourier coefficients $C_n/H_n^{(2)}(kR)$ (see (3.9)) with large numbers, are smaller than the corresponding Fourier coefficients $C_n(-i)^n$ in the series (2.9) for $\hat{F}^*(\varphi)$. This means, that the function $\hat{V}(\varphi)$ *is smoother* than $\hat{F}(\varphi)$. The smaller kR , the smaller the value of n at which the Fourier coefficients in (3.9) begin to decrease in comparison with the corresponding Fourier coefficients in (2.9). Consequently, the smaller kR , the smoother the function $\hat{V}(\varphi)$ in comparison with its asymptotics (at $R \to \infty$) $\hat{F}(\varphi)$.

If the given function $U(\sigma)$ is not very indented, then, the smoother the function $\hat{V}(\sigma)$ in (3.4), the larger is $|b_\Sigma|$ and the more essential the restriction (3.6).

In this sense, the larger the distance between $\hat{C}$ and Σ, the weaker the restriction on the approximability of the fields on $\hat{C}$. This *restriction is weakened as the contour moves away from the currents* creating the field $u(r,\varphi)$; it shows itself least of all in the pattern space. In the next section we will consider the case when this restriction exists only up to some limit distance from the currents, and at a larger distance not only does it grow feeble but even disappears completely.

We illustrate this phenomenon on the example when $\hat{C}$ is a resonant rectangle (Subsection 2.2.4). It is not a specific line in the meaning given in Section 2.2, because the integral (2.4) does not exist for the corresponding function $\hat{F}(\varphi)$. However, the integral in (3.5) exists for any bounded contour Σ, and currents on the contour $\hat{C}$ create a noncomplete set of fields on Σ.

Consider, for simplicity, the contour $\hat{C}$ as the square of the side a; let the origin of Cartesian coordinates be placed at the square center. In this case

$$\hat{u}(x,y) = \cos\frac{\pi x}{a}\cos\frac{\pi y}{a}. \tag{3.15}$$

Let Σ be the circle of radius R. The field (3.9) on Σ has the form

$$\hat{u}(R,\varphi) = \cos\left(\frac{kR}{\sqrt{2}}\cos\varphi\right)\cos\left(\frac{kR}{\sqrt{2}}\sin\varphi\right), \tag{3.16}$$

where $k = \pi\sqrt{2}/a$ is the resonant frequency. To obtain (3.7), expand this function into the Fourier series. The coefficients C_n differ from zero only if $n = 4m$ $(m = 0, 1, \ldots)$, and they do not depend on m. With an accuracy to a nonessential factor, $\hat{V}(\varphi)$ is

$$\hat{V}(\varphi) = \frac{1}{R}\sum_{m=0}^{\infty}\frac{1}{H_{4m}^{(2)}(kR)}\cos(4m\varphi). \tag{3.17}$$

This series is convergent and square-integrable at any R. Hence the fields of currents, located on the sides of the resonant square (and, in general, of a resonant rectangle), create a noncomplete set on any circle (and, in general, on any contour) encircling the currents. However, decreasing the terms of the series (3.17) begins only at $4m > kR$. At $kR >> 1$ the series (3.17) contains many high harmonics, and $\hat{V}(\varphi)$ is a fast varying function. Therefore, for any smooth function $U(\sigma)$ the residual of fields on Σ (left-hand side in (3.6)) decreases as $kR \to \infty$. The residual of patterns equals zero in this case, that is, any pattern is approximable.

3. Not only do the fields $u(\sigma)$ created by currents on $\hat{C}$, but also their normal derivatives $\partial u(r,\varphi)/\partial N|_{\Sigma}$ make up a noncomplete function sets on contour Σ. The symmetry of these properties of functions $u(r,\varphi)$ and $\partial u(r,\varphi)/\partial N|_{\Sigma}$ is obviously caused by the symmetry between electrical and magnetic fields. Remember (see Subsection 1.2.2), that the two-dimensional scalar problem arises from the three-dimensional vector one with $H_z \equiv 0$, $\partial/\partial z \equiv 0$. Then $u(r,\varphi) = E_z$, $\partial u/\partial N|_C = H_s$, where s is directed along the line C with the normal N. incompleteness of the set of functions $u(r,\varphi)$, means that an arbitrarily chosen function cannot be approximated by the fields E_z, created by the currents located on the cylinder with the director $\hat{C}$. Similarly, incompleteness of the set of functions $\partial u(r,\varphi)/\partial N|_{\Sigma}$ means, that an arbitrarily chosen function cannot be approximated by fields H_z, created by the currents of the same kind.

To obtain the conditions of type (3.3) on $\partial u/\partial N$, we construct a field in the whole plane, solving the Neumann problem (instead of the Dirichlet one), exterior with respect to Σ. Calculate the function $\partial\hat{u}/\partial N|_{\Sigma}$ by the known field $\hat{u}(r,\varphi)$ inside Σ. Then find the field outside Σ, which has the same normal derivative and satisfies the radiation condition. Denote by $\hat{w}(r,\varphi)$ the field, equal to $\hat{u}(r,\varphi)$ inside Σ, and to the solution of the above Neumann problem outside Σ. The field $\hat{w}(r,\varphi)$ has only one singularity, namely, its jump on Σ. Denote this jump by

$$\hat{W}(\sigma) = [\hat{w}(r,\varphi)]\,. \tag{3.18}$$

Apply the second Green formula to the fields $u(r,\varphi)$ and $\hat{w}(r,\varphi)$, created by the current $j(s)$ on $\hat{C}$, and by the "magnetic current" $\hat{W}(\sigma)$ on Σ, respectively. Similarly to (3.2), we obtain

$$-\int_{\Sigma} \frac{\partial u}{\partial N}\hat{W}(\sigma)d\sigma = \int_{\hat{C}} \hat{w}(s)j(s)ds, \tag{3.19}$$

where $\hat{w}(s)$ is the value of field $\hat{w}(r,\varphi)$ on $\hat{C}$. The field $\hat{w}(r,\varphi)$ coincides with $\hat{u}(r,\varphi)$ (i.e. equals zero) on $\hat{C}$. Hence the field $u(r,\varphi)$ created by a current on $\hat{C}$, should fulfil the following condition

$$\int_{\Sigma} \frac{\partial u}{\partial N}\hat{W}(\sigma)d\sigma = 0. \tag{3.20}$$

As above, it follows from this that the functions $\partial u/\partial N|_{\Sigma}$ can approximate only the given functions on Σ orthogonal to $\hat{W}(\sigma)$ in the sense of (3.20). Similarly to (3.6), the residual of the approximation of any function by the functions $\partial u/\partial N|_{\Sigma}$ is not less (with the appropriate normalization of $\hat{W}(\sigma)$) than the squared modulus of the inner product (3.20) of this function and $\hat{W}^*(\sigma)$.

If Σ is a circle of radius R, and $\hat{u}(r,\varphi)$ is given by the series (2.8) at $r \le R$, then, similarly to (3.8), the solution $\hat{w}(r,\varphi)$ to the Neumann problem at $r > R$, is the series

$$\hat{w}(r,\varphi) = \sum_{n=0}^{\infty} C_n J_n'(kR)\frac{H_n^{(2)}(kr)}{H_n^{(2)\prime}(kR)}\cos(n\varphi). \tag{3.21}$$

Its normal derivative at $r = R + 0$ equals the normal derivative of the function $\hat{u}(r,\varphi)$ at $r = R - 0$, and the jump of the function $\hat{w}(r,\varphi)$ on Σ, that is, the function $\hat{W}(\sigma)$ (3.18) is of the form

$$\hat{W}(\sigma) = -\frac{2i}{\pi kR}\sum_{n=0}^{\infty} C_n \frac{1}{H_n^{(2)\prime}(kR)}\cos(n\varphi) \tag{3.22}$$

(see (3.9)).

The asymptotics of $H_n^{(2)\prime}(kR)$ at $n >> kR$ is much the same, as of $H_n^{(2)}(kR)$, that is the multiplier $1/H_n^{(2)\prime}(kR)$ decreases rapidly with n increasing at $n >> kr$, and the Fourier coefficients of large numbers in the series (3.22) are smaller than corresponding Fourier coefficients in the series (2.9) for $\hat{F}(\varphi)$. Considerations, formulated above, concerning $\hat{V}(\varphi)$ (see (3.9)), are also true for (3.22); the restriction on the approximability decreases on going away from the currents.

The asymptotics of $H_n^{(2)\prime}(kR)$ at $kR >> n$ is also close to that of $H_n^{(2)}(kR)$. Therefore, similarly to (3.9), the series (3.22) at $kR \to \infty$ differs from that of (2.9) for $\hat{F}^*(\varphi)$, only in a multiplier proportional to $\exp(ikR)/\sqrt{kR}$. Of course, this conforms with the fact that, in the far zone, the normal derivative (i.e. $\partial/\partial r$) of any field fulfilling the radiation condition, differs from the field itself only in a multiplier.

4. The above constructions can be transferred (almost without any correction) to the fields, created by magnetic currents located on specific lines $\hat{C}^{(m)}$ of the magnetic type defined by the condition (2.34). We will not write all the formulas for this case. On any contour Σ encircling $\hat{C}^{(m)}$ one can construct the functions analogous to $\hat{V}^{(m)}$ and $\hat{W}^{(m)}$. The functions appear in the conditions imposed on the field $u^{(m)}(r,\varphi)$ created by the magnetic current, and on its normal derivative $\partial u^{(m)}/\partial N$. These formulas are similar to (2.37) for patterns of such currents.

The function $\hat{V}^{(m)}(\sigma)$ is constructed as a continuation of the auxiliary field $\hat{u}^{(m)}(r,\varphi)$ introduced in Subsection 2.1.3 into the domain, exterior to Σ, with its continuity on Σ; $\hat{V}^{(m)}(\sigma)$ is the jump of the normal derivative of the auxiliary field $\hat{v}^{(m)}(r,\varphi)$ obtained as a result of this continuation. The function $\hat{W}^{(m)}(\sigma)$ is constructed as a continuation of the same field $\hat{u}^{(m)}(r,\varphi)$, with the continuity of the normal derivative; $\hat{V}^{(m)}(\sigma)$ is the jump of $\hat{w}^{(m)}(r,\varphi)$ on Σ constructed in this way.

The conditions mentioned in the previous paragraph are of the form

$$\int_\Sigma u^{(m)}(\sigma)\hat{V}^{(m)*}(\sigma)d\sigma = 0; \qquad \int_\Sigma \frac{\partial u^{(m)}}{\partial N}\hat{W}^{(m)}(\sigma)d\sigma = 0. \tag{3.23}$$

They are completely analogous to the conditions (3.3), (3.20). Their consequences are the conditions, analogous to (3.3), and to those formulated after (3.20). These conditions, namely, the vanishing of the inner product of the given function and the orthogonal complement function, are necessary for approximability of the given function. If there exist more than one function $\hat{u}^{(m)}(r,\varphi)$ for the given line $\hat{C}^{(m)}$, then the orthogonal complement set contains more than one function both for $u^{(m)}(r,\varphi)|_\Sigma$ and $\partial u^{(m)}/\partial N|_\Sigma$. In this case the sufficient condition of approximability is the orthogonality of the given function to all the functions from the sets of orthogonal complements of functions $u^{(m)}|_\Sigma$ and $\partial u^{(m)}/\partial N|_\Sigma$.

3.2 Construction of a Field by its Zero Line. Case of the Circle Arc

1. Finding the zero line $\hat{C}$ of the given real field is not a very complicated problem. If the field $\hat{u}(r,\varphi)$ is of simple form, then the problem is solvable analytically or, at least, it is easily solvable numerically. This also relates to the specific lines of magnetic type $\hat{C}^{(m)}$, determined by the field $\hat{u}^{(m)}(r,\varphi)$ according to (2.33); their numerical construction can be made by (2.184). As a rule, we consider below the fields $\hat{u}(r,\varphi)$ and lines $\hat{C}$.

As far as we know, the inverse problem on finding solutions to the Helmholtz equation by their zero lines is not solved in general form. The sufficient conditions, under which the line is a zero line of some field $\hat{u}(r,\varphi)$, are also unknown. This problem, as well as the analogous three-dimensional vector one, is a key problem in all approximability theory. In this and the next sections, we give several examples of solving this inverse problem in some simple cases.

In the previous section it was stated that, if $\hat{C}$ is a specific line, that is, the patterns created by currents distributed on it cannot approximate all the patterns, then fields of these currents also cannot approximate all the near fields, that is, the fields given on any contour Σ, encircling $\hat{C}$ and located at a finite distance from it. By the definition given in Section 2.1 the field $\hat{u}(r,\varphi)$ has no singularity in the whole plane. We consider here the more general case when the line $\hat{C}$ is a zero line of the field $\hat{u}(r,\varphi)$ having no singularity only in a part of the plane. We will use the term "specific" for this line, either; in this way we extend the concept of the specific line and keep the notations $\hat{C}$, $\hat{u}(r,\varphi)$, and $\hat{C}^{(m)}$, $\hat{u}^{(m)}(r,\varphi)$, respectively.

The result obtained in Section 3.1 remains valid for any contour Σ, such that the field $\hat{u}(r,\varphi)$ is analytical on Σ and inside it, even if this field has some singularities outside Σ. This result consists in the fact that the fields $u(r,\varphi)$, created by a complete set of currents located on $\hat{C}$, form a noncomplete set of functions $u(\sigma)$. This follows from the constructions of the previous section, in which the properties of $\hat{u}(r,\varphi)$ outside Σ were not used. The method was also proposed for constructing the complement set of functions on Σ, orthogonal to $u(\sigma)$ (or to $\partial u/\partial N$), based on solving the Dirichlet (or Neumann) problem in the domain exterior to Σ with the Sommerfeld radiation condition, or the equations of the potential theory. This method is also applicable to our case.

2. In this section we consider the problem of constructing the field $\hat{u}(r,\varphi)$ by its zero line $\hat{C}$ in the case when $\hat{C}$ is a circle arc. Since in this case $\hat{C}$ is a coordinate line of the polar coordinate system, in which variables in the Helmholtz equation are separable, the solution is obtained, of course, in an elementary way. If our considerations were limited by seeking the field $\hat{u}(r,\varphi)$, having no singularity in the whole plane, then the problem would, in principle, be already solved in Section 2.2. Here we are interested in a wider class of the specific lines, which were defined in the last but one paragraph of the previous subsection. Just this class is essential in the problem of the near fields approximability. In this sense, the arc of any circle (not only the resonant one) is a specific line.

Let $\hat{C}$ be an arc of the circle of radius a. If for some $n = 0, 1, \dots$ we have $J_n(ka) = 0$, then $\hat{C}$ is an arc of the resonant circle, and it is a zero line of the field $\hat{u}(r,\varphi)$ given by (2.42). This field has no singularity in the whole plane. It is unique to within the polarization degeneracy, that is, to within the replacement of $\cos(m\varphi)$ by $cos(m(\varphi - \varphi_0))$ with arbitrary φ_0; we will not mention this fact below.

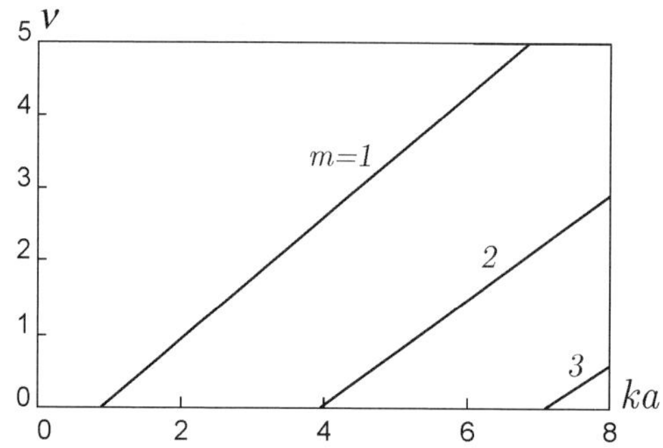

Figure 3.3:

roots at $3.96 \leq ka < 7.09$, and so on. For instance, there are two roots at $ka = 5.0$: $\nu = 0.71$ $(m = 2)$, and $\nu = 3.42$ $(m = 1)$.

At $m >> \nu$, the root of equation (3.28), treated as an equation for ka, is equal (with the same accuracy, as (3.26)) to

$$ka \simeq \pi(m + \nu/2 - 3/4). \tag{3.29}$$

For instance, at $\nu = 1/2$, the function $N_\nu(ka)$ is proportional to $\cos(ka)$, and (3.26) becomes the exact formula $ka = \pi(m - 1/2)$. Asymptotically, at $ka >> \nu$, the number m of zeros of $N_\nu(ka)$ equals $ka/\pi + 3/4$.

At noninteger ν the analyticity domain of the field (3.27) is the same as that of (3.24). However, if ka is such that the solution to (3.28) is integer $(\nu = n)$, then the condition for $\hat{C}$ to be an arc of the nonresonant circle, is not violated, because in this case the field (3.27) also has the singularity at the circle center O. Therefore, only point O does not have to be placed inside Σ. The contour Σ can be the same as in Fig. 3.2; two rays can be replaced with any curves or can consist of two closed contours, the interior of which encircles the point O.

The field, more general than (3.24) and (3.27), which equals zero on the circle of radius a, is the following linear combination of these fields:

$$\hat{u}(r, \varphi) = [N_\nu(ka)J_\nu(kr) - J_\nu(ka)N_\nu(kr)]\cos(\nu\varphi). \tag{3.30}$$

This field equals zero on $\hat{C}$ at any index ν. In particular, ν can be equal to zero; for this reason we have replaced sin with cos in (3.30). The radius a can be as small as desired.

If ν is complex, then the field (3.30) is complex, too. Since the Helmholtz equation does not contain any complex parameters and $u(r, \varphi)$ is not subject to any condition as the radiation one, then the real and imaginary parts of the field, defined by (3.30), are also solutions to the Helmholtz equation. In this case (3.30) contains two real fields equal to zero at $\hat{C}$. The analyticity domain of the field (3.30) is the same as of fields (3.24) and (3.27).

4. The orthogonal complement functions $\hat{V}(\sigma)$, corresponding to the contour Σ located in the common domain of analyticity of several fields $\hat{u}(r, \varphi)$, are different for different fields. For the fields (3.24) and (3.27) there exists a finite number of them (as well as finite numbers of the roots of equations (3.25) and (3.28)), approximately equal to $2ka/\pi$; for the fields (3.30) this number depends on the free parameter ν.

Although this parameter can be arbitrary, the class of corresponding orthogonal complement functions is, in fact, not very wide. At $ka >> 1$ and $|\nu| << ka$ the fields (3.30) are asymptotically proportional to $\sin(k(r-a))$, that is, these fields cease to depend on ν with increasing ka, and all the functions of the orthogonal complement become close to each other. At large $|\nu|$ (in the domain of the Debye asymptotics of cylindrical functions) these fields are proportional to $[(a/r)^\nu - (r/a)^\nu]$. At both $r < a$ and $r > a$ they are very fast varying on Σ. For these "exotic" functions $\hat{V}(\sigma)$ the requirement $b_\Sigma = 0$ (see (3.4)), necessary for approximability of some function $U(\sigma)$ given on Σ, plays no practical role. This requirement could be an essential restriction only if the function $U(\sigma)$ is fast varying itself. This means that even if some function $U(\sigma)$ is nonapproximable because the integral (3.4) differs from zero, but $|b_\Sigma| << 1$, then there exists an approximable function, close to $U(\sigma)$. Of course, this function is not arbitrarily close to $U(\sigma)$ (in this case $U(\sigma)$ would be approximable itself), but the residual (3.6) is small for this function.

Note that this fact weakens the difference between approximability ($b = 0$) and nonapproximability ($b \neq 0$), in the same way as the concept of approximability weakens the difference between realizability and nonrealizability (see Section 1.3). We will return to this question (methodical, in fact) in Section 4.4.

Thus, the field $\hat{u}(r, \varphi)$, having an arc of the nonresonant circle as its zero line, can be described fully enough by the above fields (3.24) and (3.27), where ν can possess only a limited set of values.

5. The larger the radius a of a circle, the more fields $\hat{u}(r, \varphi)$ (as well as functions $\hat{V}(\sigma)$ connected with them) exist. The number of such functions infinitely increases at $ka \to \infty$, even if only essentially different functions, not very fast varying on Σ, are taken into account. At $a \to \infty$ the distance between the arc $\hat{C}$ and singular points of $\hat{u}(r, \varphi)$ (the circle center and the ray outgoing from it) grows infinitely. As a result, the fields become analytical inside very large contours Σ, encircling $\hat{C}$. This is in accordance with the fact, that the arc passes at $a \to \infty$ into a straight-line segment, that is, into a specific line, such that all the solutions to the Helmholtz equation, asymmetrical with respect to the line and analytical in the whole plane, are the fields $\hat{u}(r, \varphi)$.

Remember that increasing the number of the orthogonal complement functions and extension of the analyticity domain of $\hat{u}(r, \varphi)$, means that restrictions on the fields, which can approximate the currents distributed on $\hat{C}$, become stronger. As seen in this example, the restrictions are getting stronger despite the prolongation of line $\hat{C}$. The length of this line is not important here, more important is its closeness to a specific line with very strong restrictions (i.e. with a rich set of functions $\hat{V}(\sigma)$). We will discuss this effect in the next section.

A method of constructing the functions $\hat{V}(\sigma)$ on Σ was described in Section 3.1. If Σ is a circle and $\hat{u}(r, \varphi)$ is given in the form of series (2.8), so that its Fourier series on Σ is of the form (3.8), then the function $\hat{V}(\sigma)$ has an explicit form of the Fourier series (3.9). Though in Section 3.1 the coordinate origin was assumed to coincide with the center of circle Σ, in (3.30) it lies in the center of the circle arc $\hat{C}$. To apply the results of Section 3.1, we should displace the coordinate origin, that is, express these fields in the form of a Fourier series with respect to the angle of the cylindrical coordinate set originating at the center of circle Σ. This recalculation can be easily done by the addition formula for cylindrical functions (see, e.g. [4, formula (9.1.79)]).

6. Consider a simple case, in which one can also easily construct the functions $\hat{u}^{(m)}(r, \varphi)$, the normal derivatives of which vanish on the same arc. The circle arc is simultaneously the specific line of the magnetic type.

Obviously, the fields $\hat{u}^{(m)}(r,\varphi)$ have the same form as (3.24), (3.27). Numbers ν in them satisfy one of the following equations

$$J'_\nu(ka) = 0, \tag{3.31a}$$

$$N'_\nu(ka) = 0 \tag{3.31b}$$

(instead of (3.25), (3.28)).

The dependencies ν on ka, given by each of these equations, are described by curves of the same type as in Figs. 3.1, 3.2. The curves corresponding to equation (3.31a) are similar to those in Fig.3.1; curve 1 starts at the point $ka = 0$, $\nu = 0$. For instance, equation (3.31a) has two solutions at $ka = 5.0$: $\nu = 3.7$ (curve 1) and $\nu = 0.78$ (curve 2). The curves, corresponding to (3.31b) are similar to those in Fig. 3.3, but they are shifted to the right, so that the curve 1 starts at $ka = 2.19$, $\nu = 0$. At $ka = 5.0$ equation (3.31b) has one solution, $\nu = 2$.

The analog of the field $\hat{u}(r,\varphi)$ (see (3.30)) is, obviously the field

$$\hat{u}^{(m)}(r,\varphi) = [N'_\nu(ka)J_\nu(kr) - J'_\nu(ka)N_\nu(kr)]\sin(\nu\varphi). \tag{3.32}$$

It also satisfies the condition (2.34) on the arc $r = a$ at any ν. All remarks made above concerning the field (3.30), refer to (3.32). The Debye asymptotics of (3.32) differs from that of the field (3.30) only in the sign in its second summand.

Thus, the circle arc is both a specific line $\hat{C}$, and a specific line of magnetic type $\hat{C}^{(m)}$ in a wider sense, discussed at the beginning of this section. The analyticity domains of the corresponding fields $\hat{u}$ and $\hat{u}^{(m)}$ coincide. Everything stated in Section 2.1 is true for any closed contour Σ, which contains this arc and does not contain the circle center. Both the fields created by a complete set of electrical currents on the arc, and the fields created by a complete set of magnetic currents on it, do not form complete sets apart, but they do together.

3.3 Analytical Extension of the Eigenoscillation Field outside the Boundary of the Domain

In this Section we describe a method of construction of the field $\hat{u}(r,\varphi)$ by the given zero line $\hat{C}$. It consists in *supplying* $\hat{C}$ to the resonant contour and subsequent *extension of the eigenoscillation field* into the domain, exterior to this contour. Some considerations about the general solution to such an extension problem will be given in subsections 5,6,7. At the beginning we give some examples, analogous to those from Section 2.2.

1. At first, let $\hat{C}$ be a nonclosed line. Supply it with a line C_{supp} to the closed line (contour) C_0 so that this contour is a resonant one (in the sense of the Dirichlet conditions). The eigenoscillation is described by some field equal to zero on C_0. Calculate the normal derivative of this field on C_0. The analytical extension is a field (i.e., a solution to the homogeneous Helmholtz equation) equal to zero on C_0, which has the same value of the normal derivative on C_0, as the eigenoscillation field. The field $\hat{u}(r,\varphi)$ is equal to: zero on C_0, the eigenoscillation field inside it, and the analytical extension outside. The function $\hat{u}(r,\varphi)$ is real, except for the degeneration cases. Of course, it does not satisfy the radiation condition.

The analytical extension problem can be considered as the Cauchy problem: find a field outside C_0 by its given value as well as the value of its normal derivative on C_0. In our case the Cauchy problem is stated in the a specific way: the values of the function and its normal derivative are not independent of each other. They are constrained by the condition that the same Cauchy problem in the interior to C_0 domain has a solution, analytical everywhere in the domain.

The possibility of supplying any nonclosed line C to the contour C_0, resonant at the given frequency, follows from the Courant theorem stating that the extension of the domain decreases its eigenfrequency monotonically. If C_{supp} is very close to $\hat{C}$, so that the area of the domain interior to C_0 is small, then this frequency is high. Transform continuously the contour C_{supp} in such a way, that each next contour is an extension of the previous one coinciding with it on $\hat{C}$. Then the eigenfrequency is continuously decreasing and it becomes equal to the given one at some shape of the contour C_0.

This consideration, in particular, shows that the same nonclosed line $\hat{C}$ can be a specific one not only at one frequency but in some frequency range. Thus, in the example of the previous section, where $\hat{C}$ is a circle arc of radius a, equation (3.25) is solvable with respect to ν at all $k > 2.40/a$. In general, the analyticity domains are different for different frequencies. However, it can happen that some contour Σ lies in the common analyticity domain (this fact was observed in the example); then the nonapproximability of any function given on Σ takes place in a wide frequency range.

Note that this fact permits to transfer the method of reconstruction of the scattering obstacle shape, described in Section 2.5, to the case when this obstacle is a screen (i.e., a nonclosed line), and the measurements can be made in the near zone. If the screen shape is close to an arc of an analytical curve, then the methods described in Section 2.4 can be applied at more than one frequency. This permits to raise the probability of some hypothesis on the shape of the scattering screen or on the reliability of the shape determination by measuring the near zone field at several frequencies.

2. The technique of constructing the field $\hat{u}(r,\varphi)$ can obviously be applied to construction of the field $\hat{u}^{(m)}(r,\varphi)$ by the given nonclosed specific line of magnetic type $\hat{C}^{(m)}$. The contour $C_0^{(m)}$, which contains the given line $\hat{C}^{(m)}$ as its part, should be the resonant one for the Neumann boundary condition at the given frequency. The sought analytical extension is determined from the Cauchy problem for the domain, exterior to $C_0^{(m)}$. The value of the field on $C_0^{(m)}$ in this Cauchy problem is equal to that of the eigenoscillation field inside $C_0^{(m)}$.

In this case the reference to the Courant theorem should be made in a slightly different way. Under the boundary condition of the Neumann type, the eigenfrequency of the lowest eigenoscillation in the domain of a small area can be low; it can have an order, inverse to the perimeter of contour $C_0^{(m)}$. However, in this case the eigenoscillations of high degree, having high frequencies, also exist. At least one of them will be higher than the given frequency. Moving $C_{\text{supp}}^{(m)}$ away from $\hat{C}^{(m)}$ we lessen this eigenfrequency and finally find the contour resonant at the given frequency.

This technique can be generalized for the problem of constructing the fields $\hat{u}(r,\varphi)$ and $\hat{u}^{(m)}(r,\varphi)$ by the given contours $\hat{C}$ and $\hat{C}^{(m)}$, respectively. For instance, the Neumann condition can be imposed on $\hat{C}_{\text{supp}}^{(m)}$ instead of the Dirichlet one when $\hat{u}(r,\varphi)$ is to be constructed. Then the mixed boundary conditions should be imposed on the contour C_0. For such conditions there also exist the resonant frequencies and the fields of eigenoscillations, as the eigenelements of self-adjoint problems. One can also state the third-kind boundary condition on C_{supp} with a real factor (in general, variable on C_{supp}), and so on. We will not mention these possibilities below.

It was shown in the example in the previous section that, in the general case, the circle arc is simultaneously both line $\hat{C}$ and line $\hat{C}^{(m)}$. The construction described above shows that such a situation is typical, at least in the case when we deal with the near fields. A nonclosed line can be supplied both to contour $\hat{C}_0$ and $\hat{C}_0^{(m)}$. If the fields of both eigenoscillations can be analytically extended and one can construct a contour lying in the common area of analyticity of both the obtained fields $\hat{u}(r,\varphi)$ and $\hat{u}^{(m)}(r,\varphi)$, then the sets of fields created by electrical and magnetic currents on C are separably noncomplete on Σ, but fields created by the currents of both kinds can approximate any field on Σ.

3. It can happen that the analyticity domain of the field $\hat{u}(r,\varphi)$ obtained as the analytical extension of the eigenoscillation field outside the interior domain of C_0, does not contain the whole contour C_0. We only require that this domain contains the whole line $\hat{C}$. If the singularity points of this field lie on C_{supp}, then in the space of patterns created on $\hat{C}$ there is, obviously, no orthogonal complement function $\hat{F}(\varphi)$ connected with this field; however, this can be caused by inappropriate choice of the line C_{supp}. Some other field $\hat{u}(r,\varphi)$, having no singularity in the whole plane, can exist, connected with the same specific line $\hat{C}$, but with another contour C_0. Recall that the existence of only one such field means that the pattern set is noncomplete.

However, if the points on the line C exist, in any neighborhood of which no analytical field exists, that is, the points at which any solution to the Helmholtz equation, equal to zero on C, has a singularity, then C is not a specific line. Such points can be, in particular, angular ones, at which two different analytical curves containing the line C, meet. If the angle α between two tangent lines at this point is not a rational part of π, then any solution to the Helmholtz equation has a singularity at this point. The proof, in essence, repeats the similar arguments given in Subsection 2.2.2, where it was related to the line consisting of two intersected straight lines.

In the local polar coordinate set (ρ,ψ) with the origin at the angle corner and the axis directed along one of the tangents, the field is equal to $\rho^\nu \sin(\nu\psi) + O(\rho^{\nu+1})$ at $k\rho << 1$. Since $\sin(\nu\alpha) = 0$, then $\nu = t\pi/\alpha$, $t = 1, 2, \ldots$ The same function should also describe the field at $\psi > \alpha$, otherwise there is no analyticity. Hence the equality $\sin(2\pi\nu) = 0$ should hold, that is, ν should be integer (by the way, this also follows from the fact that the field is proportional to ρ^ν at $k\rho << 1$), and then α/π is a rational fraction. Otherwise, the mentioned Cauchy problem has no solution even in the small neighborhood of the angular point (as a one-valued function).

4. Some examples of Section 3.2 can be considered as an implementation of this technique: supplying the line $\hat{C}$ to C_0 and analytically extending the field outside it. Return to the field (3.24) having a circle arc as its zero line; take, for example, $ka = 5.0$. Construct a curvilinear triangle, consisting of the arc $r = a$ and two rays: $\varphi = 0$ and $\varphi = \beta$, where the angle β is determined from the condition $\sin(\nu\beta) = 0$. Such a triangle is the contour C_0. Choose ν to be a noninteger number, for instance, $\nu = 1.9$. Then the angle β can possess one of three values: 0.53π, 1.06π, or 1.6π. The first of the triangles is presented in Fig. 3.4.

The whole triangle cannot be a line $\hat{C}$, because the angle between its sides is the irrational part of π. The circle center cannot be placed inside Σ; the field $\hat{u}(r, \varphi)$ has a singularity at the point $r = 0$ (see (3.24)). However, the segments of both rays at $r > 0$ can be included into $\hat{C}$. If the central angle of the arc of the nonresonant circle at $ka = 5.0$ is smaller than 0.53π, then the resonant triangle can have a shape shown in Fig. 3.4. If this angle is larger than 1.6π, then such a resonant contour cannot be constructed. At the same time, the arc itself remains a specific line as the zero line of the field (3.24) at $\nu = 1.9$; however, it loses this property after being supplemented with two rays. If the currents could be located not only on the arc but also on these rays, then the fields created by them on any contour Σ would form a complete set.

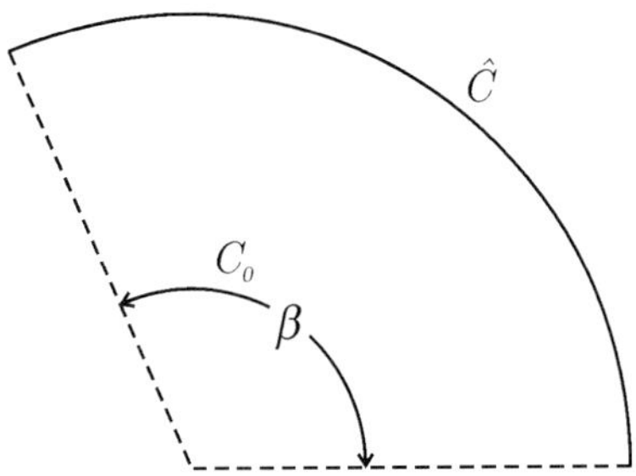

Figure 3.4:

The field $\hat{u}(r, \varphi)$, described by (3.30), can also be considered as the field of the eigenoscillation of some contour C_0 and the analytical extension of this field. Except for the line $\hat{C}$ (an arc of the nonresonant circle), the contour C_0 also contains an arc of either larger (Fig. 3.5(a)) or smaller (Fig. 3.5(b)) radius with the same center, and two segments of the rays $\varphi = 0$ and $\varphi = \beta$, where β is connected with the given value of ν by the above condition $\sin(\nu\beta) = 0$. The radius c of the second arc can be determined from the obvious equation

$$N_\nu(ka)J_\nu(kc) - J_\nu(ka)N_\nu(kc) = 0. \tag{3.33}$$

For instance, if $\nu = 0$ is chosen, then from (3.33) we have two values of kc for the arc of radius $ka = 5.0$: $kc = 1.19$ (Fig. 3.5(a)) and $kc = 8.17$ (Fig. 3.5(b)). If both radii are given in (3.33), that is, if $\hat{C}$ consists of two arcs of the concentric nonresonant circles, then (3.33) is the equation for ν. One can add two rays (at $r > 0$) to these arcs, keeping the property to be a specific one for the combined line C_0 (the curvilinear quadrangle).

The connection between the above arguments and constructions of Section 3.2 is completely trivial. It is based on the fact that the field $\hat{u}(r, \varphi)$ can have other zero lines besides $\hat{C}$. They can be added to $\hat{C}$; if they, together with the given line $\hat{C}$, form the contour C_0, then this contour will be resonant and the field $\hat{u}(r, \varphi)$ will be a field of the eigenoscillation inside C_0 and its analytical extension outside it.

Compare the conditions we have for the closed and nonclosed specific lines. It is necessary for the closed specific line to be a resonant contour. Remember that this condition is not sufficient for the line to be specific. Such a line can be specific only at a countable set of frequencies. The nonclosed specific line is part of an infinite (continual) set of resonant

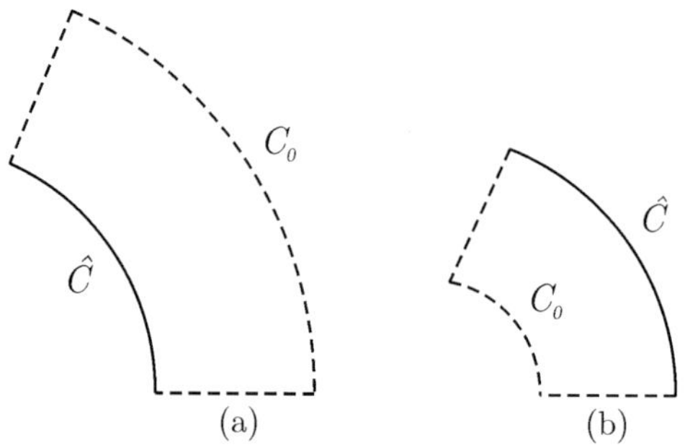

Figure 3.5:

contours. It can become a specific line in a *continuous frequency range*. Otherwise, the restrictions on the approximation of arbitrary fields and patterns are stronger for a nonclosed specific line, than for a closed one.

Remember that a similar case also occurs in the problem of realizability: the class of patterns (fields), which can be created by currents distributed on closed lines is wider than that created by currents on nonclosed ones.

5. The simplest method for the formal construction of the analytical extension of the eigenoscillation field $\hat{u}(r,\varphi)$ existing inside a resonant contour C_0 consists, probably, in the following. Let us construct some nonresonant circle with the centre at the coordinate origin and wholly lying inside C_0. Expand the field $\hat{u}(r,\varphi)$ on it into the Fourier series, leaving out again the terms with $\sin(n\varphi)$,

$$\hat{u}(a,\varphi) = \sum_{n=0}^{\infty} B_n \cos(n\varphi). \tag{3.34}$$

The coefficients B_n are connected in the obvious way with the coefficients C_n from expression (2.8): $B_n = C_n J_n(ka)$, where a is the circle radius. This expression is unique, therefore, at any r the field and its analytical extension (at the points where it exists) can be expressed in the form

$$\hat{u}(r,\varphi) = \sum_{n=0}^{\infty} B_n \frac{J_n(kr)}{J_n(ka)} \cos(n\varphi). \tag{3.35}$$

The question concerning the analytical extension and its singularities is reduced to the question concerning the convergence of series (3.35), that is, concerning the decreasing rate of the coefficients B_n in (3.34) with respect to the number n.

Using these terms, we first state the condition, under which the series (3.35) converges at infinity, that is, at $r \to \infty$. Then it will represent a function having the asymptotics (2.9). Substituting the above connection between coefficients B_n and C_n into (3.34), we have

$$\hat{F}(\varphi) = \frac{1}{\sqrt{2\pi i}} \sum_{n=0}^{\infty} (-i)^n \frac{B_n}{J_n(ka)} \cos(n\varphi), \tag{3.36}$$

and, with an accuracy to one factor,

$$\int_0^{2\pi} \left|\hat{F}(\varphi)\right|^2 d\varphi = \sum_{n=0}^{\infty} \frac{(1+\delta_{0n})B_n^2}{J_n^2(ka)}. \tag{3.37}$$

The Debye asymptotics of cylindrical functions at $n >> ka$, $n >> 1$ is of the form

$$J_n(ka) = \frac{1}{\sqrt{2\pi n}}\left(\frac{eka}{2n}\right)^n, \tag{3.38a}$$

$$H_n^{(2)}(ka) = i\sqrt{\frac{2}{\pi n}}\left(\frac{2n}{eka}\right)^n. \tag{3.38b}$$

(The second formula will be used further.) According to (3.38a), the asymptotics of the coefficient $B_n/J_n(ka)$ is the nth power of the ratio a_0/a, where

$$a_0 = \frac{2}{ek}\lim_{n\to\infty}\left(n\left|B_n\right|^{1/n}\right). \tag{3.39}$$

We already had such a type of formula in (1.21) (see Section 1.3), when the realizability of the function $F(\varphi)$ was discussed.

Therefore, to establish the convergence of the series (3.36), we should calculate a_0 by (3.39). If $a_0 > a$, then series (3.36) does not exist, that is, the analytical extension up to $r \to \infty$ of the function (3.34) without singularities is impossible. If $a_0 < a$, then the series (3.36) converges, the function $\hat{F}(\varphi)$ has the finite norm, and the field $\hat{u}(r,\varphi)$ (see (3.35)) has no singularity at any r including $r \to \infty$. The contour C_0 and any of its parts are the specific lines. Finally, if $a_0 = a$, then the function $\hat{F}(\varphi)$ exists, but it contains the δ-functions, its norm is infinite, hence the set of patterns created by currents on C_0 is complete.

Let us consider in detail the case when $a_0 > a$ (in particular, $a_0 = \infty$), that is, when the sequence $|B_n|$ tends to zero at $n \to \infty$ but not sufficiently fast for the function (3.36) to exist. At any finite distance $r = R > a$, the asymptotics of the Fourier coefficients of the function $\hat{u}(R,\varphi)$ at $n \to \infty$ (see, (3.35)) equals $B_n\,(R/a)^n$. Define the value

$$l = \lim_{n\to\infty}\left|B_n\right|^{1/n}. \tag{3.40}$$

In distinction from the expression (3.39) for a_0, this formula does not contain the factor n under the lim sign. If $l > 0$, then $a_0 = \infty$, and if $a_0 < \infty$, then $l = 0$. Of course, $l < 1$. Otherwise the series (3.34) is either not convergent (if $l > 1$, then $B_n \to \infty$ at $n \to \infty$) or convergent to a generalized function (if $l = 1$, then $B_n \to 1$ at $n \to \infty$). However, from the assumption, the circle $r = a$ wholly lies in the domain in which $\hat{u}(r,\varphi)$ has no singularity. From the notation (3.40) the Fourier coefficients of the function $\hat{u}(R,\varphi)$ have an asymptotic estimation $(Rl/a)^n$ at $n \to \infty$. At $R < a/l$ the series converges, that is, the analytical extension has no singularity inside the circle $r = a/l$. At $R > a/l$ the series diverges. The singularity point of the analytical extension of the field $\hat{u}(r,\varphi)$, closest to the coordinate origin, lies on the circle $r = a/l$.

Of course, the value a/l does not depend on the way of choice of the circle radius a, since the circle lies in the analyticity domain of the field $\hat{u}(r,\varphi)$. This fact immediately follows

Define the distance between two patterns as

$$\Delta = \left[\int_0^{2\pi} \left| F(\varphi) - \tilde{F}(\varphi) \right|^2 d\varphi \right]^{1/2} \tag{4.1}$$

(see (1.14)). The patterns $F(\varphi)$ and $\tilde{F}(\varphi)$, as well as all the patterns, mentioned below in this section, are normalized by unity:

$$\int_0^{2\pi} |F(\varphi)|^2 \, d\varphi = 1, \tag{4.2a}$$

$$\int_0^{2\pi} \left| \tilde{F}(\varphi) \right|^2 d\varphi = 1. \tag{4.2b}$$

In this case the largest value of Δ equals two, being the maximal distance between two points on the unit sphere. Practically interesting are the small values of Δ, approximately $\Delta \lesssim 0.5$.

Denote the current creating $\tilde{F}(\varphi)$, by $\tilde{j}(s)$. According to (1.11), its norm is

$$N = \left[\int_C \left| \tilde{j}(s) \right|^2 ds \right]^{1/2}. \tag{4.3}$$

It is necessary to find the minimal norm N among the norms of all currents on C, creating the patterns, for which

$$\Delta \le \delta. \tag{4.4}$$

Denote this norm by $N(\delta)$. It is the functional of $F(\varphi)$.

Instead of the optimal current synthesis of the patterns, one could formulate, in a similar way, the problem of the near-field synthesis. In this problem a closed contour Σ, encircling the line C, as well as a desired field distribution on Σ, are given. One should find the current on C with the minimal norm, which creates the field on Σ, removed from the given field not farther, than at δ. A technique will be given in Subsection 4.1.7 of the next section, based on the usage of a set of the basis functions, which permits to solve this problem for the general case. We will not write the corresponding solution: it is found by the same scheme, as for the patterns, and leads to similar results. Note only that, at the same given accuracy δ, the norm of the current required for creation of the near field, is, in general, larger than that for creation of the pattern. This fact is similar to the result, obtained in Section 3.1 for a more simpler problem on approximation. We have shown there, that the situation can occur, in which the approximation of the near field on the given line is impossible, but approximation of any pattern is possible. The problem of the optimal current synthesis can be considered as a generalization of the approximation problem.

2. First, state the qualitative behavior of the function $N(\delta)$. It is, of course, decreasing (non-increasing), because the larger the acceptable difference of the function $\tilde{F}(\varphi)$ from the given $F(\varphi)$, the wider the class of $\tilde{F}(\varphi)$, and, in general, the smaller the minimal value of $N(\delta)$ over this class.

Find the pattern normalized by (4.2b), which is created by the current on C having the minimal norm. Denote this pattern by $\tilde{F}_{\min}(\varphi)$, its norm by $N_{\min}$, and the distance between

$F(\varphi)$ and $\tilde{F}_{\min}(\varphi)$ by η:

$$\eta = \left[\int_0^{2\pi} \left| F(\varphi) - \tilde{F}_{\min}(\varphi) \right|^2 d\varphi \right]^{1/2} . \tag{4.5}$$

If there exist several patterns (normalized by (4.2b)), realizable by the currents on C with the same norm $N_{\min}$, then $\tilde{F}_{\min}(\varphi)$ is that one of them, which is the closest to $F(\varphi)$.

The form of dependence $N(\delta)$ is defined by the answer to question, whether the given function $F(\varphi)$ is approximable (or, in particular, realizable) by the currents located on C, or not. First consider the case, when $F(\varphi)$ is realizable. Then $N(\delta)$ is described by one of the curves 1-3 in Fig. 4.1(a),(b).

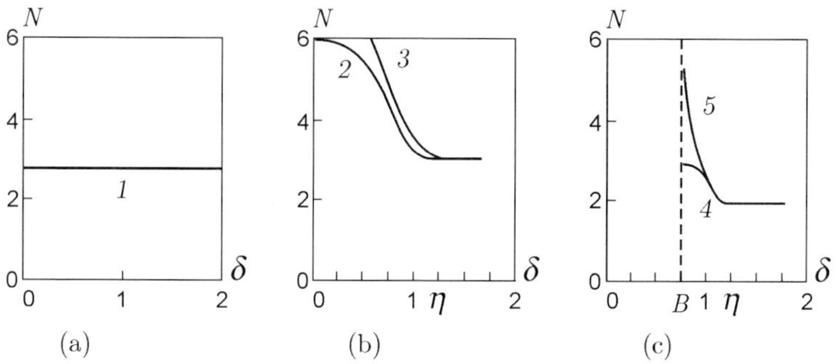

Figure 4.1:

Let $\eta \leq \delta$, which means that the pattern class defined by inequality (4.4), in which we look for the pattern with the minimal current norm, contains $\tilde{F}_{\min}(\varphi)$. Then this function is the sought pattern. Therefore, $N(\delta)$ does not decrease as δ increases, and it equals $N_{\min}$. At $\eta > \delta$, the current norm, treated as a functional of $\tilde{F}(\varphi)$, is a convex one, and its minimal value is reached on some $\tilde{F}(\varphi)$, for which

$$\Delta = \delta. \tag{4.6}$$

If $\eta = 0$ (i.e. the given pattern $F(\varphi)$ is $\tilde{F}_{\min}(\varphi)$, corresponding to this contour), then $N(\delta)$ does not vary with increasing δ. In this case, the graph $N(\delta)$ is the straight line $N(\delta) = N_{\min}$ (Fig. 4.1(a)). If $\eta \neq 0$, then the curve, describing $N(\delta)$, decreases with increasing δ up to $\delta = \eta$, and then passes into the horizontal line $N(\delta) = N_{\min}$ (curves 2,3 in Fig. 4.1(b)).

Let $\eta > 0$ and the pattern $F(\varphi)$ be realizable, it means, $N(0) < \infty$. Then $N(\delta)$ decreases monotonically as δ increases from $N(0)$ to $N_{\min}$ and then remains constant (curve 2 in Fig. 4.1(b)). If $F(\varphi)$ is nonrealizable, but approximable (i.e. $N(0) = \infty$, but $N(\delta) < \infty$ at any $\delta > 0$), then $N(\delta)$ is described by the curve with the vertical asymptote $\delta = 0$ (curve 3 in Fig. 4.1(b)).

Let $F(\varphi)$ be nonapproximable. Denote the distance between $F(\varphi)$ and the set of realizable functions $\tilde{F}(\varphi)$ by B ($B > 0$). In contrast to $f(\varphi)$ in (1.30), functions $\tilde{F}(\varphi)$ are normalized, therefore $B \geq b$. Because $\tilde{F}_{\min}(\varphi)$ belongs to this set, $\eta \geq B$. There are no realizable functions at $\delta < B$, satisfying (4.4). Then two possibilities can occur. In the first one, the lower limit of the norm of currents creating the patterns $\tilde{F}(\varphi)$ which are not nearer to $F(\varphi)$ than B, has reached, that is $N(B) < \infty$ (curve 4 in Fig. 4.1(c)). If there are several patterns with $\delta = B$, then $N(B)$ is the minimal norm of the currents corresponding to them.

The other case takes place when the pattern is nonapproximable and all the patterns with $\Delta = B$ are nonrealizable. Then the curve $N(\delta)$ has the vertical asymptote $\delta = B$, that is, $N(B) = \infty$, but $N(B+\varepsilon) < \infty$ at any $\varepsilon > 0$ (curve 5 in Fig. 4.1(c)).

3. We give some elementary examples for all five cases shown in Fig. 4.1, when the line C is a circle. Introduce the following notations. Let $\psi_m(\varphi)$, $m = 0, 1, ...$ be the orthonormal set of patterns:

$$\psi_0(\varphi) = \frac{1}{\sqrt{2\pi}}, \quad \psi_m(\varphi) = \frac{1}{\sqrt{\pi}}\cos(m\varphi), \quad m = 1, 2, ... \tag{4.7}$$

As above, we confine ourselves to the even functions of φ. Denote the orthonormal set of currents on C by $j_n(\theta)$, $n = 0, 1, ...$ For (and only for) the circle, this set coincides with (4.7):

$$j_0(\theta) = \frac{1}{\sqrt{2\pi}}, \quad j_n(\theta) = \frac{1}{\sqrt{\pi}}\cos(n\theta), \quad n = 1, 2, ... \tag{4.8}$$

It is obvious that

$$\int_0^{2\pi} \psi_m(\varphi)\psi_q(\varphi)d\varphi = \delta_{mq}, \quad \int_0^{2\pi} j_n(\theta)j_p(\theta)d\theta = \delta_{np}. \tag{4.9}$$

Expand the given function $F(\varphi)$ and the realizable one $\tilde{F}(\varphi)$ into the Fourier series by $\psi_n(\varphi)$:

$$F(\varphi) = \sum_{m=0}^{\infty} A_m \psi_m(\varphi), \tag{4.10a}$$

$$\tilde{F}(\varphi) = \sum_{m=0}^{\infty} \tilde{A}_m \psi_m(\varphi); \tag{4.10b}$$

in comparison with (1.20), we have made some changes in the notation here.

For simplicity, in this subsection we put $F(\varphi)$ as a real function, so that A_m are real numbers. Then the Fourier coefficients $\tilde{A}_m$ of the function $\tilde{F}(\varphi)$ minimizing N, are real (see (4.77) below) and $\tilde{F}(\varphi)$ can be considered as a real function (for the circle). The normalizing conditions (4.2) give

$$\sum_{m=0}^{\infty} A_m^2 = 1, \tag{4.11a}$$

$$\sum_{m=0}^{\infty} \tilde{A}_m^2 = 1. \tag{4.11b}$$

Expand the current $\tilde{j}(\theta)$ creating the pattern $\tilde{F}(\varphi)$ into the series by the functions (4.8):

$$\tilde{j}(\theta) = \sum_{n=0}^{\infty} \tilde{B}_n j_n(\theta). \tag{4.12}$$

The coefficients $\tilde{B}_n$ are easily expressed by $\tilde{A}_n$. This can be made using the general formula (1.18) (where the kernel $\mathcal{K}(\varphi, \theta)$ for a circle has the form $\exp[ika\cos(\theta - \varphi)]$) and calculating the integrals obtained, by the formula (2.15). With accuracy to a nonessential factor, we obtain the known formula

$$\tilde{B}_n = \frac{\tilde{A}_n}{J_n(ka)}, \tag{4.13}$$

where a is the radius of circle C. The squared current norm is

$$N^2 = \sum_{n=0}^{\infty} \frac{\tilde{A}_n^2}{J_n^2(ka)}. \tag{4.14}$$

Here we have also omitted a factor (this fact will not be mentioned during discussion of the numerical examples).

Begin by determining the function $\tilde{F}_{\min}(\varphi)$. To this end, it is necessary to find the maximal denominator in (4.14). For the first three examples (corresponding to Fig.4.1(a),(b)) we take the nonresonant circle at $ka = 5.0$. For this ka the maximal of $|J_n(ka)|$, $n = 0, 1, ...$ is $J_4(5.0) = 0.39$. It means that the current with minimal norm, creating a pattern $\tilde{F}(\varphi)$ normalized by (4.2b), has only one Fourier component with $n = 4$, so that $\tilde{F}(\varphi) = \psi_4(\varphi)$, $N_{\min} = 2.6$. No current on this circle, creating a normalized pattern, has a norm less than 2.6. According to (4.5), (4.11), the distance between any function $F(\varphi)$ with the Fourier coefficients A_n and the function $\tilde{F}_{\min}(\varphi)$ equals $\eta = \sqrt{2 - 2A_4}$.

Curve 1 in Fig.4.1(a) corresponds to the case when the given function $F(\varphi) = \psi_4(\varphi)$; then, at all δ, $\tilde{F}(\varphi) = \tilde{F}_{\min}(\varphi)$ and $N(\delta) = N_{\min}$.

Let $F(\varphi)$ be $\psi_0(\varphi)$. Then $\eta = \sqrt{2}$ (according to (4.5), it always equals $\sqrt{2}$ when $F(\varphi)$ is orthogonal to $\tilde{F}_{\min}(\varphi)$ corresponding to the given line C), and $N(0) = 1/|J_0(ka)|$; for the chosen value of ka, we have $N(0) = 5.6$.

At any $\delta \leq \eta$ the function $\tilde{F}(\varphi)$ corresponding to the minimal N is a linear combination of the functions $F(\varphi)$ and $\tilde{F}_{\min}(\varphi)$, that is, in our example ($A_0 = 1$) it has the form $\tilde{F}(\varphi) = \tilde{A}_0\psi_0(\varphi) + \tilde{A}_4\psi_4(\varphi)$. Indeed, let $\tilde{F}(\varphi)$ contain one more summand $\tilde{A}_m\psi_m(\varphi)$, $m \neq 0, 4$. It follows from (4.1) that

$$(\tilde{A}_0 - 1)^2 + \tilde{A}_4^2 + \tilde{A}_m^2 = \delta, \tag{4.15}$$

and from (4.11b) that

$$\tilde{A}_0^2 + \tilde{A}_4^2 + \tilde{A}_m^2 = 1. \tag{4.16}$$

Then $\tilde{A}_0^2 = 1 - \delta/2$, $\tilde{A}_4^2 + \tilde{A}_m^2 = \delta - \delta^2/4$. These two values do not depend on $\tilde{A}_m^2$, they depend only on δ. The larger $\tilde{A}_m$, the smaller $\tilde{A}_4$. In the expression for N, consisting of three summands, only the sum of the last two $\tilde{A}_4^2/J_4^2(ka) + \tilde{A}_m^2/J_m^2(ka)$ depends on $\tilde{A}_m$. Since $|J_4(ka)| > |J_m(ka)|$, the value of the above sum is minimal when $\tilde{A}_m = 0$.

It is easy to find an explicit form of the dependence $N(\delta)$. The equation of curve 2 in Fig.4.1(b) is

$$N(\delta) = \left\{ N^2(\eta) + \left(\frac{\eta^2 - \delta^2}{\eta^2} \right)^2 [N^2(0) - N^2(\eta)] \right\}^{\frac{1}{2}} \tag{4.17}$$

(this formula is a particular case of formula (4.19) below). At $\delta > \eta$ we have $N(\delta) = N(\eta)$.

Curve 3 in Fig.4.1(b) is obtained if $F(\varphi)$ differs, for instance, from $\psi_0(\varphi)$ by a summand being a nonrealizable but approximable function. Recall that nonrealizability of a function means that its Fourier coefficients A_m decrease so slowly that $\lim_{m\to\infty}(m|A_m|^{1/m}) = \infty$. Of course, in this case the series (4.10a) should converge. If the distance between $F(\varphi)$ and the realizable function ($\psi_0(\varphi)$ in our case) is small, it means the norm of nonrealizable summand equals $\varepsilon << 1$, then even at $\delta >> \varepsilon$, curve 3 in Fig.4.1(b) is near to curve 2. The corresponding value η differs from $\sqrt{2}$ by the value of the order ε, too.

The curves in Fig. 4.1(c) show the dependence $N(\delta)$ in the case when C is a resonant circle (or, in general, a specific line). Let, for instance, $ka = 2.405$ (the first root of $J_0(ka)$). The corresponding orthogonal complement function is $\psi_0(\varphi)$. If the Fourier series of $F(\varphi)$ has the coefficient $A_0 \neq 0$, then the pattern $F(\varphi)$ is nonapproximable. Among the numbers $|J_n(2.405)|$, $n = 0, 1, ...$ the maximal is $|J_1(2.405)| = 0.52$. Therefore, $\tilde{F}_{\text{min}}(\varphi) \equiv \psi_1(\varphi)$, and, according to (4.14), $N_{\text{min}} = 1.9$.

For instance, curve 4 in Fig. 4.1(c) corresponds to $F(\varphi) = A_0\psi_0(\varphi) + A_2\psi_2(\varphi)$ at $A_0 = A_2 = 1/\sqrt{2}$. The realizable normalized function $\tilde{F}(\varphi)$, closest to $F(\varphi)$, equals $\psi_2(\varphi)$. The distance between this function and $F(\varphi)$ is expressed by A_0 using (4.11a):

$$B = \sqrt{2 - 2\sqrt{1 - A_0^2}}. \tag{4.18}$$

The value B describes the *width of the "slot"*: the distance between the vertical asymptote and y-axis in Fig. 4.1(c). The chosen A_0 gives $B = 0.77$, and the norm of the current creating the pattern $\psi_2(\varphi)$ is $N(B) = 1/|J_2(ka)| = 2.3$, at $ka = 2.405$. If we do not require the norms of $\tilde{F}(\varphi)$ and $F(\varphi)$ to be equal, then the realizable pattern, nearest to $F(\varphi)$, is $A_2\psi_2(\varphi)$, and the distance between these two patterns is $A_0 = 0.71$, that is, a little smaller than by (4.18).

As in the example with the nonresonant circle, the distance between $F(\varphi)$ and $\tilde{F}_{\text{min}}(\varphi)$ is $\eta = \sqrt{2}$. Curves 2 and 4 in Figs 4.1(b),(c) coincide with an accuracy to the obvious shift B. More exactly, at $B \leq \delta \leq \eta$, the equation of curve 4 is

$$N(\delta) = \left\{ N^2(\eta) + \left(\frac{\eta^2 - \delta^2}{\eta^2 - B^2} \right)^2 [N^2(B) - N^2(\eta)] \right\}^{\frac{1}{2}}. \tag{4.19}$$

At $B = 0$ this equation coincides with (4.17).

To derive (4.19), one should at first prove that the function $\tilde{F}(\varphi)$ providing the minimum of N, has the form

$$\tilde{F}(\varphi) = \tilde{A}_1\psi_1(\varphi) + \tilde{A}_2\psi_2(\varphi). \tag{4.20}$$

The first summand in (4.20) is proportional to $\tilde{F}_{\text{min}}(\varphi)$, the second one differs from $\tilde{F}(\varphi)$ only in the absence of the summand with $\psi_0(\varphi)$, proportional to $\hat{F}(\varphi)$. In the general case, this second summand is proportional to the difference

$$F(\varphi) - \hat{F}(\varphi) \int_0^{2\pi} F(\varphi)\hat{F}^*(\varphi)d\varphi. \tag{4.21}$$

This function is orthogonal to $\hat{F}(\varphi)$. The proof of (4.20) is constructed in the same way as at $B = 0$. Namely, one should add to (4.20) the summand $\tilde{A}_m\psi_m(\varphi)$, $m \neq 1, 2$, and show that the norm of (4.14) is minimal (at fixed δ) if $A_m = 0$.

Thus $\tilde{F}(\varphi)$ has the form (4.20), so that

$$N^2(\delta) = \tilde{A}_1^2 N^2(\eta) + \tilde{A}_2^2 N^2(B). \tag{4.22}$$

The coefficients $\tilde{A}_1$ and $\tilde{A}_2$ are expressed by A_2, since $\delta^2 = 2 - 2A_2\tilde{A}_2$ and $\tilde{A}_1^2 + \tilde{A}_2^2 = 1$. In its turn, A_2 can be expressed by B using the condition $A_0^2 + A_2^2 = 1$ and the formula (4.18). After elementary (but cumbersome) derivations we have (4.19).

Curve 4 in Fig. 4.1(c) has a slightly different form if $F(\varphi)$ contains, except for $\hat{F}(\varphi)$, the term proportional to $\hat{F}_{\min}(\varphi)$, for instance, if $F(\varphi) = A_0\psi_0(\varphi) + A_1\psi_1(\varphi)$. Then any realizable function does not exist at $\delta < B$, and $\hat{F}(\varphi)$ minimizing N, coincides with $\psi_1(\varphi)$ at $\delta \geq B$. In this case, the dependence $N(\delta)$ is described by the same straight line as line 1 in Fig. 4.1(a), but beginning not at $\delta = 0$, but after the "slot" , that is, at $\delta = B$. This case is described by the same formula (4.19) at $N(B) = N(\eta)$.

Finally, curve 5 in Fig. 4.1(a) is observed if $F(\varphi)$ contains, besides $A_0\psi_0(\varphi)$, a summand, the Fourier series of which converges slowly and is a nonrealizable function. If the norm ε of this summand is small, then, at $\delta - B >> \varepsilon$, its influence is small and curve 5 tends to curve 4.

4. As has been mentioned, some *other*, different from (4.3), *norms of current* (i.e. other functionals of the realizable pattern $\tilde{F}(\varphi)$), are of practical interest. Minimal values of these norms are the functionals of given patterns $F(\varphi)$ and depend on δ. Besides (4.3), the most interesting characteristics are

$$N_1(\delta) = \left\{ \int_C \left[|\tilde{j}(s)|^2 + \left| \frac{d\tilde{j}(s)}{ds} \right|^2 \right] ds \right\}^{1/2}, \tag{4.23a}$$

$$N_2(\delta) = \left\{ \int_C \left| \frac{d\tilde{j}(s)}{ds} \right|^2 ds \right\}^{1/2}. \tag{4.23b}$$

As above, the minimum should be found among all $\tilde{F}(\varphi)$ satisfying the condition (4.4). Of course, the value (4.23b) is not the norm in the mathematical sense, because it can equal zero at $\tilde{j}(s) \neq 0$ identically. The norm N characterizes the value of current on C, N_2 describes the rate of its change, N_1 unites these two criteria. In technical problems the summands in N_1 can have different weight factors describing the relative importance of both possible deficiencies of antenna: the large current and the fast varying one.

It is clear, that the minimal $N_1^2(\delta)$ is not equal to the sum of the minimal $N^2(\delta)$ and $N_2^2(\delta)$, because their minima are reached on different functions $\tilde{F}(\varphi)$. At $\tilde{F}(\varphi)$, providing the minimum to N_1, both summands in (4.23a) are not smaller than their minimal values, so that

$$N_{1\min}^2(\delta) \geq N_{\min}^2(\delta) + N_{2\min}^2(\delta). \tag{4.24}$$

The qualitative behavior of dependencies $N_1(\delta)$ and $N_2(\delta)$ is almost the same as that of $N(\delta)$, with the obvious difference that for $N_2(\delta)$ we have $\tilde{F}_{\min}(\varphi) = const$, and the minimal value of $N_2(\delta)$ equals zero for all the lines C.

5. At the end of the section we give a brief analysis of another possible statement of the optimal current synthesis problem. In Section 2.3 we have considered the additional features arising in the approximation problem in the case when not the whole (complex) pattern $F(\varphi)$, but only its amplitude $\Phi(\varphi)$, is given. In this case the possibilities are wider than those for approximation of the whole pattern. For instance, many amplitude patterns are approximable even by the specific lines.

Similarly to the above, one can also formulate the current synthesis problem, giving only the function $\Phi(\varphi)$, which is the amplitude of the pattern $F(\varphi)$ to be approximated with the minimal current norm. In such a statement the problem consists in finding a complex realizable pattern $\tilde{F}(\varphi) = \tilde{\Phi}(\varphi)\exp[-i\tilde{\Psi}(\varphi)]$ and the phase $\Psi(\varphi)$ supplying the given amplitude pattern to $F(\varphi)$ of the same form, such that the distance Δ between $F(\varphi)$ and $\tilde{F}(\varphi)$ (see (3.33)) is equal or smaller than the given δ (condition (4.4)), and the norm of current realizing $\tilde{F}(\varphi)$ is minimal among the norms of all the currents satisfying (4.4). The distinction of this statement from that used above in this section is that, in addition to the current, the phase $\Psi(\varphi)$ is also varied. Of course, the norm of such a current is not larger than that in the problem where the whole pattern $F(\varphi)$ is given.

The direct solution of this problem is more complicated than that formulated at the beginning of the section. Calculation of the amplitude and phase of the complex function is a nonlinear operation, therefore the standard variational technique (used for the general case in Section 4.3 below) is not applicable in this case. However, a simple iterative procedure can be applied to solve this problem.

Let a function $\Phi(\varphi)$ be given. In the first step $\Phi(\varphi)$ is supplied with an arbitrary phase $\Psi_0(\varphi)$ and the optimal current synthesis problem is solved for the function $F_0(\varphi) = \Phi_0(\varphi)\exp[-i\Psi_0(\varphi)]$, that is, the pattern $\tilde{F}_0(\varphi) = \tilde{\Phi}(\varphi)\exp[-i\tilde{\Psi}_0(\varphi)]$ is found, such that the distance between $\tilde{F}_0(\varphi)$ and $F_0(\varphi)$ is Δ, and the norm of the current creating $\tilde{F}_0(\varphi)$, is minimal among the norms of all the currents creating patterns at this distance from $F_0(\varphi)$. In the second step the value Δ is minimized by choosing the phase $\Psi_1(\varphi)$ of the function $F_1(\varphi) = \Phi(\varphi)\exp[-i\Psi_1(\varphi)]$ at fixed function $\tilde{F}_0(\varphi)$. Because the squared distance between $F_1(\varphi)$ and $\tilde{F}_0(\varphi)$ is

$$\Delta^2 = 2 - 2\int_0^{2\pi} \Phi(\varphi)\tilde{\Phi}_0(\varphi)\cos[\tilde{\Psi}_0(\varphi) - \Psi_1(\varphi)]d\varphi, \tag{4.25}$$

and $\Phi(\varphi)\tilde{\Phi}_0(\varphi) \geq 0$, then Δ is minimal at $\cos[\tilde{\Psi}_0(\varphi) - \Psi_1(\varphi)] \equiv 1$, it means, $\Psi_1(\varphi) = \tilde{\Psi}_0(\varphi)$. In this step the function $\tilde{F}_0(\varphi)$ is not changed, therefore the current norm keeps its previous value. At the end of the step we have $\Delta^2 = 2 - 2\int_0^{2\pi}\Phi(\varphi)\tilde{\Phi}_0(\varphi)d\varphi$.

The two above steps are repeated in turns: at first the realized pattern $\tilde{F}_1(\varphi) = \tilde{\Phi}(\varphi)\exp[-i\tilde{\Psi}_1(\varphi)]$ is found to minimize the current norm by choosing $\tilde{\Psi}_1(\varphi)$ at fixed $\Psi_1(\varphi)$, then $\Psi_2(\varphi)$ is chosen as $\Psi_2(\varphi) = \tilde{\Psi}_1(\varphi)$ to minimize the distance Δ, and so on.

In each step either the current norm N is minimized at a fixed distance Δ, or Δ is minimized at fixed N. As these two values are limited from below, the iterative procedure converges with respect to them. At the limit we obtain the curve $N(\delta)$, lying on the left of the curves in Fig. 4.1, corresponding to the pattern with amplitude $\Phi(\varphi)$ and any phase, nonoptimal in the above sense.

4.2 Generalized Functions of Double Orthogonality. Nonapproximability and Existence of Nonradiating Currents

1. In the problem formulated and solved in the previous section, the most important was to calculate the current $\tilde{f}(s)$ on C, generating the given pattern $\tilde{F}(\varphi)$. Elementary examples illustrating the idea of the optimal current synthesis, were related to the circle. Two sets of functions (4.7), (4.8), complete and orthonormal on C and $(0, 2\pi)$, respectively, were used. The functions possessed the following property: if the current distribution is described by the function $j_n(s)$ of the set (4.7), then the pattern generated by it is described by the function $\psi_m(\varphi)$ of the set (4.8). Therefore, the expansion coefficients of the current with respect to $j_n(s)$ (see (4.12)) and the pattern with respect to $\psi_m(\varphi)$ were uniquely connected with each other by a simple formula (4.13). As it will be shown in Section 4.3, the existence of such a simple relation for the arbitrary line C allows to solve the synthesis problem in the general case. In this section the corresponding sets of the currents $j_n(s)$ and patterns $\psi_m(\varphi)$ will be constructed for the general case, which possess the same properties as the trigonometrical functions (4.7), (4.8) for the case of the circle.

The usage of such sets of functions is one of the ways to solve the integral equation of the first kind (see (1.18))

$$\int_C \mathcal{K}(s, \varphi) j(s) ds = F(\varphi) \tag{4.26}$$

with respect to $j(s)$ at given $F(\varphi)$. Under the conditions formulated in Section 1.3, at which (4.26) has the solution $j(s)$ with a finite norm (i.e. when the pattern $F(\varphi)$ is realizable by the contour C), the integral equation can be solved by expanding $F(\varphi)$ and $j(s)$ with respect to functions $\psi_m(\varphi)$ and $j_n(s)$, respectively; the solution is elementary. This approach has been used in the antenna theory for a long time. It is not that effective for solving equation (4.26), because the procedure for determining the functions $j_n(s)$, $\psi_m(\varphi)$ is complicated in the computing sense. However, its application in general constructions is expedient.

The functions $j_n(s)$, $\psi_m(\varphi)$ are determined only by the line C and kernel $\mathcal{K}$; they do not depend on the right-hand side of (4.26). We call these functions the *generalized functions of double orthogonality*, because the simpler term "functions of double orthogonality" is fixed for the set of functions which is a special case of the sets considered here, namely when C is a straight-line segment. Of course, these functions are not the *generalized functions* in the usual sense.

The kernel $\mathcal{K}(s, \varphi)$ of equation (4.26) is

$$\mathcal{K}(s, \varphi) = D e^{ikr(\theta)\cos(\theta - \varphi)}, \quad D = -\frac{\sqrt{-2i}}{4\sqrt{\pi}}. \tag{4.27}$$

Here s is the dimensionless length of the arc, $\theta = \theta(s)$ is the polar angle, the equation of the line C is assumed to have the form $r = r(\theta)$. For simplicity we put $r(\theta)$ to be a single-valued function. If C is nonclosed, then $r(\theta)$ is defined only on a part of the interval $(0, 2\pi)$.

2. In this section we use some of the simplest terms of functional analysis and its elementary facts. The physical interpretation of the intermediate and final results is given, wherever possible.

There exists an operator K: the integral operator (4.26) with the kernel (4.27) mapping any function $j(s)$ defined on C, for which the integral (4.26) exists, into a function $F(\varphi)$ defined on $(0, 2\pi)$:

$$Kj = F. \tag{4.28}$$

In general, the inverse operator mapping F into j does not exist: there is no operator mapping any function $F(\varphi)$ into a function $j(s)$ on C. If it existed, then all the patterns would be realizable, that is, the operator equation would have a solution at any $F(\varphi)$.

Introduce the inner products in the spaces of functions $j(s)$ (currents):

$$(j_1, j_2) = \int_C j_1(s) j_2^*(s) ds \tag{4.29}$$

(see (1.27)) and $F(\varphi)$ (patterns):

$$(F_1, F_2) = \int_0^{2\pi} F_1(\varphi) F_2^*(\varphi) d\varphi. \tag{4.30}$$

The values (j, j) and (F, F) are the norms of $j(s)$ (see (4.3)) and $F(\varphi)$, respectively; they are nonnegative numbers. We will only deal with functions $j(s)$, $F(\varphi)$, having finite norms.

Define the operator K^{ad} adjoint to K by the equality

$$(K^{\mathrm{ad}} F, j) = (F, Kj), \tag{4.31}$$

which is valid for any functions $j(s)$, $F(\varphi)$. It maps any function $F(\varphi)$ defined on $(0, 2\pi)$ into a function $j(s)$ defined on C. This operator is the integral one with the kernel $\mathcal{K}^*(s, \varphi)$:

$$K^{\mathrm{ad}} F = D^* \int_0^{2\pi} e^{-ikr(\theta)\cos(\theta-\varphi)} F(\varphi) d\varphi. \tag{4.32}$$

Indeed, for this operator the left and right-hand sides of (4.31) have the forms

$$(K^{\mathrm{ad}} F, j) = D^* \int_C \left[\int_0^{2\pi} e^{-ikr(\theta)\cos(\theta-\varphi)} F(\varphi) d\varphi \right] j^*(s) ds \tag{4.33}$$

and

$$(F, Kj) = D^* \int_0^{2\pi} \left[\int_C e^{-ikr(\theta)\cos(\theta-\varphi)} j^*(s) ds \right] F(\varphi) d\varphi, \tag{4.34}$$

respectively. Both these integrals are equal to each other, which proves (4.32).

Construct the sets of functions $j(s)$ and $\psi(\varphi)$, following [10]. Introduce two iterated operators $K^{\mathrm{ad}} K$ and $K K^{\mathrm{ad}}$. The first of them acts on the functions defined on C and also maps them into the functions defined on C. Similarly, the second one acts on the functions defined on $(0, 2\pi)$ and maps them into the functions defined on $(0, 2\pi)$. Both operators are self-adjoint, that is, for any two pairs of functions $j_1(s)$, $j_2(s)$ and $F_1(\varphi)$, $F_2(\varphi)$, respectively, the equalities

$$(K^{\mathrm{ad}}Kj_1, j_2) = (j_1, K^{\mathrm{ad}}Kj_2), \tag{4.35a}$$
$$(KK^{\mathrm{ad}}F_1, F_2) = (F_1, KK^{\mathrm{ad}}F_2) \tag{4.35b}$$

are valid. These equalities are easily proved by applying the definition (4.31) twice.

Both $K^{\mathrm{ad}}K$ and $K\ K^{\mathrm{ad}}$ are the integral operators. The kernel of the first one can be written as the integral

$$|D|^2 \int_0^{2\pi} e^{ik\left[r(\theta)\cos(\theta-\varphi)-r(\theta')\cos(\theta'-\varphi)\right]} d\varphi. \tag{4.36}$$

The integral is tabulated, it is expressed by the Bessel function of the argument $kR(\theta, \theta')$,where $R(\theta, \theta')$ is the distance between the points on C corresponding to the polar angles θ and θ':

$$R^2(\theta, \theta') = r^2(\theta) + r^2(\theta') - 2r(\theta)r(\theta')\cos(\theta - \theta'). \tag{4.37}$$

Thus,

$$K^{\mathrm{ad}}Kj = 2\pi|C|^2 \int_C J_0(kR(\theta, \theta'))j(\theta)ds. \tag{4.38}$$

The kernel of the second operator $K\ K^{\mathrm{ad}}$, expressed as

$$|D|^2 \int_C e^{2ikr(\theta)\sin\left((\varphi-\varphi')/2\right)\sin\left(\theta-(\varphi'+\varphi)/2\right)} ds, \tag{4.39}$$

cannot be reduced to such a simple form. As distinct from the kernel in (4.38), the last one is, in general, not real (see the note at the end of the subsection). If C is a circle, then the kernel (4.39) is real and proportional to $J_0(2ka\sin((\varphi - \varphi')/2))$.

Choose the functions $j_n(s)$ as *eigenfunctions of the operator* $K^{\mathrm{ad}}K$ and $\psi_m(\varphi)$ as *eigenfunctions of the operator* $K\ K^{\mathrm{ad}}$. Consider two homogeneous equations

$$K^{\mathrm{ad}}Kj_n = \lambda_n j_n, \tag{4.40a}$$
$$KK^{\mathrm{ad}}\psi_m = \mu_m \psi_m \tag{4.40b}$$

with eigenvalues λ_n, $(n = 0, 1, ...)$ and μ_m, $(m = 0, 1, ...)$. Since operators of the equations are self-adjoint, the eigenvalues are real and $\lim_{n\to\infty}\lambda_n = \lim_{m\to\infty}\mu_m = 0$. The eigenfunctions j_n and ψ_m make up the complete orthogonal sets. If there exist multiple eigenvalues, that is, several linearly independent eigenfunctions corresponding to one eigenvalue , then these functions can be orthogonalized by a linear transformation; we do not mention this fact again. Normalize both sets by one:

$$(j_n, j_p) = \delta_{np}, \tag{4.41a}$$
$$(\psi_m, \psi_q) = \delta_{mq}. \tag{4.41b}$$

This way we also simultaneously normalize functions Kj_n and $K^{\text{ad}}\psi_m$. Indeed, multiplying equations (4.40) scalarwise by j_n, ψ_m, respectively, and using (4.31), (4.41), we obtain

$$(Kj_n, Kj_n) = \lambda_n, \tag{4.42a}$$
$$(K^{\text{ad}}\psi_m, K^{\text{ad}}\psi_m) = \mu_m. \tag{4.42b}$$

In particular, this leads to the eigenvalues λ_n, μ_m being nonnegative.

Owing to the symmetry in the structure of operators in (4.40), there is a connection between the sets of functions j_n and ψ_m. Applying the operator K to both sides of equation (4.40a) gives

$$KK^{\text{ad}}(Kj_n) = \lambda_n(Kj_n). \tag{4.43}$$

Compare (4.43) as an equation with respect to Kj, with (4.40b). If the function Kj_n is not equal to zero identically, then it is proportional to one of the eigenfunctions ψ_m of operator $K\ K^{\text{ad}}$, namely, with that corresponding to the eigenvalue $\mu_m = \lambda_n$.

In the same way, applying the operator K^{ad} to (4.40b), we obtain

$$K^{\text{ad}}K(K^{\text{ad}}\psi_m) = \mu_m(K^{\text{ad}}\psi_m). \tag{4.44}$$

If the function $K^{\text{ad}}\psi_m$ does not equal zero identically, then it coincides with one of the eigenfunctions j_n of the operator $K^{\text{ad}}K$, namely, with that corresponding to the eigenvalue $\lambda_n = \mu_m$.

Denote the number m for which $\mu_m = \lambda_n$ by $m(n)$, and the inverse function by $n(m)$, so that, by definition,

$$\lambda_{n(m)} = \mu_m, \quad \mu_{m(n)} = \lambda_n. \tag{4.45}$$

Obviously, $n(m)$ takes on all the values corresponding to all λ_n when m possesses all values corresponding to all μ_m, and vice versa.

Determining the proportionality factors between $Kj_{n(m)}$ and ψ_m from (4.42a), (4.41b), and between $K^{\text{ad}}\psi_{m(n)}$ and j_n from (4.42b), (4.41a), we obtain

$$Kj_{n(m)} = \sqrt{\mu_m}\psi_m, \tag{4.46a}$$
$$K^{\text{ad}}\psi_{m(n)} = \sqrt{\lambda_n}j_n, \tag{4.46b}$$

respectively.

In this way we have constructed two complete orthonormal sets of functions. If all the eigenvalues λ_n and μ_m do not equal zero, then the functions j_n, ψ_m are connected by (4.46). In our problem the operator K is the integral one corresponding to (4.26).

For the case when C is a circle, the functions j_n, ψ_m are trigonometrical (4.7), (4.8). The eigenvalues are easily calculated, for instance, from (4.42b) by substitution of the explicit form (4.32) of the kernel of the integral operator K^{ad}. Using the tabulated integral for the Bessel functions once more, we obtain

$$\mu_m = |D|^2 J_m^2(ka), \tag{4.47}$$

where a is the circle radius.

Since the kernel (4.38) of equation (4.40a) is real, the eigenfunctions $j_n(\theta)$ are real, too (with accuracy to the factor $\exp(i\alpha)$, $\alpha = const$). This results, in particular, in the functions $\psi_m(\varphi)$ fulfilling the same condition (2.18), as the orthogonal complement functions $\hat{F}(\varphi)$. This is easily derived from (4.46a) and the properties of the kernel of integral operator K: the above formula (4.46a) and the reality of $j_n(s)$ give $\sqrt{\mu_m}\psi_m^*(\varphi) = K^{\text{ad}} j_n(s)$; the kernel $\mathcal{K}$ (see 4.27) possesses the obvious property $\mathcal{K}^*(s,\varphi) = \mathcal{K}(s,\varphi+\pi)$; therefore, $\psi_m^*(\varphi) = \psi_m(\varphi+\pi)$. Of course, this is in accordance with (4.46b). We will use this property in Subsection 4.

If the line C has a symmetry center, that is, if $r(\theta+\pi) = r(\theta)$, then the functions $\psi_m(\varphi)$, depending only on C, must fulfil the condition $\psi_m(\varphi+\pi) = \psi_m(\varphi)$. Together with (2.18) this means that $\psi_m(\varphi)$ are real. Of course, this result can be obtained directly from the above formulas. For instance, it is seen from the kernel (4.39) of the integral equation for $\psi_m(\varphi)$, they are real if $r(\theta+\pi) = r(\theta)$; this results in $\psi_m(\varphi)$ being real. The circle is a line with the center symmetry, and the functions $\psi_m(\varphi)$ (4.7) are real for it.

3. In order to solve equation (4.28), expand the given function $F(\varphi)$ and the sought current into the series by functions $\psi_m(\varphi)$ and $j_n(s)$:

$$F(\varphi) = \sum_{m=0}^{\infty} A_m \psi_m(\varphi), \tag{4.48a}$$

$$j(s) = \sum_{m=0}^{\infty} B_m j_{n(m)}(s). \tag{4.48b}$$

Substituting these series into (4.28) and equating the coefficients at $\psi_m(\varphi)$ gives

$$B_m = \frac{A_m}{\sqrt{\mu_m}}. \tag{4.49}$$

According to (4.48b), (4.49), the squared norm of the current is

$$N^2 = \sum_{m=0}^{\infty} \frac{|A_m|^2}{\mu_m}. \tag{4.50}$$

The solution of the form (4.48b) with coefficients (4.49) is valid if the series (4.50) converges. At first, consider the lines C for which $\mu_m \neq 0$ at all m. Then for (4.50) to be convergent it is necessary and sufficient that the coefficients $A_m = (F, \psi_m)$ of expansion of the given pattern $F(\varphi)$ decrease with growing m so fast, in comparison with $\sqrt{\mu_m}$, that B_m by (4.49) tend to zero sufficiently fast.

We have stated the conditions of realizability of a given pattern by the contour C, formulated in terms of generalized functions of double orthogonality. The statement unites the realizability conditions of the pattern by currents localized in a finite domain and those located on the line C. The first condition (existence of the limit (1.20)) depends only on $F(\varphi)$, the second one is defined by mutual location of C and the circle of radius a_0 (see (1.20)). Despite its brevity, the realizability condition in the form of the demand that the series (4.50) should converge, is practically less convenient than that in the form given in Section 1.3; to test the first one it is necessary to know the asymptotic behavior of the eigenvalues of equation (4.40b) at large indices. Of course, the new form of the condition confirms the known result formulated in Section 1.3: because the difference between a nonrealizable (but approximable) pattern and realizable ones depends on the far coefficients of the expansion (4.48a), the transfer between such patterns can be carried out by an infinitely small perturbation of those coefficients.

The conditions for the line C to be specific, also follows from (4.50). This obviously takes place when, at some $m = q$,

$$\mu_q = 0. \tag{4.51}$$

In this case, such and only such patterns $F(\varphi)$ are approximable, for which $A_q \equiv (F, \psi_q) = 0$, that is, which are orthogonal to ψ_q. Thus, if the condition (4.51) is fulfilled, then there exists the orthogonal complement function $F(\varphi) = \psi_q$. If $A_q \neq 0$, then the nearest to $F(\varphi)$ (nonnormalized) approximable function $\tilde{F}(\varphi)$ is obtained from $F(\varphi)$ by removing the summand $A_q\psi_q$ from the series (4.48a). The distance between $F(\varphi)$ and $\tilde{F}(\varphi)$ equals $|A_q|$, that is $\left|\int_0^{2\pi} F(\varphi)\hat{F}^*(\varphi)\right|$; the same result was obtained in Section 4.1 (see the text after (4.18)).

4. Usage of the generalized functions of double orthogonality allows, in particular, to analyze qualitatively the passage to the limit when the line C becomes specific. Besides, these functions demonstrate in a natural way the connection between this property and the *existence of a nonradiating current* on C. This connection is more complicated than can be expected from the example given in Subsection 1.1.2. Moreover, it turns out that there is a high degree of symmetry between these two properties.

If the nonradiating current $\hat{\jmath}(s)$ is possible on C, then, by definition,

$$K\hat{\jmath} = 0. \tag{4.52}$$

Consequently, $\hat{\jmath}(s)$ is one of the eigenfunctions of the operator (4.40a), it corresponds to the eigenvalue

$$\lambda_p = 0. \tag{4.53}$$

The nonradiating current is possible on the line C if and only if (4.53) is fulfilled at some number $n = p$.

These two properties, namely, the nonapproximability (see (4.51)) and existence of nonradiating current (see (4.53)), can be possessed by the line *independently*. This can be illustrated in the three examples considered above. The resonant circle ($J_0(ka) = 0$) possesses both these properties: the pattern $\cos(q\varphi)$ is the orthogonal complement function $\hat{F}(\varphi)$, so that the circle is a specific line, and the current $\cos(q\theta)$ on it does not radiate. Coincidence of these two functions of φ and θ (i.e. s) is inherent in the circle and the simultaneous existence of the two above properties is not one-to-one connected with this coincidence. There are "continually many" contours possessing both these properties. The example of the line for which $\hat{F}(\varphi)$ exists but $\hat{\jmath}(s)$ does not, is an arc of resonant circle (the nonclosed specific line): a nonclosed current always radiates. The example of the line, on which nonradiating current $\hat{\jmath}(s)$ exists, but there is no orthogonal complement function in the pattern space associated with the line, is an arbitrary non-specific closed resonant contour, for example, the curvilinear quadrangle (see Fig. 3.5). It was proved that the field $\hat{u}(r, \varphi)$, without singularities in the whole plane, does not exist for this contour.

To explain how the independence of conditions (4.51) and (4.53) conforms with (4.45), arrange the functions $j_n(s)$ and $\psi_m(\varphi)$ in the order of their complication, for instance, by the increase in the rate of their varying with s and φ. One can arrange the functions $j_n(s)$ by the increase in the ratio of two norms N_1/N, where N and N_1 are defined in (4.3), (4.23), respectively. Under the condition (4.41a) this means that the functions $j_n(s)$ are arranged by increasing the norm N_1. Similarly, arrange $\psi_m(\varphi)$ in order of increasing the value

$$\int_0^{2\pi} \left[|\psi_m(\varphi)|^2 + |d\psi_m/d\varphi|^2 \right] d\varphi. \tag{4.54}$$

Then the functions $j_n(s)$, fast varying along the line C, correspond to large n, and the functions $\psi_m(\varphi)$, fast varying with φ, correspond to large m. The usual arrangement of eigenfunctions by decreasing the eigenvalues would leave the functions, corresponding to the zero eigenvalues, "without numbers", that is they should be placed last, after all the other functions. In our problem just such functions are of special interest.

Fig.4.2 shows the dependencies of μ_m and λ_n on the numbers m and n, respectively; to demonstrate it, the dependencies are shown as continuous curves, although only their values at the integer m and n make a sense. Curve 1 in Fig. 4.2(a) corresponds to the line C which is not specific and not close to such a line; curve 2 corresponds to the specific line, that is to the line for which the condition (4.51) is fulfilled. Curve 1 in Fig. 4.2(b) corresponds to the line C having no nonradiating currents, curve 2 corresponds to that having a current $j_p(s)$ which does not radiate, $j_p(s) = \hat{j}(s)$.

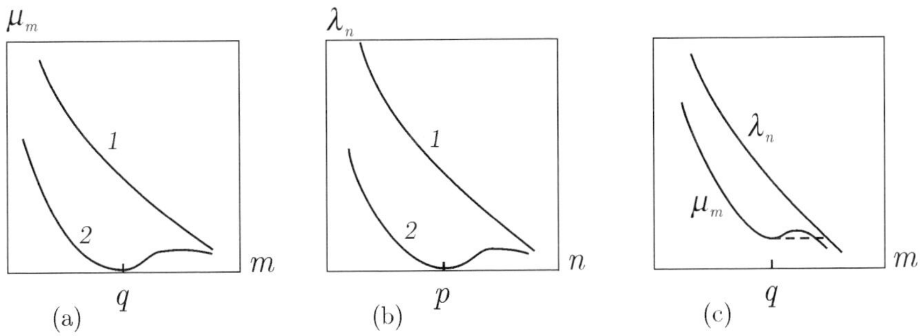

Figure 4.2:

Based on the conditions (4.45), consider the situation when the frequency k passes through the value at which C becomes the specific line, or there exists a nonradiating current on C. In the first case $k \to k_q$, such that $\mu_q(k_q) = 0$, in the second one $k \to k_p$, $\lambda_p(k_p) = 0$. For lines like a circle, the values p and q are equal to each other, these lines are resonant contours and specific lines simultaneously. As was mentioned, in this case the nonapproximability is an obvious consequence of the existence of a nonradiating current.

Assume that at $k \to k_q$ we have $\mu_q \to 0$, that is, the line C becomes specific, but (as in the second example) it is nonclosed, a nonradiating current does not exist on it, $\lambda_n \neq 0$ for all n. Dependencies of μ_m and λ_n on their indices are given in Fig. 4.2(c) simultaneously. The horizontal (dashed) line corresponds to the condition $\lambda_{n(q)} = \mu_q$ (see (4.45)); $n(q) \to \infty$

at $\mu_q \to 0$. In this case the nonapproximability can be treated also as a consequence of the existence of a nonradiating current, but this "current" is infinitely fast varying along C. Of course, such a "current" does not radiate, the fields generated by its neighboring segments compensate each other completely. Remember that the norm of current $j_n(s)$ does not depend on the frequency, it is equal to unity at all n.

A similar situation occurs if (as in the third example) the closed contour C becomes resonant at $k \to k_p$, that is, a nonradiating current $\hat{\jmath}(s) = j_p(s)$ can be created on it, but the contour does not become the specific line. Then at $\lambda_p \to 0$ we have $m(p) \to \infty$ (see curves 1 in Fig. 4.2(a) and 2 in Fig. 4.2(b). The "pattern" $\psi_{m(p)}(\varphi)$ corresponding to the current $j_p(s)$ becomes infinitely fast varying at the limit. As in the previous paragraph, one can conventionally say that the orthogonal complement function $\hat{F}(\varphi)$, corresponding to the nonradiating current, exists also for this (nonspecific) line, but this "function" is infinitely fast varying with the angle φ. Of course, any function $f(\varphi)$ is orthogonal to this "function" in the sense that the condition (1.23) is fulfilled for any $f(\varphi)$. Only if the infinitely fast varying "functions" $\hat{\jmath}(s)$ and $\hat{F}(\varphi)$ are considered as usual ones, can one state that nonapproximability is a consequence of the existence of a nonradiating current .

Previous considerations are not only of methodical interest. They suggest the following numerical way to find out if the given resonant contour C is a specific line, and if so, then they allow to find $\hat{F}(\varphi)$ approximately. First, we should find the eigenoscillation, eigenfrequency k_p and the current $\hat{\jmath}(s)$ (see (4.58) below). Then the eigenfunction $j_p(s)$ of equation (4.40a), close to $\hat{\jmath}(s)$, should be determined at $k \neq k_p$, $|k - k_p| << k_p$. This function corresponds to a small eigenvalue λ_p and the function Kj_p. If the limit of $Kj_p/\sqrt{\lambda_p}$ at $k \to k_p$ exists, then according to (4.46a) (rewritten as $Kj_p = \sqrt{\lambda_p}\psi_{m(n)}$), it equals $\psi_{m(n)}(\varphi)$, that is, the line becomes specific, and $\hat{F}(\varphi) = \psi_{m(p)}(\varphi)$. If the above limit does not exist, then, although the contour C becomes resonant at $k = k_p$, it does not become specific and all patterns can be approximated by the currents on C.

In fact, the above limit should not be calculated. If at $k \to k_p$ the calculation stops being stable and the function $\psi_{m(p)}(\varphi)$ is fast varying, then the limit does not exist and the contour is not specific. If the $\psi_{m(p)}(\varphi)$ is smooth, then one can stop at some frequency $k \neq k_p$ corresponding to small λ_p and consider the obtained $\psi_{m(p)}(\varphi)$ as $\hat{F}(\varphi)$ for this frequency. According to the note at the end of Subsection 2, the function $\psi_m(\varphi)$ fulfils the condition (2.18), necessary for this. The field $\hat{u}(r, \varphi)$ can be easily constructed by $\hat{F}(\varphi)$. The field has a zero line close to C. After it is found, the problem of determination of $\hat{F}(\varphi)$ for a contour close to the given one, and the resonant frequency corresponding to this contour, is solved.

We have shown that any line C is one of the four types; they can be characterized by the conditions (4.51), (4.53). If both conditions are not fulfilled, then neither the current $\hat{\jmath}$, nor the pattern $\hat{F}$ exist. If only (4.53) is fulfilled, then C is a nonspecific resonant contour. If only (4.51) is fulfilled, then the line is specific, but the current $\hat{\jmath}$ does not exist on it. Finally, if both the conditions are fulfilled, then C is a specific resonant contour. In any locality of the line of one of these types, there exist other lines of the same type.

5. If C is a resonant contour, then the solution to equation (4.28) is nonunique: adding the summand $B_p\hat{\jmath}(s)$ with any coefficient B_p to the solution $j(s)$ does not violate the equation. This nonuniqueness is analogous to that which happens spuriously in the diffraction problem on the metallic body with the same boundary C, while reducing it to the integral equation with respect to the current on the boundary. This integral equation has the same form (4.28) with the right-hand side, defined by the incident field. Although the kernel of this equation is different from (4.27), the corresponding integral operator maps (similarly to K) the nonradiating current into zero. Therefore, adding a term proportional to $\hat{\jmath}(s)$, to the solution, does not violate the integral equation. In the diffraction problem there is no nonuniqueness: when C is a resonant contour, then the integral equation is not equivalent to this problem. In the problem of the current creating the given exterior field, *nonuniqueness actually exists* for such contours.

Thus, two cases can take place while determining the current from equation (4.28). If the line C is specific, then one can approximate (in particular, realize) only the patterns, orthogonal to the function $\hat{F}(\varphi)$. If C is a resonant contour, then the solution is nonunique, it is determined with accuracy to the summand $\hat{\jmath}(s)$ with arbitrary coefficient B_p.

Consider briefly a problem arising in the receiving antenna theory. One should create the field $J(s)$ on the contour C, illuminating it by a convergent cylindrical wave. The pattern $F^*(\varphi)$ of this wave is to be found.

The function $F(\varphi)$ satisfies the equation

$$K^{\mathrm{ad}}F = J. \tag{4.55}$$

There are also two cases within this problem. If the contour C is resonant, then the field $J(s)$ can be approximated (and even realized) if the condition $(J, \hat{\jmath}) = 0$ is fulfilled, where $\hat{\jmath}(s)$ is a nonradiating current, this means, $K\hat{\jmath} \equiv 0$. We give the proof similar to that in Subsection 1.1.3. Consider the product $(J - K^{\mathrm{ad}}F, \hat{\jmath})$. It equals the difference $(J, \hat{\jmath}) - (K^{\mathrm{ad}}F, \hat{\jmath})$. According to (4.31), the second summand is $(F, K\hat{\jmath})$, that means it is zero for any pattern $F(\varphi)$. Therefore,

$$(J - K^{\mathrm{ad}}F, \hat{\jmath}) = (J, \hat{\jmath}). \tag{4.56}$$

Applying the Cauchy inequality (1.28) to (4.56) gives

$$(J - K^{\mathrm{ad}}F, J - K^{\mathrm{ad}}F)(\hat{\jmath}, \hat{\jmath}) \geq |(J, \hat{\jmath})|^2 . \tag{4.57}$$

The value $(J - K^{\mathrm{ad}}F, J - K^{\mathrm{ad}}F)$ is the squared distance between the given field J and a realized one $K^{\mathrm{ad}}F$. Normalize $\hat{\jmath}(s)$ by the condition $(\hat{\jmath}, \hat{\jmath}) = 1$. Then the above distance is larger than (or equal to) $|(J, \hat{\jmath})|$. Therefore, only such fields J on C can be approximated, which are orthogonal to the nonradiating current $\hat{\jmath}(s)$.

If the contour is specific, then the problem of determination of $F(\varphi)$ from equation (4.55) has many solutions. By definition, there exists the pattern $\hat{F}(\varphi)$ such that $(\hat{F}, Kj_n) = 0$, $n = 1, 2, \ldots$ According to (4.31), we have $(K^{\mathrm{ad}}\hat{F}, j_n) = 0$ at all n, which means, $K^{\mathrm{ad}}\hat{F} \equiv 0$. In terms of functional analysis this is a known result: if the set of functions $j_n(s)$, $n = 1, 2, \ldots$ is complete, but the set Kj_n is noncomplete, then the kernel of adjoint operator K^{ad} is not empty; namely, this operator transforms the orthogonal complement function $\hat{F}(\varphi)$ into zero. One can add the function $\hat{F}(\varphi)$ with any coefficient to the solution of equation (4.55); this does not change its right-hand side. If we have several functions $\hat{F}(\varphi)$, then any linear combination of them can be added to the solution.

The symmetrical position of the functions $\hat{F}(\varphi)$ and $\hat{\jmath}(s)$ can also be emphasized in the analysis method used above, which is based on investigation of the function $\hat{u}(r, \varphi)$. By definition, the line is specific if some field $\hat{u}(r, \varphi)$, satisfying the homogeneous Helmholtz equation and having no singularities in the whole plane, and not equal to zero identically (such a field has the asymptotics of the type (2.3) with nonzero $F(\varphi)$), is equal to zero on this line. On the other hand, the nonradiating current exists on the resonant contour, that is, on the line where the field is zero (it can also be denoted by $\hat{u}(r, \varphi)$), which satisfies the homogeneous Helmholtz equation, does not equal zero identically, and has no singularities inside C. The current is connected with $\hat{u}(r, \varphi)$ by the obvious formula

$$\hat{\jmath}(s) = \left.\frac{\partial \hat{u}}{\partial N}\right|_C, \tag{4.58}$$

which means, $\hat{\jmath}(s)$ is equal to the current on the walls of the resonator, coinciding with C, at its eigenfrequency. Such a current does not create any field outside C.

6. The generalized functions of double orthogonality can be connected with the above auxiliary field $\hat{u}(r, \varphi)$ created by the cylindrical wave incoming from infinity and having the angle dependence $\hat{F}^*(\varphi)$, and with the field $u(r, \varphi)$ creating a current on C. To this end, consider first the operator, obtained from the integral one K^{ad} (see (4.32)) after replacing the function $r(\theta)$ (which describes the line C by the equation $r = r(\theta)$) in its kernel with the free variable r. This expanded operator does not depend on the line C. Omitting the nonessential factor D^*, we have

$$K_{\text{ex}}^{\text{ad}} F = \int_0^{2\pi} e^{-ikr\cos(\theta-\varphi)} F(\varphi) d\varphi. \tag{4.59}$$

This operator maps the function $F(\varphi)$ into the function of point (r, φ). If this point lies on C, then (4.59) coincides with K^*.

The left-hand side of (4.59) is the function of r, φ. It has no singularities and satisfies the homogeneous Helmholtz equation (2.2) in the whole plane. The latter is easy to check, writing the kernel in the form $\exp[-ik(x\cos(\varphi) + y\sin(\varphi)]$, where $x = r\cos(\theta)$, $y = r\sin(\theta)$. The function (4.59) coincides with (2.12) by the form and, according to (2.13), its asymptotics is

$$\frac{1}{\sqrt{kr}}\left[e^{ikr}F(\varphi+\pi) + e^{-ikr}F(\varphi)\right]; \tag{4.60}$$

here a nonessential factor is also omitted.

According to (4.60), for the field (4.59) to be real, the function $F(\varphi)$ must fulfil the condition

$$F(\varphi + \pi) = F^*(\varphi). \tag{4.61}$$

Substituting (4.61) into (4.59), we see that the latter is real; this means that the condition (4.61) is necessary and sufficient for the field (4.59) to be real. In fact, this condition coincides with (2.16) which is fulfilled by the orthogonal complement function $\hat{F}(\varphi)$. Some nonessential difference in the coefficients is caused by the omitted constant factors in the two formulas above.

Thus, if $F(\varphi)$ fulfils the condition (4.61), then the field (4.59) is the real solution to the Helmholtz equation. If it equals zero on C, then it is the field $\hat{u}(r, \varphi)$, C is the specific line $\hat{C}$, and $K^{\text{ad}}\hat{F} = 0$, which means, operator $K\ K^{\text{ad}}$ has the zero eigenvalue μ_q.

Extension of the operator K^{ad} to $K_{\text{ex}}^{\text{ad}}$ (see (4.59)) connects the technique of this section with that based on investigation of the field $\hat{u}(r, \varphi)$. The formula (2.6) which was considered as a version of the reciprocity theorem, is equivalent to formula (4.31). In fact, the proof of (2.6), based on the Green formula, is contained in the proof of identity of the integrals (4.33), (4.34). In some aspects the methods of Sections 2.1, 4.2 are mutually complementary, and each of them has its own area of application. So, limiting ourselves by the operator K^{ad} (without introducing $K_{\text{ex}}^{\text{ad}}$) one might have difficulty in proving that equation (4.40b) for the curvilinear quadrangle of Fig. 3.5, has no zero eigenvalue , and for two straight lines crossing at the angle (2.47), this equation has the zero eigenvalue of the infinite multiplicity. This fact was proved while investigating the field $\hat{u}(r, \varphi)$ (see (2.42)).

Similarly to (4.59), one can introduce the operator K_{ex} acting onto $j(s)$, by the formula

$$K_{\text{ex}} j = \frac{i}{2} \int_C H_0^{(2)}(k\rho) j(s) ds, \tag{4.62}$$

where ρ is the distance between the integrating point on C and the point (r, θ) in the plane. This operator maps the current $j(s)$ into the field $u(r, \theta)$ created by this current at an arbitrary point (r, θ). If the point lies at infinity $(r \to \infty)$, then the operator K_{ex} differs from K by the factor $\exp(-ikr)/\sqrt{kr}$, only. Just this fact allows us to say that the operator $K^{\text{ad}} K$ has the zero eigenvalue $\lambda_p = 0$ $(j_p = \hat{\jmath})$ for any resonant contour; the multiplicity of the eigenvalue is finite, because the eigenvalues of the Dirichlet problem (i.e., the numbers k_p) can have only the finite-multiple degeneration.

Note a distinction, disturbing the symmetry a little between the operators K and K^{ad}, as well as between the functions $\hat{\jmath}(s)$ and $\hat{F}(\varphi)$, respectively. It follows from $K\hat{\jmath} \equiv 0$ that $K_{\text{ex}}\hat{\jmath} \equiv 0$ in the whole infinite domain, exterior with respect to the contour C, but from $K^{\text{ad}}\hat{F} \equiv 0$ it does not follow that $K_{\text{ex}}^{\text{ad}}\hat{F} \equiv 0$. The distinction is connected with the analytical properties of solutions to the Helmholtz equation, namely, the fields $u(r, \varphi)$ and $\hat{u}(r, \varphi)$. The first one is the complex travelling wave, it fulfils the radiation condition. By the Rellich theorem, if $u(r, \varphi)$ equals zero at infinity, more specifically, if $K\hat{\jmath} \equiv 0$ (i.e., $j = \hat{\jmath}$), then it equals zero everywhere (except for, perhaps, the domain interior with respect to C). It cannot have zero lines and be nonzero outside these lines. Unlike $u(r, \varphi)$, the field $\hat{u}(r, \varphi) = K_{\text{ex}}^{\text{ad}} F$ is created by the convergent cylindrical wave incoming from infinity. It is the standing wave, it can have zero lines $\hat{C}$ (the specific ones) at $F = \hat{F}$. One can conventionally say that the whole plane is the interior domain with respect to the line (infinitely remote circle) where the sources of the field $\hat{u}(r, \varphi)$ are distributed. In fact, all the constructions of Chapter 3 are based on this concept.

We complete this subsection by a remark concerning the formula (2.183), which means that for any line C one can point out such a sequence $u_m(r, \varphi)$ of solutions to the Helmholtz equation, that the integral of $|u_m(r, \varphi)|^2$ over C tends to zero. An example of such a sequence is the fields $K_{\text{ex}}^{\text{ad}}\psi_m$. The functions $K^{\text{ad}}\psi_m$ are the values of these fields on C, and the integral in (2.183) is the norm of these functions. According to (4.42b), this integral equals μ_m and tends to zero at $m \to \infty$. The field $u_m(r, \varphi)$ is created by the convergent cylindrical wave having the form $\psi_m^*(\varphi) \exp(ikr)/\sqrt{kr}$ at infinity. At large m the function $\psi_m(\varphi)$ is fast varying and the field $u_m(r, \varphi)$ is small in any finite domain. This does not contradict the fact that the functions $\psi_m(\varphi)$ themselves do not decrease with increasing m; according to (4.41b), they are normalized by unity. We repeat, in new terms, the remark made below (2.183): of course, it does not follow from $\mu_m \to 0$ that there exists a limiting function to which the functions $\psi_m(\varphi)$ tend, and that the operator (4.40b) has zero eigenvalue .

7. The sets of currents $j_n(s)$ and patterns $\psi_m(\varphi)$ were constructed above, using the operator K mapping the currents into the patterns. Similar complete orthonormal sets of functions, connected by equations (4.45), can be constructed on C and on any contour Σ encircling C, using, instead of K, the operator K_Σ, which maps the current $j(s)$ on C into the field $u(\sigma)$ on Σ (see [11]). The operator K_Σ can be obtained from (4.62) by putting $\rho = \rho(s,\sigma)$ as the distance between the points s on C and σ on Σ

$$K_\Sigma j = \frac{i}{2}\int_C H_0^{(2)}(k\rho(s,\sigma))j(s)ds. \tag{4.63}$$

This operator maps $j(s)$ into $u(\sigma)$.

The inner product over C is defined by formula (4.29). Instead of the product (4.30) in the pattern space, we introduce the inner product in the space of fields on Σ:

$$(u_1,u_2) = \int_\Sigma u_1(\sigma)u_2^*(\sigma)d\sigma. \tag{4.64}$$

The adjoint operator K_Σ^{ad} is defined by the condition

$$(K_\Sigma^{\text{ad}}u_1, j_2) = (u_1, K_\Sigma j_2) \tag{4.65}$$

and has the form

$$K_\Sigma^{\text{ad}}u = -\frac{i}{2}\int_\Sigma H_0^{(1)}(k\rho(s,\sigma))u(\sigma)d\sigma. \tag{4.66}$$

This formula is proved in the same way as (4.32). Both sides of (4.65) are the double integrals over C and Σ and differ from each other by an order of integration.

Like the symmetry (4.33), (4.34) between the operators K and K^{ad}, the symmetry between K_Σ and K_Σ^{ad} is caused by that of the kernel in the integral transformations (4.63) and (4.66) with respect to s and σ, correspondingly. In its turn, this is a consequence of the reciprocity property, as was mentioned above (4.62).

All further constructions with the operators K_Σ and K_Σ^{ad} can be made in the same way as with K and K^{ad}. Two sets of the generalized functions of double orthogonality can be obtained as the eigenfunctions of the equations of type (4.40). Similarly to the above, one can solve the problem of the current which creates the given field $U(\sigma)$ on Σ. In general, these two sets differ from the function sets generated by the operators (4.40).

The same specific situations occur when the operators $K_\Sigma K_\Sigma^{\text{ad}}$ or (and) $K_\Sigma^{\text{ad}} K_\Sigma$ have zero eigenvalues. The first of them has the zero eigenvalue $\mu_q = 0$, when the operator KK^{ad} has the same eigenvalue . The multiplicity of zero eigenvalue of the operator $K_\Sigma K_\Sigma^{\text{ad}}$ can be larger than that of KK^{ad}. When the contour Σ is sufficiently close to C, then the operator $K_\Sigma K_\Sigma^{\text{ad}}$ can have zero eigenvalue , even if KK^{ad} does not. The two last sentences formulate, in operator terms, the results obtained in Chapter 3 where the fields $\hat{u}(r,\varphi)$ were investigated: many lines C are zero lines of the fields having no singularities only in the finite part of the plane, close to C. On these lines one can create the currents which approximate any pattern, but no currents can approximate some fields on the contours Σ, sufficiently close to C.

The operator $K_\Sigma^{\text{ad}} K_\Sigma$ has zero eigenvalue if and only if the operator $K^{\text{ad}} K$ also has it. The multiplicity of these eigenvalues is the same, and the same nonradiating currents (4.52) correspond to them. This follows from the fact that, if a current on C does not create any field on Σ, then it does not create any pattern either, and vice versa.

8. Finally, let us mention a possibility to construct the sets of generalized functions of double orthogonality, using the operators mapping the magnetic fields $j^{(m)}$ located on C into the patterns or into the fields on Σ. Denote the first of these operators by $K^{(m)}$ and the operator mapping the current $j^{(m)}$ into the field at any point (r, θ), by $K_{\text{ex}}^{(m)}$. Introduce the magnetic current as a jump of the field u on the line C, that is, put $j^{(m)} = [u]$ (see (1.6b)). The operator $K_{\text{ex}}^{(m)}$ is also integral, similar to (4.62):

$$K_{\text{ex}}^{(m)} j^{(m)} = \frac{i}{2} \int_C j^{(m)}(s) \frac{\partial}{\partial N} H_0^{(2)}(k\rho) ds, \tag{4.67}$$

where $\partial / \partial N$ is the normal derivative on C. If the point (r, θ) lies on the contour Σ, then (4.67) is transformed (in obvious notation) into $K_\Sigma^{(m)}$; at $r \to \infty$ (4.67) differs from $K^{(m)}$ only by the factor $\exp(-ikr)/\sqrt{kr}$.

All the constructions made for K, and for the operators obtained from it, can be wholly carried over to $K^{(m)}$, $K_{\text{ex}}^{(m)}$, and so on. The concept of the resonant contour is different from above only in that: inside this contour the solution to the Helmholtz equation with zero normal derivative on C exists. The nonradiating current $\hat{j}^{(m)}$ equals the value of this field on C. If $C$is the circle, then $j_n^{(m)}(\theta)$ and $\psi_m^{(m)}(\varphi)$ are the same trigonometrical functions, as $j_n(\theta)$ and $\psi_m(\varphi)$, and the eigenvalues are proportional to $[J_n'(ka)]^2$.

4.3 Optimal Current Synthesis. The General Case

1. In this section the general formal solution to the problem stated in Subsection 4.1.1, is given. Some numerical examples are presented.

The problem consists in finding a realizable pattern $\tilde{F}(\varphi)$ at a distance δ from the given one $F(\varphi)$ and created by the current with minimal norm. Both patterns are normalized by the condition (4.2). The current norm is calculated by (4.3). Other norm definitions will also be considered further.

The solution is based on the expansion of the current and pattern by the generalized functions of double orthogonality, introduced in Section 4.2. Then the Lagrange multiplier method is applied and the Ritz technique, in which the varied variables are the Fourier coefficients of the function $\tilde{F}(\varphi)$, is used. We will use the following expressions, concerning the patterns and currents:

a. The expansions of $F(\varphi)$, $\tilde{F}(\varphi)$ by the generalized functions of double orthogonality:

$$F(\varphi) = \sum_{m=1}^{\infty} A_m \psi_m(\varphi), \tag{4.68a}$$

$$\tilde{F}(\varphi) = \sum_{m=1}^{\infty} \tilde{A}_m \psi_m(\varphi); \tag{4.68b}$$

b. The squared norm of current creating the pattern $\tilde{F}(\varphi)$, expressed by $\tilde{A}_m$:

$$N^2 = \sum_{m=1}^{\infty} \frac{\left|\tilde{A}_m\right|^2}{\mu_m}; \tag{4.69}$$

c. The normalizing conditions for the patterns:

$$\sum_{m=1}^{\infty} |A_m|^2 = 1; \tag{4.70a}$$

$$\sum_{m=1}^{\infty} \left|\tilde{A}_m\right|^2 = 1; \tag{4.70b}$$

d. The squared distance between the functions $F(\varphi)$ and $\tilde{F}(\varphi)$:

$$\Delta^2 = \sum_{m=1}^{\infty} \left|A_m - \tilde{A}_m\right|^2 . \tag{4.71}$$

The numbers μ_m in (4.69) are real and nonnegative, $\mu_m \to 0$ at $m \to \infty$.

The expression (4.69) permits to answer the general question on the current which has the minimal norm among all the currents distributed along the given line C creating the patterns with the unit norm. If μ_t is maximal among all μ_m, then the minimal value of the norm calculated by (4.69) using the condition (4.70b) is $N_{\min} = 1/\sqrt{\mu_t}$; it is reached at $|\tilde{A}_t| = 1$ and $\tilde{A}_m = 0$, $m \neq t$. Indeed, if some other coefficients $\tilde{A}_m$, $m \neq t$, were different from zero, then N^2 could be expressed as

$$N^2 = \frac{1}{\mu_t} + \sum_{m=1}^{\infty} \left|\tilde{A}_m\right|^2 \left(\frac{1}{\mu_m} - \frac{1}{\mu_t}\right). \tag{4.72}$$

Since $\mu_t \geq \mu_m$, all the terms in the sum are positive (nonnegative), which gives $N^2 \geq 1/\mu_t = N^2_{\min}$.

The current with the minimal norm is determined with accuracy to the constant phase α of the coefficient $\tilde{A}_t$. The function $\tilde{F}_{\min}(\varphi)$ corresponding to this current is $\tilde{F}_{\min}(\varphi) = \exp(i\alpha)\psi_t(\varphi)$. The distance η between $F(\varphi)$ and $\tilde{F}_{\min}(\varphi)$ is minimal if α is equal to the phase of the coefficient A_t. Then

$$\eta^2 = 2 - 2|A_t| \tag{4.73}$$

(see Subsection 4.1.3). Consequently, $\tilde{F}_{\min}(\varphi) = \exp(i \arg A_t)\psi_t(\varphi)$.

As it was shown in Section 4.1, at $\delta \geq \eta$ the optimal current synthesis problem has the obvious solution

$$\tilde{F}(\varphi) \equiv \tilde{F}_{\min}(\varphi), \quad N(\delta) = N_{\min}. \tag{4.74}$$

Below we solve this problem for $\delta < \eta$. Then the minimum is reached on the function $\tilde{F}(\varphi)$, for which equality (4.71) holds after replacing Δ^2 with δ^2. Remember that the minimal value of N, at the given δ, is denoted by $N(\delta)$. It is the functional of $F(\varphi)$.

2. The conditional extremum problem for the functional (4.69), where the variables $\tilde{A}_m$ are subject to equalities (4.70b), (4.71), is reduced by the Lagrange multipliers method to the unconditional extremum problem for the functional

$$L(\tilde{A}_1, \tilde{A}_2, ...) = \sum_{m=1}^{\infty} \frac{\left|\tilde{A}_m\right|^2}{\mu_m} + l_1 \sum_{m=1}^{\infty} \left|\tilde{A}_m\right|^2 + l_2 \sum_{m=1}^{\infty} \left|\tilde{A}_m - A_m\right|^2 . \tag{4.75}$$

The necessary conditions of the extremum are

$$\frac{\partial L}{\partial \tilde{A}_m} = 0, \qquad \frac{\partial L}{\partial \tilde{A}_m^*} = 0, \quad m = 1, 2, ... \tag{4.76}$$

They give

$$\tilde{A}_m = \frac{\alpha A_m}{1 + \beta/\mu_m}, \quad \tilde{A}_m^* = \frac{\alpha A_m^*}{1 + \beta/\mu_m}, \tag{4.77}$$

where

$$\alpha = \frac{l_2}{l_1 + l_2}, \quad \beta = \frac{1}{l_1 + l_2}. \tag{4.78}$$

It follows from (4.77), that the values $\alpha/\,(1 + \beta/\mu_m)$ are real for all m. This is possible only if α and β are real. Then the Lagrange multipliers l_1, l_2 are real, too.

According to (4.77), the larger μ_m, the smaller the relative difference between the coefficients $\tilde{A}_m$ and A_m. The difference $A_t - \tilde{A}_t$ corresponding to the largest μ_m is the smallest. The terms of the series (4.68a), (4.68b), which correspond to small μ_m, (in particular, very far-standing terms) differ from each other most of all. If there is $\mu_q = 0$ among the numbers μ_m, then $\tilde{A}_q = 0$. Since at small μ_m the coefficient $\tilde{A}_m$ is approximately proportional to μ_m, we have $\tilde{A}_q \to 0$ at $\mu_q \to 0$, and the corresponding summand in the norm N (4.69) also tends to zero (see (4.83) below). This result is obvious, since the pattern $\tilde{F}(\varphi)$ is realizable, and at the limit it is orthogonal to the orthogonal complement function $\hat{F}(\varphi) = \psi_q(\varphi)$. This means, that the coefficient $\tilde{A}_q$, equal to $(\tilde{F}, \psi_q)$ by (4.68a), is zero.

The equation system (4.77) is solved together with (4.70b) and (4.71) (at $\Delta = \delta$). Using (4.70), the latter can be rewritten in the form

$$\sum_{m=1}^{\infty} (A_m^* \tilde{A}_m + A_m \tilde{A}_m^*) = 2 - \delta^2, \tag{4.79}$$

which, after substituting (4.77), gives

$$\alpha \sum_{m=1}^{\infty} \frac{|A_m|^2}{1 + \beta/\mu_m} = 1 - \frac{\delta^2}{2}. \tag{4.80}$$

On the other hand, (4.70b) together with (4.77) yields

$$\alpha^2 \sum_{m=1}^{\infty} \frac{|A_m|^2}{(1 + \beta/\mu_m)^2} = 1. \tag{4.81}$$

Since A_m are given and they depend only on $F(\varphi)$, the two last formulas make up the equation system in unknown α and β. Eliminating α in these equations leads to the equation for β:

$$\left[\sum_{m=1}^{\infty} \frac{|A_m|^2}{1+\beta/\mu_m}\right]^2 = \left(1-\frac{\delta^2}{2}\right)^2 \sum_{m=1}^{\infty} \frac{|A_m|^2}{(1+\beta/\mu_m)^2}. \tag{4.82}$$

After α and β are found, one can calculate the coefficients $\tilde{A}_m$ by (4.77) and obtain the function $\tilde{F}(\varphi)$. The norm N can be calculated either in the form of the sum

$$N^2(\delta) = \alpha^2 \sum_{m=1}^{\infty} \frac{|A_m|^2 \mu_m}{(\mu_m+\beta)^2}, \tag{4.83}$$

obtained from (4.69), (4.77), or in the form containing only α and β:

$$N^2(\delta) = \frac{\alpha-1}{\beta} - \frac{\alpha\delta^2}{2\beta}. \tag{4.84}$$

To obtain (4.84), one should divide both sides of equalities (4.80), (4.81) by α and α^2, respectively, and subtract the results. As follows from (4.69), (4.77), this difference is $-\beta N^2(\delta)/\alpha$, which gives (4.84). Below, in Subsection 4, we obtain this formula in a simpler way.

3. Expressions for the solution of the optimal current synthesis problem by the given whole (complex) pattern, obtained above, allow progress in one particular case of the problem when only the amplitude pattern is given. A corresponding formulation of the problem was given in Subsection 4.1.5. In this problem the functions to be varied were not only the realizable pattern $\tilde{F}(\varphi)$, but also the phase of the function $F(\varphi)$ to be approximated by $\tilde{F}(\varphi)$ with a given distance δ from it and the minimal current norm $N(\delta)$.

According to (4.80), (4.81), the coefficients α, β depend only on the squared moduli of the Fourier coefficients of $F(\varphi)$, that is, on the numbers $|A_m|^2$. It follows from (4.83) that the norm $N(\delta)$ depends only on $|A_m|^2$; it is independent of the phases of A_m. The numbers

$$|A_m|^2 = \left| \int_0^{2\pi} \Phi(\varphi)\psi_m^*(\varphi) e^{-i\Psi(\varphi)} d\varphi \right| \tag{4.85}$$

can be considered as functionals of the phase $\Psi(\varphi)$. If the functions $\psi_m(\varphi)$ are real, then these functionals are stationary at $\Psi(\varphi) \equiv 0$. Indeed, one can show that at $\Psi(\varphi) = \varepsilon\gamma(\varphi)$, where $\gamma(\varphi)$ is an arbitrary function and $\varepsilon << 1$, $|A_m|^2$ differ from those at $\Psi(\varphi) \equiv 0$ by a value of order ε^2at all m. Therefore, $N(\varphi)$ is also stationary at $\Psi(\varphi) \equiv 0$ with respect to variations of the phase $\Psi(\varphi)$. In this stationary point all A_m are real.

In Subsection 4.2.2 it was shown that all the functions $\psi_m(\varphi)$ are real when the line C has central symmetry. Consequently, if equation $r = r(\theta)$ of this line possesses the property $r(\theta+\pi) = r(\theta)$, then one can approximate the pattern $F(\varphi)$ with the given amplitude $\Phi(\varphi)$ by a function $\tilde{F}(\varphi)$ corresponding to a current with the stationary (with respect to small variations of $\Psi(\varphi)$) norm when $F(\varphi)$ is real ($\Psi(\varphi) \equiv 0$). Then, according to (4.77), the Fourier coefficients $\tilde{A}_m$ of the function $\tilde{F}(\varphi)$, on which the current norm is stationary, are real, and therefore the function $\tilde{F}(\varphi)$ is real itself.

Note that we have shown only stationarity, but not minimality of the current norm at the point $\Psi(\varphi) \equiv 0$. As is seen in the examples of other functionals having stationary points (see, e.g. [44]) *stationarity need not be minimality* in such problems.

If all the functions $\psi_m(\varphi)$ are not real, but have the same phase, then one can easily point out the phase $\Psi(\varphi)$ of the pattern $F(\varphi)$, on which the current norm $N(\delta)$is stationary with respect to small variations of $\Psi(\varphi)$. According to (4.85), all the coefficients A_m are real (the condition of stationarity of $N(\delta)$) if $F(\varphi)$has the same phase as all $\psi_m(\varphi)$. In this case the pattern $\tilde{F}(\varphi)$ also has the same phase.

4. The above results can be obtained without using the expansions (4.68). To this end, one should express the current norm $N(\delta)$ and the additional conditions (4.70b), (4.71) directly by the current $\tilde{j}(s)$ creating the pattern $\tilde{F}(\varphi)$:

$$N^2(\delta) = (\tilde{j}, \tilde{j}), \tag{4.86a}$$

$$(K\tilde{j}, K\tilde{j}) = 1, \tag{4.86b}$$

$$(K\tilde{j} - F, K\tilde{j} - F) = \delta^2. \tag{4.86c}$$

The functional (4.75) expressed by $\tilde{j}(s)$, is

$$L(\tilde{j}) = (\tilde{j}, \tilde{j}) + l_1(K\tilde{j}, K\tilde{j}) + l_2(K\tilde{j} - F, K\tilde{j} - F). \tag{4.87}$$

The current $\tilde{j}(s)$ on which this functional is stationary, satisfies the Euler equation obtained from equating to zero the first variation of $L(\tilde{j})$, that is, the linear (with respect to ε) term in $L(\tilde{j} + \varepsilon\gamma)$ at arbitrary function $\gamma(\varphi)$. This variation is

$$\delta L = 2\,\mathrm{Re}\left[(\gamma, \tilde{j}) + l_1(K\gamma, K\tilde{j}) + l_2(K\gamma, K\tilde{j} - F)\right]. \tag{4.88}$$

Using the adjoint operator K^{ad} introduced by (4.31), we obtain

$$\delta L = 2\,\mathrm{Re}\left[(\gamma, \tilde{j} + l_1 K^{\mathrm{ad}} K\tilde{j} + l_2 K^{\mathrm{ad}}\left[K\tilde{j} - F\right])\right]. \tag{4.89}$$

Equating δL to zero, the complex function γ being chosen as an arbitrary real and imaginary function independently, we obtain the sought Euler equation [45], [46]

$$\tilde{j} + l_1 K^{\mathrm{ad}} K\tilde{j} + l_2 K^{\mathrm{ad}}\left[K\tilde{j} - F\right] = 0. \tag{4.90}$$

Substitute the expansions $\tilde{j}$ and F of the forms (4.12), (4.68a) in (4.90) and make use of equation (4.40a). Then we obtain

$$\tilde{B}_n = \frac{\alpha A_n}{\sqrt{\lambda_n}(1 + \beta/\lambda_n)}, \tag{4.91}$$

where α and β are expressed by l_1, l_2 according to (4.78). In this way we have got the above sum (4.83) for the current norm (4.86a), and equations (4.80), (4.81) for the additional conditions (4.70b), (4.71).

One can also easily obtain the expression (4.84) for the current norm immediately, by α, β, that is, according to (4.78), by the Lagrange multipliers. To this end, one should multiply the Euler equation (4.90) by $\tilde{j}(s)$ and integrate over C. Then the first term gives $N^2(\delta)$ (see

(4.86a)); the second and third ones (using (4.31) and (4.86b)) are equal to l_1 and $l_2[1-(\tilde{F}, F)]$, respectively. Since the normalizing conditions for $\tilde{F}(\varphi)$ and $F(\varphi)$ together with (4.86a) give $(F, \tilde{F}) = 1 - \delta^2/2$ (see (4.79)), we obtain the expression

$$N^2(\delta) = -l_1 - l_2 \frac{\delta^2}{2}, \tag{4.92}$$

which coincides, according to (4.78), with (4.84).

One drawback of the method for deriving the main formulas using the Euler equation is the absence of explicit formula (4.77) for the Fourier coefficients of the function $\tilde{F}(\varphi)$ minimizing the functional L. However, (4.90), (4.46) give $\tilde{A}_n = \tilde{B}_n/\sqrt{\lambda_n}$, from which, using (4.91), one can easily obtain (4.77). On the other hand, application of the technique based on (4.87), (4.88) shows, how in a natural way, the operators K^{ad} and $K^{\text{ad}}K$ are introduced in the optimal current synthesis theory. Of course, the Euler equation (4.90) together with the conditions (4.86b), (4.86c), can be solved in another way, without expanding $\tilde{j}(s)$ by eigenfunctions of the operator $K^{\text{ad}}K$.

Beside $N(\delta)$, in Subsection 4.1.4 two other current norms $N_1(\delta)$, $N_2(\delta)$ were introduced, which contain the derivative $d\tilde{j}/ds$. The functionals, corresponding to these norms, are similar to L (see (4.87)). For instance, for the norm $N_1(\delta)$ one can construct the functional

$$L_1(\tilde{j}) = (\tilde{j}, \tilde{j}) + \left(\frac{d\tilde{j}}{ds}, \frac{d\tilde{j}}{ds}\right) + l_1(K\tilde{j}, K\tilde{j}) + l_2(K\tilde{j} - F, K\tilde{j} - F). \tag{4.93}$$

The Euler equation for the current $\tilde{j}(s)$, which provides the stationarity of L_1, can be obtained in the same way as equation (4.90) for the functional (4.87). The summand

$$\int_C \frac{d\tilde{j}}{ds}\frac{d\gamma}{ds} ds \tag{4.94}$$

in the first variation of (4.93) is reduced to the form $-\int_C \gamma \left(d^2\tilde{j}/ds^2\right) ds$ by the partial integration. For simplicity, we suppose C to be a closed contour so that the end terms obtained while integrating $d(\gamma d\tilde{j}/ds)/ds$ diminish. Thus, the Euler equation for the functional $L_1(\tilde{j})$ is

$$\tilde{j} - \frac{d^2\tilde{j}}{ds^2} + l_1 K^{\text{ad}} K\tilde{j} + l_2 K^{\text{ad}}\left[K\tilde{j} - F\right] = 0. \tag{4.95}$$

This equation should be solved together with the conditions (4.86b), (4.86c).

The numerical solution of the optimal current synthesis problem for the norms $N_1(\delta)$ and $N_2(\delta)$ by expanding the function $\tilde{j}(s)$ into the series by $\tilde{j}_n(s)$ is very laborious, because it requires differentiating the functions $\tilde{j}_n(s)$. In the examples given below we confine ourselves to the case of a circle, then $\tilde{j}_n(s)$ are the simple trigonometric functions (4.8). Here we only note that (4.92) is valid also for the norm $N_1(\delta)$ and the Euler equation (4.95) corresponding to it. Indeed, multiplying (4.95) by $\tilde{j}(s)$ and integrating over C using the partial integration technique, we obtain the value $N_1(\delta)$ from the first two summands, and, finally, the expression (4.92) for $N_1^2(\delta)$. Of course, this formula is valid also for $N_2^2(\delta)$ (see (4.23b)) and, in general, for any norm containing $|\tilde{j}(s)|^2$ and $|d\tilde{j}/ds|^2$ under the integral with arbitrary weight factors. However, the Lagrange multipliers l_1, l_2 are different for each of the norms.

5. The connection between β and δ is given by equation (4.82). It is more convenient to determine the dependence $\delta(\beta)$ from this equation. Similarly, the parameter α can be considered as a function of β, given by (4.81). Remember that β is real.

At first assume that all the eigenvalues $\mu_m \neq 0$, which means that the line C is not specific. Then, at $\beta = 0$ both sums in (4.82) equal unity according to (4.70a), hence $\delta = 0$. The sum in (4.81) is also equal to unity, so that $\alpha = 1$. It is impossible to determine $N(\delta)$ directly from (4.83). To open the indefiniteness, one should use equations (4.83), (4.81) at $\beta \to +0$. In fact, we can avoid this difficulty, using the formula (4.50), which is, the series (4.83) at $\beta = 0$. As was shown in Section 1.3, the value $N(\delta)$ at $\delta \to 0$ is determined by the rate of decreasing $|A_m|$ at $m \to \infty$. If $|A_m|$ decrease sufficiently fast, then $N(0) < \infty$ and the pattern $F(\varphi)$ is realizable. In the opposite case $N(0) = \infty$. If δ grows, then the values β and $\alpha - 1$ grow as well, and the formulas (4.83), (4.84) can be used.

There exist functions $F(\varphi)$, for which β is *negative* at some values of δ. Let, for instance, the series (4.68a) for $F(\varphi)$ not contain the term with $\psi_t(\varphi)$ (i.e. $A_t = 0$), where t is the index corresponding to the maximal μ_m ($\mu_t \geq \mu_m$, $m \neq t$). As we have ascertained in Section 4.1 while discussing Fig. 4.1(b), in this case the term with $\psi_t(\varphi)$ should be presented in $\tilde{F}(\varphi)$, that is, $\tilde{A}_t \neq 0$. According to (4.77), this is possible only if $\beta = -\mu_t$. Other cases can also take place when β is negative. Probably, in each of these cases, it is convenient to apply some simpler method, like that used in Section 4.1.

To analyze equation (4.82) for the specific line (i.e. at $\mu_q = 0$), we assume at first, that $\mu_q \neq 0$, but $\mu_q << \mu_m$, $m \neq q$, and find δ for values of β, small in comparison with μ_m, $m \neq q$, but large in comparison with μ_q. At $\beta << \mu_m$, $\beta >> \mu_q$ one can neglect the summand β/μ_m in the denominator of all the terms of both series (4.82), except for $m = q$, and omit the term $m = q$. Using the condition (4.70a), equation (4.82) takes the form $1 - |A_q|^2 = (1 - \delta^2/2)^2$, from which $\delta = B$, where

$$B = \sqrt{2 - 2\sqrt{1 - |A_q|^2}}. \tag{4.96}$$

Consequently, the dependence $\delta(\beta)$ is described by the curve given in Fig. 4.3(a): it starts at the point $\beta = 0$, $\delta = 0$ and goes onto the horizontal line $\delta = B$ at $\beta >> \mu_q$. When β becomes comparable with other eigenvalues, then the curve begins to rise again.

It is obvious that the curve has the form presented in Fig. 4.3(b) at $\mu_q = 0$, that is, when C is a specific line. The inverse function is given in Fig. 4.3(c). The value of $\beta(\delta)$ is equal to zero in the interval $0 \leq \delta \leq B$. As it follows from (4.81), at these δ, the value $\alpha(\delta)$ is different from the unit: $\alpha = (1 - |A_q|^2)^{-1/2}$. According to (4.84), the current norm $N(\delta)$ is infinite for these δ.

This result, obtained from the analysis of the formulas (4.81), (4.82), (4.84), repeats the property of the specific lines: the patterns, nonorthogonal to the orthogonal complement functions, that is, represented by the series (4.68a) with $A_q \neq 0$, are nonapproximable; not only these patterns themselves, but also all those close to them are nonrealizable. There exists the nonzero *lower limit B of the distance* between $F(\varphi)$ and a realizable patterns $\tilde{F}(\varphi)$. This limit can be either reached (Fig. 4.1(c), curve 4) or not (Fig. 4.1(b), curve 5); this depends on the decrease rate of the terms in the series (4.68a).

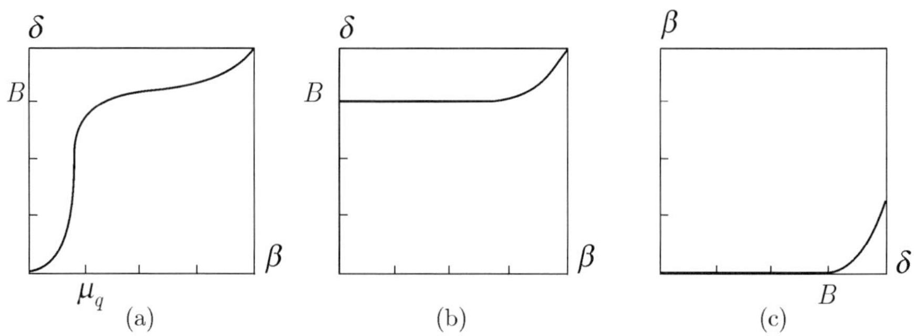

Figure 4.3:

The formula (4.96) for the slot width B can be easily obtained without using the technique applied in this section. Its particular form was given in (4.18). The value B^2 is equal to the minimum of the squared distance between $F(\varphi)$ and $\tilde{F}(\varphi)$, calculated over all the functions $\tilde{F}(\varphi)$ with $\tilde{A}_q = 0$, normalized by unity according to (4.70b). The coefficients $\tilde{A}_m$ are calculated from the requirement of the unconditional minimum of the functional

$$L = \sum_{m \neq q}^{\infty} |\tilde{A}_m - A_m|^2 + A_q^2 + l \sum_{m \neq q}^{\infty} |\tilde{A}_m|^2. \tag{4.97}$$

Equating the partial derivatives of L with respect to $\tilde{A}_m$ and $\tilde{A}_m^*$ $(m \neq q)$ to zero, we obtain, similarly to (4.76) that $\tilde{A}_m = A_m/(1+l)$, $m \neq q$, that is, $\tilde{F}(\varphi)$ differs from $F(\varphi)$ only in the absence of the term $m = q$ and in the proportional change of all other coefficients, caused by the normalizing condition. Substituting $\tilde{A}_m$ into (4.70b) and taking into account that $\sum_{m \neq q}^{\infty} |\tilde{A}_m|^2 = 1 - |\tilde{A}_q|^2$, one can calculate l and then determine B^2 as the sum of two first summands in L. Under the notation (1.26), $B^2 = 2 - 2\sqrt{1 - |b|^2}$ and it is larger than $|b|^2$, because the patterns $f(\varphi)$, in contrast to $\tilde{F}(\varphi)$, are not normalized.

6. In this subsection we give the numerical results [12], [17], [47] regarding the dependence of the current norms on δ. The examples for the flat-topped (Π-shaped), Gaussian and cosecant patterns are considered. All the patterns are real. Almost all calculations are made for the lines C in the form of circle or its arc. For the nonresonant circle it is taken $ka = 1$ (Figs. 4.4, 4.5) and $ka = 6$ (Fig. 4.6, curve 1, and Fig. 4.7). For the resonant one (Fig. 4.8) $ka = 3.83$, which means that it is a zero of the Bessel function $J_1(ka)$, so that the orthogonal complement function $\hat{F}(\varphi)$ is $\pi^{-1/2}\cos\varphi$. Fig. 4.9 corresponds to C in the form of a circle arc of the angle 229° (4 radians); for the nonresonant circle $ka = 1$, and for the resonant one $ka = 3.83$. In all figures (except Fig. 4.7) the abscissa corresponds to the value δ^2 (instead of δ, as in Fig. 4.1).

Fig. 4.4 shows the norms of currents to be distributed on the nonresonant circle ($ka = 1$), to create the pattern close to the flat-topped one of the width 2γ radians:

$$F(\varphi) = \begin{cases} 1/\sqrt{2\gamma}, & |\varphi| \leq \gamma, \\ 0, & |\varphi| > \gamma. \end{cases} \tag{4.98}$$

Figures (a), (b) and (c) show different current norms.

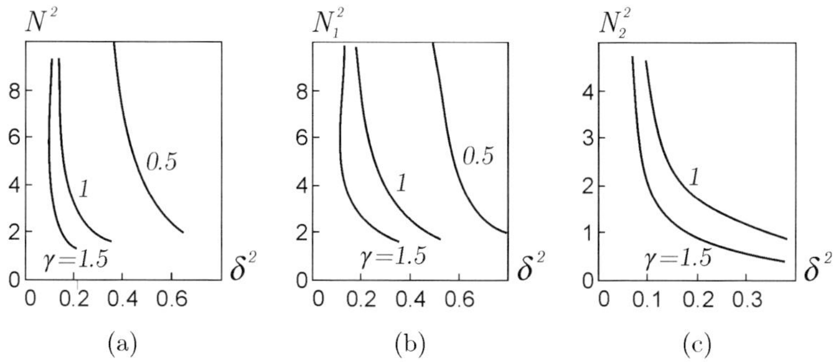

Figure 4.4:

Of course, the flat-topped pattern is nonrealizable ($N(0) = \infty$); however, the line C is nonspecific, so that there is no finite "slot" between $F(\varphi)$ and $\tilde{F}(\varphi)$, that is, $N(\delta) < \infty$ at any $\delta > 0$. The wider the pattern (i.e. the larger γ), the smaller the norm of the current that is required to approximate it with the given accuracy δ. One can approximate the wide pattern ($\gamma = 1.5$) by the currents with small norms.

The comparison of different norms shows that the current oscillates very fast at this degree of approximation. It is practically impossible to approximate the sharper patterns ($\gamma = 1.0$, or, moreover, $\gamma = 0.5$) with $\delta^2 < 0.3 \div 0.4$ without the large and fast varying currents. At smaller δ the current norms become very large.

In Fig. 4.5(a),(b) the squared current norms N^2, N_1^2 are presented for the same circle with $ka = 1.0$. The pattern to be approximated is the Gaussian one (1.22):

$$F(\varphi) = \frac{e^{A/2}}{\sqrt{2\pi I_0(A)}} e^{-A\sin^2(\varphi/2)}. \qquad (4.99)$$

Here I_0 is the modified Bessel function, the constant factor is determined from the normalizing condition (4.2a).

The parameter A defines the pattern width; the larger A, the sharper $F(\varphi)$. The pattern half-width equals $2\arcsin\sqrt{\ln 2/A}$ (see Section 2.3). Three of four values A, namely, 0.75, 1.51, 5.67, for which the graphs are given in Fig. 4.5, are chosen in such a way, that the half-width of the patterns (4.99) coincides with the parameter γ of the above flat-topped patterns (4.98). The corresponding values γ are 1.5, 1.0, 0.5. Moreover, the curves $N^2(\delta)$, $N_1^2(\delta)$ are given for $A = 2.2$, which will be discussed further.

The Gaussian pattern (4.99) is realizable by the current on the circle with $ka = 1$ at $A < 2$ and nonrealizable at $A > 2$. This follows from the result, given in Section 1.3, according to which the patterns with $a_0 < a$, where $a_0 = A/2k$, are realizable by the currents on the circle of radius a; the patterns with $A < 2ka$ fulfil this condition.

Therefore, for the wide patterns, namely, at $A < 2$ the norms $N(\delta)$ and $N_1(\delta)$ are finite even at $\delta = 0$. Replacing $F(\varphi)$ with the pattern $\tilde{F}(\varphi)$ close to it, having smaller norm, decreases the norm only a little. The curves in Fig. 4.5, corresponding to $A = 1.51$ and $A = 0.75$ fall very slowly with increasing δ^2. To obtain not too large norms for the nonrealizable

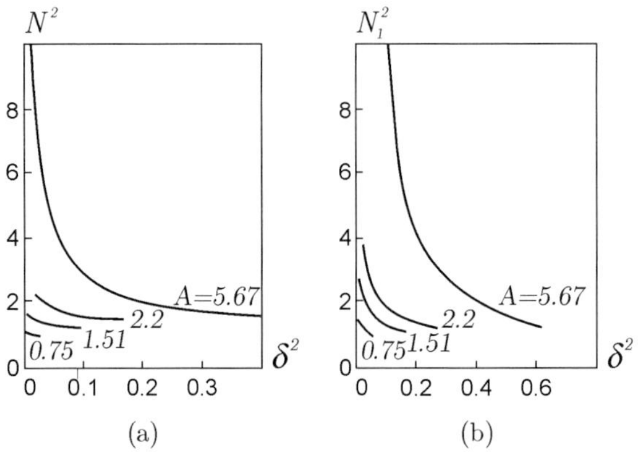

Figure 4.5:

(too sharp) pattern with $A = 5.67$ one should permit the deviation $\delta^2 > 0.1$. For instance, at $\delta^2 = 0.1$ we have $N^2 = 2.7$, $N_1^2 = 7.5$.

The curves for $A = 2.2$ in Fig. 4.5 allow to compare the results of the optimal current synthesis for two patterns close to each other, one of which is realizable ($A = 1.51$) and the other nonrealizable ($A = 2.2$). The results illustrate the idea stated above that the difference between realizable and nonrealizable patterns has no special practical signification. These patterns are especially different from the point of view of the exact realization problem, because $N(0) < \infty$, $N_1(0) < \infty$ at $A = 1.51$ and $N(0) = N_1(0) = \infty$ at $A = 2.2$. However, this difference quickly diminishes with δ increasing: just at $\delta^2 = 0.05$ the values $N^2(\delta)$ for both patterns are close to each other, and at $\delta^2 = 0.2$ the values $N_1^2(\delta)$ are also close.

Curve 1 in Fig. 4.6 shows the dependence of the squared norm N^2 on δ^2 for the cosecant pattern

$$F(\varphi) = \begin{cases} 1/\cos\varphi, & 105^o \le \varphi \le 165^o, \\ 0, & 0 \le \varphi < 105^o; \quad 165^o < \varphi < 360^o; \end{cases} \tag{4.100}$$

the normalizing factor is omitted here. The current is distributed on the circle with $ka = 5.0$. Of course, this pattern is nonrealizable, but the optimal approximation is reached at a not too large current norm even at $\delta^2 = 0.1$; better results are obtained at $\delta^2 = 0.2$. In Fig. 4.7 the given pattern $F(\varphi)$ (dashed line) and realizable ones (solid lines) obtained in the optimal current synthesis problem at the above values of δ^2, are compared. At a larger acceptable residual ($\delta^2 = 0.2$) the optimal pattern is a little farther from the given one, but it is created by the current with essentially smaller norm, than at the smaller residual.

Curve 2 in Fig. 4.6 is taken from paper [47]. It describes the value N^2 for the current distributed along the *straight-line segment* of the length, equal to the circle length with $ka = 6.0$, for the same pattern (4.100). At not too large δ (about $\delta^2 < 0.3$) the cosecant pattern can be approximated by currents on the circle with the norm smaller than on the straight-line segment, equal to the circle diameter.

All the curves $N^2(\delta)$, $N_1^2(\delta)$ in Figs.4.4 – 4.7 correspond to the curves in Fig. 4.1(b). The approximation of the realizable patterns (e.g. the Gaussian one with $A < 2$) gives curves like

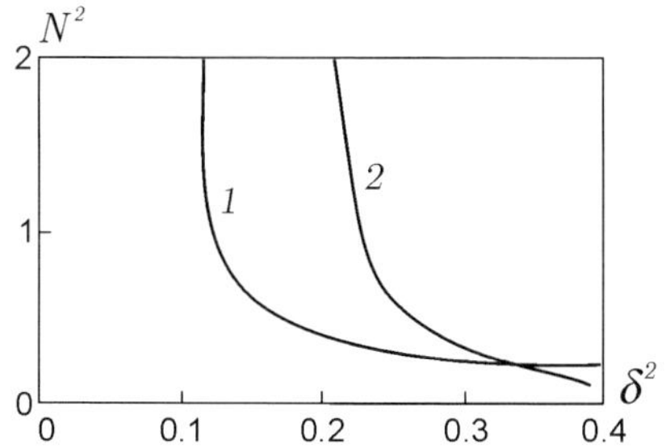

Figure 4.6:

curve 2 in Fig. 4.1(b). If $F(\varphi)$ is nonrealizable (the Gaussian ones with $A > 2$, nonrealizable by currents on the circle with $ka = 1.0$; or the patterns (4.98), (4.100), nonrealizable by any current), then we have curves like curve 3 in Fig. 4.1(b), which are the curves with the vertical asymptote $\delta = 0$.

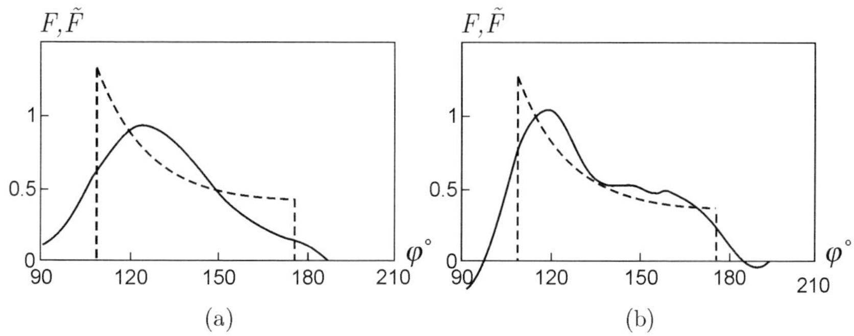

Figure 4.7:

Consider the currents located on the circle with $ka = 3.83$. In Fig. 4.8 the dependence N^2 on δ^2 for the flat-topped pattern (4.98), is shown. Since this circle is resonant and the pattern is nonrealizable, the curves in Fig. 4.8 are similar to curve 5 in Fig. 4.1(c), namely, they have vertical asymptotes $\delta^2 = B^2$, $B > 0$ (dashed lines).

Unlike the case of the nonresonant circle, presented in Fig. 4.69(a), for the resonant one, the *widening* of the given *pattern* can lead not to a decrease, but to an *increase* in the current norm, necessary for its approximation with the given accuracy. The curve for $\gamma = 0.5$ related to the sharpest of all considered flat-topped patterns, lies in Fig. 4.8 to the left of the curves for $\gamma = 1.5$ and $\gamma = 1.0$. For the nonresonant circle (Fig. 4.69(a)) the first curve lies to the right of the two last curves. This seeming paradox in location of the curves is explained by the dependence of the coefficient A_1 at the orthogonal complement function $\hat{F}(\varphi) = \pi^{-1/2} \cos \varphi$ in the expansion (4.68) on γ. As follows from (4.98), this dependence is $A_1(\gamma) = \sqrt{2/\pi} \sin \gamma / \sqrt{\gamma}$, so that A_1 grows as $\sqrt{\gamma}$ at small γ. For chosen γ the values of A_1 are: $A_1(0.5) = 0.54$, $A_1(1.0) = 0.67$, $A_1(1.5) = 0.65$. The least of these values is at $\gamma = 0.5$. The width and form of the pattern $F(\varphi)$ influence the rate of decrease of $N^2(\delta)$ with increasing δ ($\delta > B$), but,

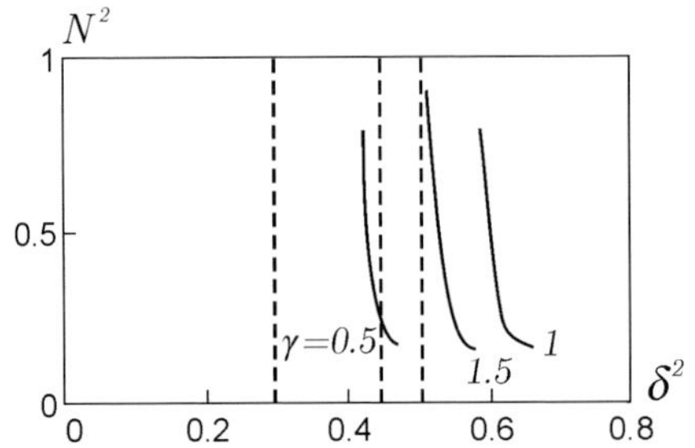

Figure 4.8:

according to (4.96), the asymptote $\delta = B$ depends only on A_q (in our case $q = 1$). Among the patterns of the type (4.98), narrower patterns have smaller coefficient A_1. The values $B^2(\gamma)$ for three considered patterns are $B^2(0.5) = 0.32$, $B^2(1.0) = 0.52$, $B^2(1.5) = 0.48$. At $\gamma << 1$ we have $B^2 = 2\gamma/\pi$, and the effect of nonapproximability is practically insignificant.

Consider the numerical results related to the optimal current synthesis problem for the case when C is the circle arc. The same flat-topped pattern is given. Begin from the nonresonant circle ($ka = 1.0$); the arc size is 4 radians. The curves $N^2(\delta^2)$ for three values of γ are shown in Fig. 4.9(a). The curves have the same behavior as in Fig. 4.4(a) where the same patterns were synthesized by the currents on the whole circle. Just as would be expected, the current norm needed for the approximation of some pattern with a given accuracy, by currents on the circle arc is larger than in the case of the whole circle. Since we seek the current having the minimal norm at the given approximation accuracy, the lengthening of the line where the current is located, can only diminish (more exactly, not increase) this norm. At least, the current on the additional segment equals zero, then the norm has the same value.

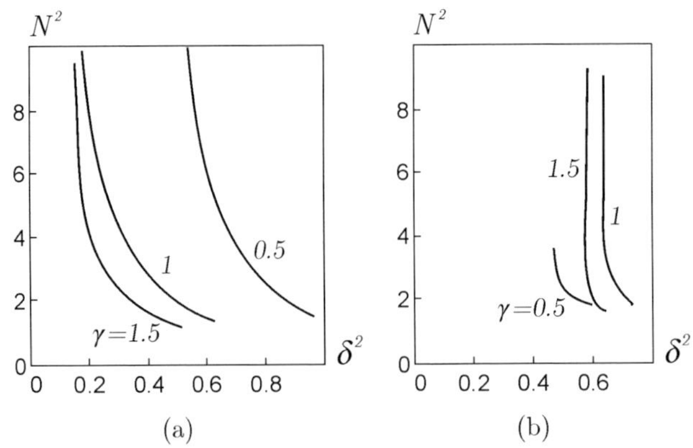

Figure 4.9:

In Fig. 4.9(b) the current norm is presented for the same flat-topped patterns, but the current is distributed on the arc of a resonant circle ($ka = 3.53$), which means, on the specific line. Therefore the curves in Fig. 4.9(b) have vertical asymptotes similar to those in Fig. 4.8. For these patterns, the same *inversion* takes place: at the given accuracy of approximation, the sharper pattern ($\gamma = 0.5$) requires a smaller current norm than the wider ones. As in the case of the whole circle, it is possible to approach the sharper pattern at less distance (in the least-squares sense), than the wider one, because the product $(F, \hat{F})$ is smaller in the first case. Repeating the arguments given at the end of the previous paragraph, we can say that, for the same pattern, the value B is larger (not smaller) in the case of the circle arc, than in the case of the whole circle.

4.4 Domain of Specific Line Influence

1. If C is a nonspecific line, then any function can be approximated by patterns of the currents on C with finite norm N, that is, as we have agreed to say, the function can be approximated by the line C. In this sense all nonspecific lines are equivalent, for them $N < \infty$. Even a small difference of C from $\hat{C}$ makes any pattern approximable on C. It is clear that the closeness C to some line $\hat{C}$ causes the function, nonapproximable by the line $\hat{C}$, to be approximable by C with a large norm; the smaller the distance between C and $\hat{C}$, the larger this norm. Further, we will specify what we mean by the distance between two close lines in this context.

We explain this by an elementary example. Let the current on circle of radius a realize the pattern $F(\varphi) = 1/\sqrt{2\pi}$; it has the norm $N = 1/|J_0(ka)|$ (see (4.14)). For the specific line $r = \hat{a}$, $J_0(k\hat{a}) = 0$, so that if the circle is close to the specific one (i.e. $k|a - \hat{a}| << 1$), then $N \approx \{k|(a - \hat{a})J_0'(k\hat{a})|\}^{-1}$. In the general case, the answer to the question of how far from the specific line its influence is observed (i.e., where it manifests itself in the current norm), can be based on the formula (4.50).

For the specific line we have $\mu_q = 0$, where μ_q is one of the eigenvalues of operator KK^{ad} (see (4.40b)). The kernel of this integral operator contains the integral over C and this fact defines the dependence μ_q on C. However, this dependence is difficult to use practically; moreover, changing C changes the set of functions $\psi_m(\varphi)$ and the coefficients $A_m = (F, \psi_m)$ contained in (4.50).

Before describing another technique for estimating the dependence N on C, we make one general remark. Compare two properties: the nonrealizability of the approximable pattern, and nonapproximability of the pattern itself. The first one can be eliminated by a small perturbation of the line C only in especial cases, say, if the pattern is realizable in general, but nonrealizable by a line C, close to the circle of radius a_0 (see (1.21)). However, there always exists a small as desired perturbation of the pattern, which makes it realizable. In the second case any small perturbation of the pattern cannot make it approximable, but there exist small perturbations of the line, making the pattern approximable. In the first case the norm of the current realizing the perturbed pattern is, in general, not large. In the second one the norm of the current on the perturbed contour, approximating the pattern, is, in general, large. This follows, for instance, from (4.50). Only in exclusive cases, for the patterns with $|A_q| << 1$, can the small perturbation of $\hat{C}$ cause μ_q to be large in comparison with $|A_q|^2$, remaining small in comparison with the unit, so that the norm (4.50) will be large.

Therefore, there is no close analogy between these two properties: nonrealizability and nonapproximability. Although the nonapproximability (as well as nonrealizability) can be eliminated by a small perturbation of the line or pattern, it is a more "rough" property, namely, it is *stable with respect to perturbations.*

2. We describe a technique allowing to estimate the norm of current satisfying equation (4.26) without solving this equation, that is, without determining the current $j(s)$. Assume that this equation has a solution. Multiply the equation

$$\int_C K^*(s,\varphi)j^*(s)ds = F^*(\varphi) \tag{4.101}$$

complex conjugated to (4.26), by some function $\Phi(\varphi)$ which will be chosen further, and integrate from 0 to 2π with respect to φ. Exchange the order of integration in the left-hand side. According to (4.32), the inner integral is $K^{\text{ad}}\Phi$, which gives

$$\int_C K^{\text{ad}}\Phi{\cdot}j^*(s)ds = \int_0^{2\pi} F^*(\varphi)\Phi(\varphi)d\varphi. \tag{4.102}$$

The Cauchy inequality, together with the definition of N^2, gives

$$N^2 \int_C \left|K^{\text{ad}}\Phi\right|^2 ds \geq \left|\int_C K^{\text{ad}}\Phi{\cdot}j^*(s)ds\right|^2, \tag{4.103}$$

so that

$$N^2 \geq \frac{\left|\int_0^{2\pi} F^*(\varphi)\Phi(\varphi)d\varphi\right|^2}{\int_C \left|K^{\text{ad}}\Phi\right|^2 ds}. \tag{4.104}$$

Remember that the two last inequalities become equalities if the function Φ is chosen in such a way that $K^{\text{ad}}\Phi$ differs from $j(s)$ only by a constant factor.

Show that the equality holds in (4.104) when the function Φ is chosen from an wide enough class of functions according to the requirement that the right-hand side of (4.104) is maximal, that is,

$$N^2 = \max_{\Phi} \frac{\left|\int_0^{2\pi} F^*(\varphi)\Phi(\varphi)d\varphi\right|^2}{\int_C \left|K^{\text{ad}}\Phi\right|^2 ds}. \tag{4.105}$$

To prove this, use again the generalized functions of double orthogonality $\psi_m(\varphi)$. Choose Φ in the form of the finite sum by $\psi_m(\varphi)$:

$$\Phi(\varphi) = \sum_{m=1}^{M} C_m \psi_m(\varphi). \tag{4.106}$$

Then, according to (4.46b),

$$K^{\text{ad}}\Phi = \sum_{m=1}^{M} C_m \sqrt{\mu_m} j_m(s), \tag{4.107}$$

and inequality (4.104) has the form

$$N^2 \geq \frac{\sum_{n=1}^{M} \sum_{m=1}^{M} A_n A_m^* C_n C_m^*}{\sum_{m=1}^{M} \mu_m |C_m|^2}. \tag{4.108}$$

Remember that A_m are the Fourier coefficients in the expansion (4.68a) of $F(\varphi)$ in the same functions $\psi_m(\varphi)$.

Introduce the notations $c_m = C_m \sqrt{\mu_m}$, $a_m = A_m / \sqrt{\mu_m}$, $\vec{c} = \{c_m\}_{m=1}^{M}$, $\vec{a} = \{a_m\}_{m=1}^{M}$. Then (4.108) can be rewritten in the form

$$N^2 \geq \frac{|(\vec{a}, \vec{c})|^2}{(\vec{c}, \vec{c})}, \tag{4.109}$$

where $(\vec{a}, \vec{c})$ is the scalar product of two M-dimensional complex vectors. According to the Cauchy inequality, the right-hand side of (4.109) is maximal when $\vec{c} = \vec{a}$, that is, $C_m = A_m / \mu_m$, and this maximum is

$$X^{(M)} = \sum_{m=1}^{M} \frac{|A_m|^2}{\mu_m}. \tag{4.110}$$

Consequently, the lower estimate of the current norm is

$$N^2 \geq X^{(M)}. \tag{4.111}$$

With growing M the class of functions $\Phi(\varphi)$ is widened, and $X^{(M)}$ can only increase: $X^{(M+1)} \geq X^{(M)}$. It is obvious that $X^{(M)}$ is bounded by the squared norm of the solution to (4.26). Hence, there exists $\lim_{M\to\infty} X^{(M)} = X^{(\infty)}$, $N^2 \geq X^{(\infty)}$. Comparing (4.110) with (4.50) we obtain the sought result:

$$N^2 = X^{(\infty)}. \tag{4.112}$$

Consequently, substituting different functions $\Phi(\varphi)$ into (4.104), we obtain the lower estimations of N^2; while widening the function set from which Φ is chosen, and transforming it into the complete set, the maximum of the right-hand side of (4.104) not only grows, but also becomes more informative. At the limit this inequality transforms to the equality. Then one can calculate N^2 without using (4.50) and without calculating the numbers μ_m and coefficients A_m.

3. As it was shown in Section 4.2, a line C is specific if (and only if) there are $\mu_q = 0$ among the numbers μ_m. This also follows from (4.110): if $\mu_q = 0$ and $A_q \neq 0$, then, at $M \geq q$, $X^{(M)} = \infty$.

It is easy to find an analogy of the condition $\mu_q = 0$, $q \leq M$, applicable also in the case when some other set of orthogonal functions $\psi_m(\varphi)$, different from the generalized functions of double orthogonality, are used in (4.106). If, for such a set, the functions $K^{\text{ad}}\psi_m$ are not orthogonal, then the denominator of (4.104) is the double sum

$$\sum_{n=1}^{M} \sum_{m=1}^{M} C_n C_m^* \beta_{nm}, \quad \beta_{nm} = \int_C K^{\text{ad}}\psi_m \left(K^{\text{ad}}\psi_n\right)^* ds, \tag{4.113}$$

and $X^{(M)}$ can be calculated from the equation

$$\det\left\{A_m A_n^* - X^{(M)}\beta_{nm}\right\}_M = 0, \tag{4.114}$$

where the index M is the determinant order and $X^{(M)}$ is presented not only in diagonal terms. If the trigonometric function set (4.7) is used as the functions $\psi(\varphi)$, then

$$\beta_{nm} = \frac{4\pi|D|^2}{\sqrt{(1+\delta_{0n})(1+\delta_{0m})}}\int_C J_n(kr)J_m(kr)\cos(n\varphi)\cos(m\varphi)ds, \tag{4.115}$$

where $r = r(\varphi)$ is the equation of the line C, the coefficient D is given in (4.32).

In general, the root of equation (4.114) differs from the value (4.110) at the same M. The M-dimensional functional spaces, constructed on different bases, are different and the extreme values of the functional are also different. These values coincide only at the limit ($M \to \infty$), because the maximum, reached in any complete function set, does not depend on this set. Since the maximum equals N^2 for the set of generalized functions of double orthogonality, it equals N^2 for any other complete set. This means that the maximal root of equation (4.114) tends to N^2 at $M \to \infty$.

If equation (4.114) has the root $X^{(M)} = \infty$ at a finite M, which can occur if C is a specific line, then, dividing the left-hand side of (4.114) by $\left[X^{(M)}\right]^{M^2}$ we have

$$\det\{\beta_{mn}\}_M = 0. \tag{4.116}$$

If $\psi_m(\varphi)$ are the general functions of double orthogonality, then $\beta_{nm} = \mu_m\delta_{nm}$ and the left-hand side of (4.116) is the product of the M first eigenvalues μ_m. The condition (4.116) is a formal generalization of the condition of existence $\mu_q = 0$ among M first numbers μ_m ($q \leq M$).

At $M \to \infty$ the condition (4.116) is fulfilled for any chosen set of functions, because β_{nm} (see 4.113) are fast decreasing with increasing index, owing to $K^{\mathrm{ad}}\psi_m$ becoming fast varying along C. For instance, for the trigonometrical function set (4.7) β_{nm} decrease by the Debye asymptotics (3.38a) of the Bessel functions (see (4.115) below). Of course, it does not follow from this, that any line is specific (see corresponding note in Subsection 4.2.4). Only if calculation of the determinant is stable with growing M and it is close to zero, can one say that the line C is close to some specific line $\hat{C}$. In this case the Fourier coefficients of the function $\Phi(\varphi)$ maximizing the right-hand side of (4.104) are determined from a well-posed system of equations, and their limits at $M \to \infty$ exist. The limiting values are the Fourier coefficients of the orthogonal complement function $\hat{F}(\varphi)$. This numerical technique, permitting to ascertain whether the line C is specific, can turn out to be simpler than those proposed in Sections 3.3, 4.2.

If the line C is nonspecific, then (4.113) gives

$$\det\{\beta_{mn}\}_M > 0. \tag{4.117}$$

Indeed, at arbitrary numbers C_n, which are not equal to zero simultaneously, the double sum

$$\sum_{n=1}^{M}\sum_{m=1}^{M}\beta_{nm}C_nC_m^* \tag{4.118}$$

is positive, because it equals the integral $\int_C \left|\sum_{n=1}^{M} C_n K^{\text{ad}} \psi_n\right|^2$ (see (4.104)). This integral equals zero only if all the functions $K^{\text{ad}}\psi_n$ are equal to zero identically. This is possible only in the case when C is a specific line. Then all the coefficients C_n except for $n = q$ can be chosen to be equal to zero, and since the term with $n = q$ equals zero even at $C_q \neq 0$, the double sum equals zero (although not all the C_n are equal to zero). The positive definiteness of the above sum results in holding inequality (4.117).

Note that, at $M = 2$, the condition (4.117) is the Cauchy inequality; at larger M it can be considered as a generalization of this inequality.

Since the left-hand side of equation (4.114) is real at real X, and it is positive at $X = \infty$ under (4.117), to determine the maximal root of this equation, it is sufficient to calculate its left-hand side with decreasing X. The first value of X at which it changes its sign, is the sought root, that is, the lower estimate of N^2 (see (16.9)). The larger M, the more informative this estimate is, and, simultaneously, the more laborious are the calculations.

4. If C is a specific line, and the function $F(\varphi)$ is not orthogonal to the appropriate orthogonal complement function, then the current with infinite norm is needed on C to approximate $F(\varphi)$ by its pattern. If C is a nonspecific line, then approximation is possible by the current with finite norm, and, according to (4.111), this norm is larger than the largest root $X^{(M)}$ of equation (4.114). Here M is the number of significant Fourier coefficients of the function $F(\varphi)$. The more oscillating is this function, the larger M must be taken for an acceptable lower estimate of the norm, and the larger is this estimate $X^{(M)}$. The currents needed for creation or approximation of oscillating patterns are larger than for smooth ones.

Let the function $F(\varphi)$ be normalized to unity by (1.12), so that in (4.111) N is the norm of the current creating the normalized pattern according to (4.101). Then the number $X^{(M)}$ can be considered as a measure of the closeness of C to a specific line. This measure depends on the function $F(\varphi)$ to be approximated. In this subsection we give another characteristic of this closeness. It does not depend on $F(\varphi)$, but the results described here are qualitative only.

The numerator of the fraction in (4.104) is smaller than $\int_0^{2\pi} |\Phi(\varphi)|^2 d\varphi$, because, according to the Cauchy inequality and condition (1.12), we have

$$\left|\int_0^{2\pi} F^*(\varphi)\Phi(\varphi)d\varphi\right|^2 \leq \int_0^{2\pi} |\Phi(\varphi)|^2 d\varphi. \tag{4.119}$$

Increase the right-hand side of condition (4.104) and consider the inequality obtained

$$N^2 \geq \frac{\int_0^{2\pi} |\Phi(\varphi)|^2 d\varphi}{\int_C |K^{\text{ad}}\Phi|^2 \, ds}. \tag{4.120}$$

This inequality is sufficient (but, of course, not necessary) for (4.104) to hold. If the current norm fulfils (4.120), then it fulfils (4.104) as well.

The inequality (4.120) does not contain the function $F(\varphi)$, it characterizes only the line C and the function $\Phi(\varphi)$. If the rate of variation of $\Phi(\varphi)$ with respect to φ is not limited, then (4.120) contains no information, because one can put $\Phi(\varphi) = \psi_m(\varphi)$ and the right-hand side of (4.120) (equal to μ_m^{-1} in this case) can be made infinitely large, and (4.120) means only that if the value of N^2 is very large, then the inequality (4.104) holds. However, if the rate of variation of the function $F(\varphi)$ (say, the maximal index of the significant Fourier coefficients in its expansion) is known, then one can take the functions $\Phi(\varphi)$ having approximately the same (at least, not larger) rate of variation. Then (4.120) gives an acceptable estimate of the norm of the current, necessary for approximation of any pattern of unit norm with the same order of oscillations. This estimate depends only on C and not on $F(\varphi)$.

The further analysis in this subsection carries over (in a simplified form) the results, obtained in the preceding section for (4.104), to (4.120). If the set of numbers μ_m corresponding to the given line C, is known, then the minimum of the right-hand side of (4.120) is easy to find. It equals μ_q^{-1}, where μ_q is minimal among the M first eigenvalues, which gives the estimate $N^2 \geq \mu_q^{-1}$. The exact formula (4.50) after using the normalization (1.12) results, of course, in the analogous estimate $N^2 \geq |A_q|^2 / \mu_q$. Similarly to the preceding subsection, we find an analog of this result, which does not contain the number μ_q.

Let $\Phi(\varphi)$ be expressible as the finite sum (4.106) by the orthonormal functions $\psi_m(\varphi)$, which in general are not assumed to be the generalized functions of double orthogonality; for instance, they can be the trigonometrical functions (4.7). The denominator in the inequality (4.120) is the double sum $\sum_{n=1}^{M}\sum_{m=1}^{M} C_n C_m^* \beta_{nm}$ (see (4.113)), where β_{nm} are given in (4.115). Then (4.120) is of the form

$$N^2 \geq \frac{\sum_{n=1}^{M} |C_n|^2}{\sum_{n=1}^{M}\sum_{m=1}^{M} C_n C_m^* \beta_{nm}}. \tag{4.121}$$

At any C_n, the right-hand side of (4.121) is less than or equal to the maximal root $Y^{(M)}$ of the equation

$$\det\left\{\delta_{nm} - Y^{(M)}\beta_{nm}\right\}_M = 0, \tag{4.122}$$

analogous to equation (4.114) but not containing the coefficients A_m. The number M in (4.122) should be chosen so that the Fourier series of the patterns $F(\varphi)$, which is of interest in the concrete problem, by the functions (4.7), can be cut at the Mth term without a large error. The wider (in this sense) the pattern class, the larger M, and the larger the estimate $Y^{(M)}$ for the norm of the currents approximating the patterns of this class.

Fixing M, we obtain from (4.122) the estimate of the *closeness of the line C to the nearest specific one* $\hat{C}$. The estimate is somewhat more exact than that given by the left-hand side of the inequality (4.117), that is, the value $\det\{\beta_{nm}\}_M$. The estimates coincide when C is a specific line, more exactly, if $\det\{\beta_{nm}\}_M = 0$ at some finite M; then $Y^{(M)} = \infty$. If the line C is close to a specific one (i.e. , at $Y^{(M)} >> 1$), then it is more convenient to write (4.122) in the form

$$\det\left\{\beta_{nm} - \delta_{nm} y^{(M)}\right\}_M = 0, \tag{4.123}$$

where $y^{(M)} = 1/Y^{(M)}$. At $y^{(M)} << 1$ this root can be easily found in an explicit form, keeping only its zeroth and first degrees in equation (4.123). According to the above, for the patterns, expressed as a finite sum of M terms of the Fourier series, we have $N^2 \geq 1/y^{(M)}$. Similarly to $X^{(M)}$, this number depends on the class of functions from which M functions ψ_m are chosen, the linear combinations of which approximate the functions $F(\varphi)$, which we are interested in. The number $y^{(M)}$ is the sought generalization of the number μ_q for the case of an arbitrary system of functions $\psi_m(\varphi)$, but this inequality, as well as (4.120), is not a necessary condition for N^2.

The physical sense of the results obtained in the last two subsections, is obvious. The more oscillating are the functions $F(\varphi)$, the larger should be the norm of the currents generating the patterns which approximate $F(\varphi)$, and the wider is the domain of influence of the specific line closest to C. This elementary concept can be illustrated in an example similar to that given at the beginning of the section. Consider the pattern $F(\varphi) = \pi^{-1/2}\cos(n\varphi)$ which should be approximated by a current on the circle of radius a. The current norm for this pattern will be $N = |J_n(ka)|^{-1}$, which is approximately equal to $||a - \hat{a}|J_n'(k\hat{a})|^{-1}$, where $J_n(k\hat{a}) = 0$, that is, $\hat{a}$ is the radius of the nearest resonant circle. The width of the domain of its influence at given N is $|a - \hat{a}|$, which is proportional to $|J_n'(k\hat{a})|^{-1}$. It increases quickly with increasing n, that is, as the function $F(\varphi)$ is more complicated.

In the general case, the approximative estimate of N^2 is $y^{(M)}$, and the domain of influence of the specific line (for which $y^{(M)} = 0$) is determined by the condition that for the lines on the boundary of the domain we have $y^{(M)} \approx N^{-2}$, where N is an "allowable" current norm.

Therefore, using an arbitrary complete set of functions $\psi_m(\varphi)$ one can estimate the closeness C to $\hat{C}$ by calculating the values β_{nm} by (4.32), (4.113) and investigating the dependence of the determinant in (4.117) and the root of equations (4.122), (4.123) on M.

5. The inequality (4.104) can be used not only for determination of a lower bound of N^2 but also for the rough estimate of N^2 itself. If the function $\Phi(\varphi)$ is $F(\varphi)$, then (4.104) has the simple form

$$\mathcal{R} \leq \int_C \left|K^{\text{ad}}F\right|^2 ds, \tag{4.124}$$

where $\mathcal{R}$ is the value having the nature of the *radiation resistance*:

$$\mathcal{R} = \frac{\int_0^{2\pi} |F(\varphi|^2\, d\varphi}{\int_C |j(s)|^2\, ds}, \tag{4.125}$$

which means that it is equal to the ratio of the radiated power to the squared current norm. For a good antenna, $\mathcal{R}$ should be large. If C is a specific line and the function $F(\varphi)$ is nonapproximable, then $\mathcal{R} = 0$.

Consider an elementary example showing by how much the right-hand side of (4.124) exceeds $\mathcal{R}$ in different cases. Let the current be located on the circle of radius a, and the pattern be $F(\varphi) = A_0 + A_1\cos\varphi$. Then equation (4.26) can be easily solved and the exact value of $\mathcal{R}$ can be calculated by (4.125)).

The ratio of $\mathcal{R}$ to the right-hand side of (4.124) is

$$\frac{\mathcal{R}}{\int_C |K^{\text{ad}}F|^2\, ds} = \frac{(a_0 + a_1)^2\, p}{(a_0 p + a_1)(a_0 + a_1 p)}, \tag{4.126}$$

where the following notations are introduced:

$$a_0 = |A_0|^2, \quad a_1 = |A_1|^2, \quad p = \frac{J_1^2(ka)}{J_0^2(ka)}. \tag{4.127}$$

The nearer this ratio to unity, the more exactly the right-hand side of (4.124) describes the radiation resistance. Of course, the ratio (4.126) does not exceed unity. It equals unity at $p = 1$ and either $A_0 = 0$ or $A_1 = 0$, and then the radiation resistance is equal to the right-hand side of (4.124).

If $J_0(ka) = 0$ or $J_1(ka) = 0$, then the circle of radius a is the specific line for which the given pattern is nonapproximable and $p = \infty$ or $p = 0$. The estimate (4.124) is useless, because $\mathcal{R} = 0$ and the right-hand side of (4.124) is finite.

At intermediate values of p, that is, when the circle is not close to a specific line, the right-hand side of (4.124) gives (*in this example*) a good approximation of $\mathcal{R}$. We show the numerical results corresponding to the pattern $F(\varphi) = 1 + \cos\varphi$ (i.e. $A_0 = A_1 = 1$). If $0.36 < p < 2.76$, then the ratio (4.126) is larger than 0.8, which means that the error obtained while calculating the value $\mathcal{R}$ from the right-hand side of (4.124) is smaller than 20 percent. An error of 10 percent is obtained in a narrower interval of values p, namely, at $0.5 < p < 2.0$. An accuracy of 20 percent is reached at $1.0 < ka < 1.8$; an accuracy of 10 percent is reached on the smaller interval of the radius values: $1.2 < ka < 1.7$. At $ka = 1.4$, we have $p = 1$ and the equality is reached in (4.124).

In this example the radiation resistance $\mathcal{R}$ reaches its maximal value at $ka = 1.1$: such a circle is maximally distanced from the specific lines ($ka = 2.4$ and $ka = 3.83$) for which $\mathcal{R} = 0$.

We give another elementary example characterizing the influence of the specific line onto the nearest domain. Let the current be distributed with surface density $j(r)$ in the ring $a < r < b$ and generate the pattern $F(\varphi) = 1$. The function $j(r)$ should be found, which minimizes the integral $N^2 = \int_a^b j^2(r) r dr$, at the given value of $\int_a^b j(r) J_0(kr) dr$. It is easy to show that the minimum is reached when $j(r)$ is proportional to $J_0(kr)$. The current density, providing the minimum, is zero on the specific line $r = \hat{a}$, $(J_0(k\hat{a}) = 0)$, because the current on this line does not radiate. In the neighborhood of this line $j(r)$ is small and grows approximately linearly with distance from the line.

6. In this subsection we give a *more rough estimate of the influence* provided by the neighborhood of a specific line $\hat{C}$ to the line C. Determine the *order of value* of the current $j(s)$ to be created on a line C ("mirror") in order that the field of this current (reflected or diffracted) approximates the function $U(\sigma)$ given on a contour Σ encircling C and $\hat{C}$ (Fig. 4.10). This function is not orthogonal to the orthogonal complement function $\hat{V}(\sigma)$ on Σ, which corresponds to the specific line $\hat{C}$, which means that

$$\int_\Sigma U(\sigma) \hat{V}^*(\sigma) d\sigma \neq 0 \tag{4.128}$$

and therefore an infinite current on $\hat{C}$ would be needed to approximate $U(\sigma)$. The functions $U(\sigma)$ and $\hat{V}(\sigma)$ are assumed to be normalized to the unit:

$$\int_\Sigma |U(\sigma)|^2 d\sigma = 1, \tag{4.129a}$$

$$\int_\Sigma \left|\hat{V}(\sigma)\right|^2 d\sigma = 1. \tag{4.129b}$$

We use the constructions of Subsection 3.1.1. Denote by $\hat{v}(r,\varphi)$ an auxiliary field, being equal to zero on $\hat{C}$ and having the jump of the normal derivative on Σ, which is equal to $\hat{V}(\sigma)$.

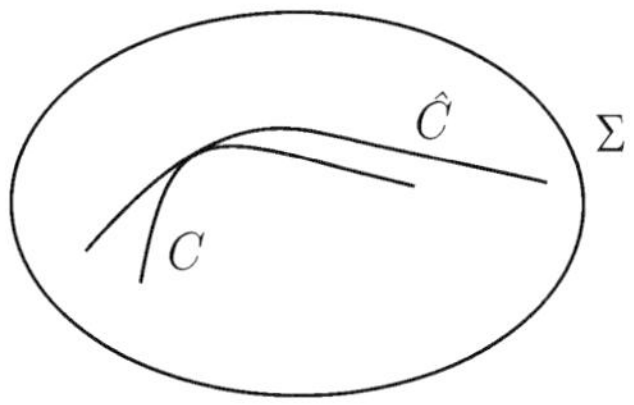

Figure 4.10:

The field $\hat{v}(r,\varphi)$ is created by the "current" $\hat{V}(\sigma)$. Denote by $u(r,\varphi)$ the field, created by the current $j(s)$; it has the jump of the normal derivative on C, equal to $j(s)$ (see (2.1)). The fields $\hat{v}(r,\varphi)$ and $u(r,\varphi)$ satisfy the homogeneous Helmholtz equation in the whole plane and the radiation condition at infinity; Green's formula applied to these function leads to

$$\int_\Sigma \hat{V}^*(\sigma)u(\sigma)d\sigma = \int_C j^*(s)\hat{v}(s)ds. \tag{4.130}$$

This formula is analogous to (3.2), but the integral on the right is taken over the line C (not over $\hat{C}$).

Estimate the function $\hat{v}(s) = \hat{v}(r,\varphi)|_C$ in the right integral. It equals zero on the line $\hat{C}$. If the distance $\delta(s)$ between C and $\hat{C}$ is small, then $\hat{v}|_C = \partial u/dN|_C\,\delta$. Assume that $\partial\hat{v}/dN|_C$ has the same order as $\partial\hat{v}/dN|_\Sigma$ and this derivative has the order $\hat{V}$. Remember that $\hat{V}$ is the difference between the normal derivatives calculated on both sides of the line Σ. According to (4.129b), we have the estimate $\hat{V} \approx L_\Sigma^{-1/2}$, where L_Σ is the length of the contour Σ.

Hence, the estimate of $j(s)$, to be obtained below, is based on the assumption (of course, *very rough*) that

$$\hat{v}|_C = \frac{\delta}{\sqrt{L_\Sigma}}. \tag{4.131}$$

It is not possible to calculate $\hat{v}|_C$ more exactly in a general form, that is, without concretization of the lines C, $\hat{C}$, and Σ.

Replacing $u(\sigma)$ in (4.130) with the close function $U(\sigma)$, we obtain the following value of the average current $j(s)$:

$$\bar{j}(s) = \frac{\sqrt{L_\Sigma}}{L_C\bar{\delta}}\int_\Sigma U(\sigma)\hat{V}(\sigma)d\sigma, \tag{4.132}$$

where the barred symbols denote the average values on the mirror.

Compare the estimate (4.132) with that for the current located on the line C, which create a field $U^0(\sigma)$ on Σ in the case when C is not placed in the domain of influence of any specific line $\hat{C}$. Denote this current by $j^0(s)$. It is equal (in order of value) to kU^0. With the same accuracy as (4.132), we have

$$\frac{\bar{j}}{j^0} = \frac{1}{k\bar{\delta}}\frac{L_\Sigma}{L_C}\int_\Sigma \frac{U(\sigma)}{U^0(\sigma)}\frac{\hat{V}(\sigma)}{1/\sqrt{L_\Sigma}}\frac{d\sigma}{L_\Sigma}. \tag{4.133}$$

When $k\bar{\delta} << 1$, then the closeness of a specific line enlarges the current on C, needed for creation of a given field, by the multiplier $L_\Sigma/L_C \cdot \left(k\bar{\delta}\right)^{-1}$, and the proportionality factor is the integral (4.130), i.e., the value characterizing the "order of nonapproximability" of the given field by fields of the currents located on the specific line.

We consider a numerical example. The "mirror" C is the arc of the circle of radius R (see Subsection 1.1.1). The nearest specific line to it is the resonant ellipse, tangent to C at the end of its small axis and having, at this point, the radius of curvature R. The function $\hat{V}(\sigma)$ is the Mathieu function. However, the curvature radius of the ellipse at this point is a^2/b, where a, b are its semi-axes ($a \geq b$). Then the distance δ between C and $\hat{C}$ at the point s is proportional to the fourth degree of the distance from s to the tangent point. The average distance is $\bar{\delta} = L_C^4 \left(1/\left(Ra^2\right) - 1/R^3\right)/640$. Let $kR = 10$. Then the nearest resonant ellipse with $a^2/b = R$ has the parameters $ka = 4.14$, $kb = 1.73$, so that $\bar{\delta}/\lambda \approx 10^{-3}\left(L_C/\lambda\right)^4$ in this example. The larger the mirror length L_C, the relatively smaller part of it lies in the domain of influence of the specific line, and a smaller current can be distributed on the mirror, creating the field prohibited for this specific line.

If the line C is a part of the analytical curve, then it is a part of a specific line (and not only one). This specific line can be constructed supplying C with the segments of analytical curves up to a closed resonant contour (see Section 3.3). This means, that the fields of currents distributed on C can approximate only the functions $U(\sigma)$ on Σ orthogonal to all the orthogonal complement functions $\hat{V}(\sigma)$, associated with the constructed specific line. We do not discuss restrictions on $j(s)$, following from this. This question was considered in a general form in Section 4.3. To obtain the concrete results for the given lines C and Σ one should solve the problem on the construction of this resonant contour and on the analytical continuation of the field of its eigenoscillation into the domain between C and Σ, and then solve the Cauchy problem on a continuation of this field outside the contour Σ.

5 Electromagnetic Field. The Maxwell Equations

5.1 Trivial Generalizations

1. All the constructions in the previous four chapters were based on the following property of *any field* being a real solution to the scalar Helmholtz equation and having no singularity in the whole plane: *any current* located on a zero line of such a field, creates the pattern orthogonal to a function describing the asymptotical angular dependence of the field at infinity. *Any* real *solution* to the homogeneous Maxwell equations, analytical in the whole space, possesses an analogous property: *any current*, located on the surface where the tangential components of the electrical field are zero, generates the pattern, orthogonal to the function describing the angular dependence of the asymptotics of this field. The orthogonality of two patterns should have the form of condition (5.10) below, where $f_{1,2}$ and $F_{1,2}$ describe the angular dependencies of two components of the fields at infinity (see (2.3)).

The property follows from the Green formula (for the scalar field) and the Lorentz lemma (for the vector field). In fact, it is the formulation of the reciprocity theorem. For instance, the pattern generated in any direction by the source located at any point on C, is equal at this point (with an accuracy to a nonessential factor) to the value of the field of a plane wave, incoming from the same direction. The proof of this statement is quite elementary. The consequences of the properties are very significant, they are applicable to any fields and currents. The first five chapters of the book consist of discussions about these consequences, on the whole.

In this section the main result, previously obtained for the two-dimensional scalar model are transferred to the three-dimensional vector problem. Only one essential difference between these two problems will be discussed in the next section.

All the constructions given in this section are the simple rephrasing of the previous results. This means, in fact, that these results have a very wide application. If the sources of the field are located in the space having a dimension less than that of the space where the field is formed, then the field can possess the property of nonapproximability (see the note after (5.8) below). This case is not exceptional. This affirmation is based only on the linearity of the equation to be held, namely, on the existence of a version of the reciprocity theorem. Therefore, it is valid for the fields of different nature (acoustical, seismic etc.), for which the concept "surface sources" makes sense.

2. We write the Maxwell equations for the monochromatic field (1.4) in terms of the complex magnitudes $\mathbf{E}(\vec{r})$, $\mathbf{H}(\vec{r})$ (the field $(\mathbf{E}, \mathbf{H})$ below) in the homogeneous form

$$\mathrm{rot}\mathbf{H} - ik\mathbf{E} = 0, \tag{5.1a}$$

$$\mathrm{rot}\mathbf{E} + ik\mathbf{H} = 0, \tag{5.1b}$$

analogous to the homogeneous Helmholtz equation (1.5). We will consider only the case, when the currents radiating the field are located on the surface C. They are given not as the right-hand side of the Maxwell equations, but as the jump of the tangential components of the magnetic field on C, that is, as a two-dimensional vector $\vec{j}$ with components

$$[H_{t_1}]|_C = j_{t_2}, \quad [H_{t_2}]|_C = -j_{t_1}. \tag{5.2}$$

Here $\vec{t_1}$, $\vec{t_2}$ are two orthogonal vectors, tangential to the surface C. These vectors, together with the outward normal $\vec{N}$ to the surface, complete the right vector triple $(\vec{t_1}, \vec{t_2}, \vec{N})$. We do not introduce the magnetic currents, therefore the field $\mathbf{E}$ has no jumps (see, however, the note at the end of Subsection 3 below).

The radiation (Sommerfeld) conditions at infinity, written in the spherical coordinate system (ρ, θ, φ), are

$$E_\theta \simeq \frac{e^{-ik\rho}}{k\rho} f_1(\theta, \varphi); \quad E_\varphi \simeq \frac{e^{-ik\rho}}{k\rho} f_2(\theta, \varphi); \tag{5.3a}$$

$$H_\theta = \frac{1}{ik}\frac{\partial E_\varphi}{\partial \rho} \simeq -\frac{e^{-ik\rho}}{k\rho} f_2(\theta, \varphi); \quad H_\varphi = -\frac{1}{ik}\frac{\partial E_\theta}{\partial \rho} \simeq \frac{e^{-ik\rho}}{k\rho} f_1(\theta, \varphi). \tag{5.3b}$$

The functions $f_1(\theta, \varphi)$, $f_2(\theta, \varphi)$ are not given here; only the structure of the first term in the expansion of the field components by $1/(k\rho)$, is determined by these conditions. The pair of functions f_1, f_2 is denoted as (f).

Similarly to the function $u(r, \varphi)$ in the scalar case, the field $(\mathbf{E}, \mathbf{H})$ is generated only by the current $\vec{j}$. It is not the total field in the space. The vectors $\mathbf{E}, \mathbf{H}$ should not satisfy any conditions on the surface C, except for (5.2). The field inducing the current $\vec{j}$ (the "incident field" in terms of the diffraction theory) is not included in $(\mathbf{E}, \mathbf{H})$ (the "diffraction field"). For instance, the condition that the electrical field must be zero on the metallic surface does not relate to the field $(\mathbf{E}, \mathbf{H})$ in (5.1), (5.3).

We do not introduce the permittivity ε and permeability μ into the Maxwell equations here. If these values are constant, then their introduction is equivalent to the frequency changing. If an inhomogeneous dielectric body is present in the field, then the function $k\varepsilon(x, y, z)$ should be substituted in equations (5.1) for the frequency k. The auxiliary field $(\hat{\mathbf{E}}, \hat{\mathbf{H}})$ introduced below should satisfy the same equations as does $(\mathbf{E}, \mathbf{H})$. It is the solution to the problem on diffraction of the spherical wave, incoming from infinity, on the dielectric body. Nothing is changed in the constructions below, especially in the formula (5.4), on which these constructions are based. We confine ourselves to this note and use the Maxwell equations in the form (5.1), that is, at $\varepsilon \equiv 1$, $\mu \equiv 1$.

The analog to the Green formula (1.10) is the simplified form of the Lorentz lemma for two fields $(\mathbf{E}_1, \mathbf{H}_1)$, $(\mathbf{E}_2, \mathbf{H}_2)$:

$$\int_{\mathcal{L}} \{[\mathbf{E}_1 \mathbf{H}_2]_N - [\mathbf{E}_2 \mathbf{H}_2]_N\}\, dS = 0, . \tag{5.4}$$

following from the Maxwell equations (5.1). Here $\mathcal{L}$ is any closed surface, N is the normal to it, the square brackets denote the vector product of two vectors. The formula (5.4) contains only the tangential field components.

Formula (5.4) is valid only in the case when both fields have no singularities, including jumps, inside $\mathcal{L}$. An additional integral appears in the right-hand side of (5.4) if the tangential components of one of fields, say, $\mathbf{H}_1$ have a jump on a surface C inside $\mathcal{L}$. This integral is taken over C and it contains the normal component of the vector product of this jump (equal to the current) and the field $\mathbf{E}_2$. The integral is

$$\int_C [\mathbf{E}_2(\mathbf{H}_1^+ - \mathbf{H}_1^-)]_N dS = \int_C \mathbf{E}_2 \cdot \vec{j} dS, \tag{5.5}$$

where (and below) an expression such as $\mathbf{E}_2 \cdot \vec{j}$ means the scalar product of two vectors. Since the current $\vec{j}$ is tangential to C, only the tangential components of $\mathbf{E}_2$ take part in the integral over C.

3. The auxiliary field $(\hat{\mathbf{E}}, \hat{\mathbf{H}})$ is determined as the field satisfying the homogeneous Maxwell equations and the following conditions at infinity

$$\hat{E}_\theta = \hat{F}_1^*(\theta,\varphi)\frac{e^{ik\rho}}{k\rho} + \hat{F}_1(\theta,\varphi)\frac{e^{-ik\rho}}{k\rho}, \tag{5.6a}$$

$$\hat{E}_\varphi = \hat{F}_2^*(\theta,\varphi)\frac{e^{ik\rho}}{k\rho} + \hat{F}_2(\theta,\varphi)\frac{e^{-ik\rho}}{k\rho}. \tag{5.6b}$$

The magnetic field components $\hat{H}_\theta$, $\hat{H}_\varphi$ are obtained (to the first order with respect to $1/k\rho$) from $\hat{E}_\theta$, $\hat{E}_\varphi$ by the formulas

$$H_\theta = \frac{\partial E_\varphi}{\partial\,(ik\rho)}, \quad H_\varphi = -\frac{\partial E_\theta}{\partial\,(ik\rho)}, \tag{5.7}$$

where only the exponent factor should be differentiated. The first summands in (5.6) and in the analogous formulas for $\hat{H}_\theta$, $\hat{H}_\varphi$ are the *nonsommerfeld* terms; they describe the spherical wave *incoming from infinity* and generating the field $(\hat{\mathbf{E}}, \hat{\mathbf{H}})$ in the whole plane. In contrast to (2.3), we assume that the field $\hat{\mathbf{E}}$ is real. Then it follows from (5.1) together with (5.6), (5.3b), that the field $\hat{\mathbf{H}}$ is imaginary in the whole space. The complex auxiliary field $(\hat{\mathbf{E}}, \hat{\mathbf{H}})$, in which neither $\hat{\mathbf{E}}$ nor $\hat{\mathbf{H}}$ are in-phase, will be considered in the next section. In this case two angular factors in each summand of (5.6a), (5.6b) are independent, as in (2.3).

If a field $\hat{\mathbf{E}}$ exists which fulfils the condition

$$\hat{E}_t\Big|_{\hat{C}} = 0, \tag{5.8}$$

on some surface $\hat{C}$, then $\hat{C}$ is called the specific surface. In (5.8) t is an arbitrary tangential direction to $\hat{C}$. In fact, this formula can be rewritten as two conditions for two orthogonal tangential directions.

Note that the field, which does not equal zero identically in the whole three-dimensional space, can be equal to zero only on a surface, that is, in a space of smaller dimension. Remember that, in the two-dimensional case, the field $\hat{u}(r,\varphi)$, which is not equal to zero in the whole plane, could equal zero on a line only. The nonapproximability following from (5.8) is applied only to a current on the surface $\hat{C}$. This is impossible if the current is located in a finite three-dimensional volume. However, the approximation can require very large currents when all points of the volume are close to the specific surface (see Section 4.4).

Let us first assume that the field $(\hat{\mathbf{E}}, \hat{\mathbf{H}})$ has no singularities in the whole space. The formula (5.4) with $\mathcal{L}$ as the infinitely remote sphere (where the expressions (5.3), (5.6) are valid) can be applied to the fields $(\mathbf{E}, \mathbf{H})$ and $(\hat{\mathbf{E}}, \hat{\mathbf{H}})$ as $(\mathbf{E}_1, \mathbf{H}_1)$, $(\mathbf{E}_2, \mathbf{H}_2)$, respectively. The integral (5.5), taken over $\hat{C}$, is caused by the field $\mathbf{H}$ jump. According to (5.8), it equals zero at any currents located on $\hat{C}$. The outward normal to $\mathcal{L}$ coincides with $\vec{\rho}$, so that the integrand in (5.4) is

$$E_\theta \hat{H}_\varphi - E_\varphi \hat{H}_\theta - \hat{E}_\theta H_\varphi + \hat{E}_\theta H_\varphi. \tag{5.9}$$

It contains only the tangential components of the fields. Introduce the components of $(\mathbf{E}, \mathbf{H})$ and $(\hat{\mathbf{E}}, \hat{\mathbf{H}})$ by (5.3), (5.6), respectively. As usual in such constructions, the terms originating from the waves of the same direction (outgoing spherical waves, i.e. the terms with the factor $\exp(-ik\rho)$) cancel out. Only the first (nonsommerfeld) terms in (5.6) remain. The analogous consideration led to the left-hand side in (2.6) in the scalar case. The formulas (5.6) contain only tangential components of the fields. As a result, we obtain

$$\int_\Omega \left[f_1(\theta,\varphi)\hat{F}_1^*(\theta,\varphi) + f_2(\theta,\varphi)\hat{F}_2^*(\theta,\varphi) \right] d\Omega = 0, \tag{5.10}$$

where Ω is the complete solid angle 4π, $d\Omega = \sin\theta d\theta d\varphi$ is its element. This formula is analogous to (1.23).

Denote by

$$(F^{(1)}, F^{(2)}) = \int_\Omega \left[F_1^{(1)}(\theta,\varphi) F_1^{(2)*}(\theta,\varphi) + F_2^{(1)}(\theta,\varphi) F_2^{(2)*}(\theta,\varphi) \right] d\Omega \tag{5.11}$$

the inner product of two pairs of the complex functions $F^{(1)}$, $F^{(2)}$ in the *function pair space*. The formula (5.10) means that any pattern (f) generated by the current located on a specific surface $\hat{C}$, is orthogonal to the pattern $(\hat{F})$, it means, $(f, \hat{F}) = 0$. The element $(\hat{F}) = \{\hat{F}_1, \hat{F}_2\}$ of the function pair space, is the orthogonal complement in the pattern space. It follows from this that any function pair (F) can be approximated by the patterns (f) with no greater precision than that given by the inequality

$$(F - f, F - f) \geq |b|^2, \tag{5.12a}$$

$$b = (F, \hat{F}); \tag{5.12b}$$

$(\hat{F})$ is assumed to be normalized to unity:

$$(\hat{F}, \hat{F}) = 1. \tag{5.13}$$

If there exist several fields $(\hat{\mathbf{E}}_p, \hat{\mathbf{H}}_p)$, $p = 1, 2, ..., P$, for which the conditions (5.8) are valid, then there exist P function pairs $(\hat{F}_p)$ orthogonal to any pattern created by a current on $\hat{C}$. Then the inequality (5.12a) contains the sum $\sum_{p=1}^{P} |b_p|^2$ on the right-hand side, where $b_p = (F, \hat{F}_p)$. The necessary conditions of approximability are the conditions (1.41); if there are no other elements of the orthogonal complement space, then these conditions are also sufficient.

By definition, for any specific surface $\hat{C}$, a solution $(\hat{\mathbf{E}}, \hat{\mathbf{H}})$ to the Maxwell equations exists, in which the vector $\hat{\mathbf{E}}$ has the property (5.8). Simultaneously, *another solution* exists, in which the vector $\hat{\mathbf{H}}$ has a similar property, namely, its tangential components are zero on the *same surface* $\hat{C}$. This follows from the invariance of the equations under the substitution $\mathbf{E} \to \mathbf{H}, \mathbf{H} \to -\mathbf{E}$. The orthogonal complement functions corresponding to these two fields are mutually orthogonal, as follows from (5.6), (5.3b).

Let, for instance, $\hat{C}$ be a cylindrical surface, parallel to the z-axis, and let none of the fields depend on z. The field $(\hat{\mathbf{E}}, \hat{\mathbf{H}})$ can be expressed by two scalar functions $\psi_1(x, y)$, $\psi_2(x, y)$, as follows: $\hat{E}_z = \psi_1(x, y)$, $\hat{H}_z = \psi_2(x, y)$. Then $\hat{E}_s$ is proportional to $\partial\psi_2/\partial N$, where s is the direction tangential to $\hat{C}$ in the plane $z = const$. In this case the conditions (5.8) are of the form:

$$\psi_1|_{\hat{C}} = 0, \qquad \left.\frac{\partial\psi_2}{\partial N}\right|_{\hat{C}} = 0. \tag{5.14}$$

Hence the functions ψ_1, ψ_2 are solutions to the Helmholtz equation in the plane $z = const$ with the Dirichlet or Neumann boundary conditions on $\hat{C}$, respectively. Another field $(\hat{\mathbf{E}}, \hat{\mathbf{H}})$ exists, in which $\hat{H}_z = \psi_1(x, y)$, $\hat{E}_z = \psi_2(x, y)$, so that $\hat{H}_s$ is proportional to $\partial\psi_2/\partial N$. The condition $\left.\hat{H}_t\right|_{\hat{C}} = 0$ is fulfilled for this field. Each specific surface $\hat{C}$ is *simultaneously a specific surface of the magnetic type.*

Therefore, if some surface $\hat{C}$ has the property that the currents located on it generate a noncomplete system of patterns, then the magnetic currents on $\hat{C}$ generate the noncomplete pattern system, too. The orthogonal complement functions of both these systems do not coincide (in terms of (2.41) $J = 0$). As in Subsection 2.1.6, this means that if the currents of both types can be located on some surface, then any function can be approximated by the patterns of all these currents together.

4. Widen the concept of the specific surface $\hat{C}$, requiring (similarly to Chapter 3) that the vector $\hat{\mathbf{E}}$ of a solution to the Maxwell equations, being real and satisfying the condition (5.8) on $\hat{C}$, has no singularities at least in a *finite part* of the space, encircled by a closed surface Σ containing $\hat{C}$. Then there exists a function on Σ, being the function of the orthogonal complement to the tangential components of the field $\mathbf{E}$ generated on Σ by a current located on $\hat{C}$.

To construct such a function, one should solve the "*first boundary value problem*" outside Σ, namely, to find a field satisfying the Maxwell equations, the Sommerfeld condition at infinity and also having on Σ the same tangential components of the electric field as does $\mathbf{E}$. Then one has to consider the field, equal to $(\hat{\mathbf{E}}, \hat{\mathbf{H}})$ inside Σ and to the found solution outside it. This field has a jump of the tangential components of the magnetic field on Σ. The jump is "current" generating this field. Denote this jump (a two-dimensional vector defined on Σ) by $\hat{V}(\sigma)$. This notation coincides with that given in (3.1) for the analogous scalar function. The field considered here also has a jump of the normal components of the electrical field on Σ, but this jump will not appear in the formulas below.

Apply formula (5.4) to the field $(\mathbf{E}, \mathbf{H})$, generated by a current located on $\hat{C}$, and to the field defined as $(\hat{\mathbf{E}}, \hat{\mathbf{H}})$ inside Σ and, as the above prolongation of $(\hat{\mathbf{E}}, \hat{\mathbf{H}})$, through Σ. The integral over the infinitely remote sphere disappears, because both fields satisfy the same radiation condition (see the text before (5.10). The term (5.5) does not arise because the second field satisfies the condition (5.8). Only the integral over Σ, containing the jump of the magnetic field, remains, which gives

$$\int_{\Sigma} \mathbf{E}(\sigma) \cdot \hat{V}(\sigma) d\sigma = 0. \tag{5.15}$$

The integrand is the scalar product of two-dimensional vectors; it contains only the tangential components of the field $\mathbf{E}$ (see the similar formula (5.10). The two-dimensional vector $\hat{V}(\sigma)$ plays the role of the orthogonal complement for the tangential to Σ components of the electric field generated by the currents located on the specific surface $\hat{C}$.

The approximability restrictions on the electric fields in the near zone are stronger than on the patterns. First of all, it can result that the field $(\hat{\mathbf{E}}, \hat{\mathbf{H}})$ for the given specific surface has no singularities except in a part of the space (inside Σ), close to $\hat{C}$. Besides, the vector $\hat{V}(\sigma)$ is, generally speaking, smoother than the function $(\hat{F})$. Here we do not verify this statement in detail. To do this, it would be necessary to repeat the considerations from Subsection 3.1.2. We should imply under Σ a spherical surface and represent the field $(\hat{\mathbf{E}}, \hat{\mathbf{H}})$ inside Σ by the Debye potentials in the form of the Fourier series by the spherical harmonics $P_n^{(m)}(\cos\theta)\cos(m\varphi)$ and the Bessel functions of the half-integer indices (see (5.18), (5.19) below). Then the mentioned first boundary value problem, exterior with respect to Σ, can be solved in the form of a series containing the half-integer Hankel functions. As a result, we would obtain an expression for the Fourier coefficients of the vector $\hat{V}(\sigma)$. The comparison of the result with the analogous Fourier series of the function $(\hat{F})$ shows that high harmonics in $\hat{V}(\sigma)$ are smaller than in $(\hat{F})$. This means that $\hat{V}(\sigma)$ is smoother than $(\hat{F})$, and, therefore, the approximability conditions for the fields in the near zone are stronger than for the patterns. We omit these derivations, which are cumbersome, but, in fact, very simple, because they duplicate similar derivations for the scalar case, given in Subsection 3.1.2.

It is easily shown that, in the case when $\hat{C}$ is a specific surface, there exists a two-dimensional vector $\hat{W}(\sigma)$, tangential to Σ, which is orthogonal to the magnetic field $\mathbf{H}$ generated by some current located on $\hat{C}$, such that

$$\int_{\Sigma} \mathbf{H}(\sigma) \cdot \hat{W}(\sigma) d\sigma. \tag{5.16}$$

This formula is analogous to (5.15), and the vector $\hat{W}(\sigma)$ can be found in the same way as $\hat{V}(\sigma)$. The field prolonging $(\hat{\mathbf{E}}, \hat{\mathbf{H}})$ into the space outside Σ should solve the second boundary value problem, which means that it has the same tangential components of the magnetic field as $\hat{\mathbf{H}}$ has. The field, equal to $(\hat{\mathbf{E}}, \hat{\mathbf{H}})$ inside Σ and to the (just mentioned) one outside it, has only one singularity, namely, the jump of tangential components of the electric field (and normal components of the magnetic one, which is not essential). This jump was denoted above by $\hat{W}(\sigma)$. The formula (5.16) follows immediately from (5.4).

5. The antenna surface must be not a specific one or close to it. The apical angle of the *sectorial antenna* must not be equal or close to the angle mentioned in (2.47).

Let us obtain analogous results for the *cone antenna*. Consider the circular cone antenna with the axis coinciding with that of the spherical coordinate system (ρ, θ, φ). The apex angle is 2α, so that the equation of the antenna surface is $\theta = \alpha$. This surface is a specific one when the field $(\hat{\mathbf{E}}, \hat{\mathbf{H}})$ exists, for which

$$\hat{E}_\rho\Big|_{\theta=\alpha} = 0, \tag{5.17a}$$

$$\hat{E}_\varphi\Big|_{\theta=\alpha} = 0. \tag{5.17b}$$

It is convenient to express the components of $(\hat{\mathbf{E}}, \hat{\mathbf{H}})$ by two scalar functions (the electrical and magnetic Debye potentials) $\hat{U}(\rho,\theta,\varphi)$, $\hat{V}(\rho,\theta,\varphi)$:

$$\hat{E}_\rho = \left(\frac{\partial^2}{\partial\rho^2} + k^2\right)\left(\rho\hat{U}\right), \tag{5.18a}$$

$$\hat{E}_\varphi = \frac{1}{\rho\sin\theta}\frac{\partial^2\left(\rho\hat{U}\right)}{\partial\rho\partial\varphi} + ik\frac{\partial\hat{V}}{\partial\theta} \tag{5.18b}$$

$$\hat{E}_\theta = \frac{1}{\rho}\frac{\partial^2\left(\rho\hat{U}\right)}{\partial\rho\partial\theta} - \frac{ik}{\sin\theta}\frac{\partial\hat{V}}{\partial\varphi}. \tag{5.18c}$$

These potentials satisfy the three-dimensional Helmholtz equation. Each of them can be written in the form

$$\frac{1}{\sqrt{k\rho}}J_{n+1/2}(k\rho)P_n^m(\cos\theta)\left\{\begin{array}{c}\sin(m\varphi)\\ \cos(m\varphi)\end{array}\right\} \tag{5.19}$$

for $\hat{U}$ and $\hat{V}$, respectively. Generally speaking, any cylindrical function could be in (5.19) instead of $J_{n+1/2}(k\rho)$. However, only the Bessel function has no singularity at $\rho = 0$. It follows from (5.18) that the conditions (5.17) can be fulfilled if either $\hat{V} \equiv 0$ or $\hat{U} \equiv 0$. Consider these two cases separately.

Assume at first that $\hat{V} \equiv 0$. Then α is a root of the equation

$$P_n^m(\cos\alpha) = 0. \tag{5.20}$$

Indeed, two components of $\hat{\mathbf{E}}$ in (5.17) do not contain $\partial\hat{U}/\partial\theta$, and therefore they are proportional to the function $P_n^m(\cos\alpha)$. For (5.17) to be valid, equation (5.20) must hold.

The orthogonal complement $(\hat{F})$ to the patterns of currents located on the cone $\theta = \alpha$ is the angular factor in the nonsommerfeld term of the asymptotics of the tangential components of $(\hat{\mathbf{E}}, \hat{\mathbf{H}})$ at $\rho \to \infty$.

According to (5.18), (5.19), the functions $\hat{F}_1^*(\theta,\varphi)$ and $\hat{F}_2^*(\theta,\varphi)$ are of the form

$$\hat{F}_1^*(\theta,\varphi) = \frac{d}{d\theta}P_n^m(\cos\theta)\sin(m\varphi), \tag{5.21a}$$

$$\hat{F}_2^*(\theta,\varphi) = -mP_n^m(\cos\theta)\frac{\cos(m\varphi)}{\sin\theta}, \tag{5.21b}$$

where the normalized factor following from (5.13) is omitted.

The numbers n and m should be integer, $n \geq m$; otherwise the function $P_n^m(\cos\theta)\cos(m\varphi)$ will have a singularity. For each pair of these numbers equation (5.20) determines the values α at which the cone $\theta = \alpha$ is the specific surface. In fact, the restrictions on the approximability of the patterns are essential only when $(\hat{F})$ is smooth, i.e., when n, m are small numbers.

For instance, for the pattern independent of φ $(m = 0)$ the cone is the specific surface at, approximately, $\alpha = 55^o$ $(n = 2)$, $\alpha = 39^o$ $(n = 3)$, $\alpha = 31^o$ $(n = 4)$. For the approximation of the patterns, proportional to $\cos\varphi$ $(m = 1)$, it is necessary to avoid the angles $\alpha = 63^o$ $(n = 3)$ and $\alpha = 49^o$ $(n = 4)$.

If $\hat{U} \equiv 0$ is assumed, then $\hat{E}_\rho \equiv 0$, $\hat{E}_\varphi = ik\partial\hat{V}/\partial\theta$. The condition (5.17a) is fulfilled automatically, and (5.17b) leads to the following equation in unknown α:

$$\frac{d}{d\theta}P_n^m(\cos\theta)\bigg|_{\theta=\alpha} = 0. \tag{5.22}$$

The orthogonal complement functions for this case are expressed by the same functions as in (5.21):

$$\hat{F}_1^*(\theta,\varphi) = -mP_n^m(\cos\theta)\frac{\sin(m\varphi)}{\sin\theta}, \tag{5.23a}$$

$$\hat{F}_2^*(\theta,\varphi) = -\frac{d}{d\theta}P_n^m(\cos\theta)\cos(m\varphi). \tag{5.23b}$$

At $m = 0$ equation (5.22) gives $\alpha = 63^o$ $(n = 3)$, $\alpha = 49^o$ $(n = 4)$. At $m = 1$, correspondingly, $\alpha = 31^o$ $(n = 3)$, $\alpha = 24^o$ and $\alpha = 69^o$ $(n = 4)$.

Consider the more general case of the conic (not necessarily circular) surface. Assume that a current located on this surface has only the radial component j_ρ. Then the pattern (f) generated by this current is orthogonal to the function $(\hat{F})$ corresponding to any field $(\hat{\mathbf{E}}, \hat{\mathbf{H}})$, which does not contain the component $\hat{E}_\rho$, that is, to a field expressed by the potential $\hat{V}$ only. This follows from the fact that the integral (5.5) equals zero for any radial current, what leads to (5.10). For such a specific surface all the patterns generated by the radial surface current are approximable.

The orthogonality of the pattern generated by a radial current to the function $(\hat{F})$ associated with such a specific surface can easily be proved for the case when the current is located on the axis. It is seen from the symmetry that such a current has only the θ-component of the electrical field, i.e., $f_2 \equiv 0$, $f_1 = f_1(\theta)$ (see (17.3)). The product $(f, \hat{F})$ contains only the integral $\int_0^{2\pi} f_1(\theta)\partial\hat{V}/\partial\varphi d\varphi$ which equals zero because f_1 does not depend on φ and $\hat{V}$ is a periodical function of φ. This fact is valid also in the case when j_ρ is not located on a straight line; the proof is quite cumbersome and so we do not give it here.

This consideration shows once more that the approximability theory can be considered as a *generalization of the elementary properties of the symmetry of fields* generated by the currents located on a straight line or on a plane (see Subsections 2.2.4, 2.3.2).

Let us show that the functions $\hat{F}_1(\theta,\varphi)$, $\hat{F}_2(\theta,\varphi)$ satisfy the conditions

$$\hat{F}_1(\pi-\theta,\pi-\varphi) = \hat{F}_1^*(\theta,\varphi); \quad \hat{F}_2(\pi-\theta,\pi-\varphi) = \hat{F}_2^*(\theta,\varphi) \tag{5.24}$$

if the auxiliary field $(\hat{\mathbf{E}}, \hat{\mathbf{H}})$ is introduced by (5.6), it means, if the field $\hat{\mathbf{E}}$ is real. These conditions are analogous to (2.18) for the two-dimensional scalar case. In general case the potentials $\hat{U}$, $\hat{V}$, as well as the total fields $\hat{\mathbf{E}}$, $\hat{\mathbf{H}}$ can be represented as the sums (with respect to n and m) of the functions (5.19). The nonsommerfeld part of $\hat{\mathbf{E}}$ is the sum of the functions $H_{n+1/2}^{(1)}(k\rho)P_n^m(\cos\theta)\cos(m\varphi)$. The asymptotics of the function $H_{n+1/2}^{(1)}(k\rho)$ contains the factor i^n. Substituting $\theta,\varphi \to \pi-\theta, \pi-\varphi$ in the product $P_n^m(\cos\theta)\cos(m\varphi)$ gives only the factor $(-1)^n$. The same factor is obtained by the Hankel function after the complex conjugation. Hence the condition (5.24) is fulfilled for all the terms of the sum and therefore for the whole function $(\hat{F})$.

This result can be obtained directly from (5.6) after applying the formula (5.4) to the field $(\hat{\mathbf{E}}, \hat{\mathbf{H}})$ and the plane wave coming from the direction (θ, φ) (see the note after (2.18)). However, the asymptotic integration in the two-dimensional case (with respect to θ and φ) is more complicated than in the one-dimensional case (see (2.18), (2.58)), and this way is more laborious than that used above.

6. The *amplitude nonapproximability* of the patterns by a specific surface is possible if infinitely many functions $\hat{F}_p$, $(p = 1, 2, ...)$ correspond to this surface. The functions

$$F_1(\theta,\varphi) = \Phi_1(\theta,\varphi)e^{-i\Psi_1(\theta,\varphi)}, \tag{5.25a}$$

$$F_2(\theta,\varphi) = \Phi_2(\theta,\varphi)e^{-i\Psi_2(\theta,\varphi)} \tag{5.25b}$$

can exist with the given amplitude pattern (Φ) and arbitrary phase (Ψ), which are not approximable by the surface. The combination of three intercrossed planes is an example of such a surface if the angles between them satisfy some conditions. The situation is like that in the two-dimensional case, where $\hat{C}$ consisted of two intercrossed straight lines (see Section 2.3). We confine ourselves to considering the partial case of such a surface, namely the *three mutually orthogonal planes*.

Let the equations of these planes in the Cartesian system be: $y = 0$, $x = 0$, $z = 0$. In the spherical coordinate system with the polar axis oriented into the z-direction and the angle φ measured from the plane $y = 0$ these equations are of the form: $\varphi = 0$, $\varphi = \pi/2$, $\theta = \pi/2$.

To find the field $(\hat{\mathbf{E}}, \hat{\mathbf{H}})$ satisfying the condition (5.8) on these planes, express it by the Debye potentials $\hat{U}(\rho,\varphi,\theta)$, $\hat{V}(\rho,\varphi,\theta)$, using the formulas (5.18). We seek these potentials in the form (see (5.19))

$$\hat{U}(\rho,\varphi,\theta) = R_n(\rho)P_n^m(\cos\theta)\sin(m\varphi), \tag{5.26a}$$

$$\hat{V}(\rho,\varphi,\theta) = R_\nu(\rho)P_\nu^\mu(\cos\theta)\cos(\mu\varphi) \tag{5.26b}$$

(the obvious indices n, m at $\hat{U}, \hat{V}$ are omitted). The functions $R_n(\rho) = J_{n+1/2}(k\rho)/\sqrt{k\rho}$ will not be needed further. For the functions (5.26) to have no singularities, the numbers n, m, ν, μ should be integer, $n \geq m$, $\nu \geq \mu$.

The components $\hat{E}_\rho$, $\hat{E}_\theta$ are the tangential ones on the planes $\varphi = 0$, $\varphi = \pi/2$, and $\hat{E}_\rho$, $\hat{E}_\varphi$ on the plane $\theta = \pi/2$. Such components should be zero (see (5.8)) and this fact imposes conditions on the Debye potentials, following from (5.18):

$$\hat{U} = 0, \quad \frac{\partial \hat{U}}{\partial \theta} = 0, \quad \frac{\partial \hat{V}}{\partial \rho} = 0 \quad \text{at} \quad \varphi = 0, \quad \varphi = \pi/2; \tag{5.27a}$$

$$\hat{U} = 0, \quad \frac{\partial \hat{U}}{\partial \varphi} = 0, \quad \frac{\partial \hat{V}}{\partial \theta} = 0 \quad \text{at} \quad \theta = \pi/2. \tag{5.27b}$$

If m, μ are even numbers, then the conditions (5.27a) are fulfilled by the potentials (5.26). The conditions on the plane $\theta = \pi/2$ require that

$$P_n^m(0) = 0, \qquad \left.\frac{\partial P_\nu^\mu(\cos\theta)}{\partial \theta}\right|_{\theta=\pi/2} = 0. \tag{5.28}$$

The first of these conditions is fulfilled at even μ for odd n $(n > m)$, the second one is fulfilled at even μ for ν $(\nu > \mu)$. At $\mu = \nu = 0$ the potential $\hat{V}$ (5.26b) does not depend on θ and φ, and the corresponding field is zero. Hence, the numbers μ and m must be even and positive.

Each of the potentials (5.26) describes the auxiliary field $(\hat{\mathbf{E}}, \hat{\mathbf{H}})$ independently at any even $m, \mu = 2, 4, ...$ and at any odd both n $(n > m)$ and ν $(\nu > \mu)$.

The orthogonal complement functions (nonnormalized) in the pattern space are

$$\hat{F}_1(\theta,\varphi) = (P_n^m(\cos\theta))' \sin(m\varphi), \quad \hat{F}_2(\theta,\varphi) = \frac{1}{\sin\theta} P_n^m(\cos\theta)\cos(m\varphi); \quad (5.29a)$$

$$\hat{F}_1(\theta,\varphi) = \frac{1}{\sin\theta} P_\nu^\mu(\cos\theta)\cos(\mu\varphi); \quad \hat{F}_2(\theta,\varphi) = (P_\nu^\mu(\cos\theta))' \sin(\mu\varphi). \quad (5.29b)$$

The functions (5.29a) arise from the potential $\hat{U}$ (5.26a), and (5.29b) from $\hat{V}$ (5.26b). The infinite countable set of the orthogonal complement functions $(\hat{F}_p)$ corresponds to the specific surface in the form of three orthogonal planes.

Any pattern (F) is approximable if and only if it is orthogonal to all these functions $(\hat{F}_p)$, that is, if $(F, \hat{F}_p) = 0$, $(p = 1, 2, ...)$, where the inner product is calculated as an integral of the type (5.10). One can multiply each of these equations by $\hat{F}_p(\theta', \varphi')$, summarize over p and exchange the order of the summation and integration. Then one obtains an equivalent condition

$$(F, \mathcal{F}) \equiv 0, \quad (5.30)$$

where $\mathcal{F} = \mathcal{F}(\theta, \varphi, \theta', \varphi')$ is a function being the sum of products of all the functions of type (5.29) for all allowed indices n, m, ν, μ (see (2.84)). The condition (5.30) must hold for all $0 \le \theta' < \pi$, $0 \le \varphi' < 2\pi$.

It is possible to reduce the above sum to a finite one of δ-functions, analogous to (2.88). Then the condition (5.30) becomes the connection between the values of (F) at some different angles, similar to (2.89). This connection will be used while investigating the amplitude approximability below.

The above way of obtaining the connection is too cumbersome. We will use a simpler one based on the symmetry properties, which are obvious in the case of three mutually orthogonal planes. This way was used in the scalar problem, according to the formulas (2.92).

First, consider the potential $V(\rho, \theta, \varphi)$ for the field generated by any current on $\hat{C}$. It can be expressed as the sum of three potentials

$$V(\theta,\varphi) = V_1(\theta,\varphi) + V_2(\theta,\varphi) + V_3(\theta,\varphi) \quad (5.31)$$

of the fields generated by the currents located on the planes $\varphi = 0$, $\varphi = \pi/2$, and $\theta = \pi/2$, respectively. We do not show the dependence of the potentials on ρ here. The field generated by the current located on a plane is symmetric with respect to this plane, so that

$$V_1(\theta,\varphi) = V_1(\theta,-\varphi), \quad V_2(\theta,\varphi) = V_2(\theta,\pi-\varphi), \quad V_3(\theta,\varphi) = V_3(\pi-\theta,\varphi). \quad (5.32)$$

Apply (5.31) to eight angles

$$\begin{gathered}(\theta,\varphi), \quad (\theta,-\varphi), \quad (\theta,\pi-\varphi), \quad (\pi-\theta,\varphi), \\ (\pi-\theta,-\varphi), \quad (\pi-\theta,\pi-\varphi), \quad (\theta,\varphi-\pi), \quad (\pi-\theta,\varphi-\pi)\end{gathered} \quad (5.33)$$

and sum the results. The last seven angles (5.33) are obtained after the imaginary "reflections" of the point (θ, φ) in one, two and three planes. The value of ρ is not changed at these reflections. According to (5.32), all summands in the left-hand side are cancelled, and we obtain

$$V(\theta, \varphi) - [V(\theta, -\varphi) + V(\theta, \pi - \varphi) + V(\pi - \theta, \varphi)] + \\ [V(\pi - \theta, -\varphi) + V(\pi - \theta, \pi - \varphi) + V(\theta, \varphi - \pi)] - V(\pi - \theta, \varphi - \pi) = 0. \quad (5.34)$$

The first square brackets contain the values of V at the points obtained after the single reflections. The second brackets describe the points obtained after the double ones, and the last term corresponds to the triple reflection. The analogous result is valid also for the potential U. Unlike V, the potential U is asymmetrical with respect to the plane on which the current is located, so that, for instance, the first formula of type (5.32) is for U of the form $U(\theta, \varphi) = -U(\theta, -\varphi)$. This difference changes only the signs in the formula analogous to (5.34), which is not essential for the considerations below.

Returning from the potentials to the patterns, we obtain the key result of this consideration: the sum of values of the patterns at eight angles participating in (5.34) equals zero for any pattern created by the current on $\hat{C}$. This means that the function $\mathcal{F}$ contains eight summands, being the products of two δ-functions of the arguments like $(\theta - \theta' + \theta_s)$ and $(\varphi - \varphi' + \varphi_s)$, $s = 1, 2, ...8$, where θ_s, φ_s are some angles (see (2.88)).

The sought necessary and sufficient condition for approximability of the pattern (F) is the equality of the sum of its values at eight angles to zero. Similarly as in Section 2.3 one can obtain the condition of the amplitude approximability of the pattern with the given amplitude (Φ): for all angles θ, φ, the largest of the values Φ_j, $j = 1, 2$, at the points (5.33) must not be larger than the sum of values at seven other points at the same j.

Consider the example when the given amplitude pattern is similar to (2.107). Let both components of (Φ) be the same: $\Phi_1(\theta, \varphi) = \Phi_2(\theta, \varphi) = \Phi(\theta, \varphi)$, and the maximum of $\Phi(\theta, \varphi)$ is reached in the direction $\theta_0 = \arcsin\sqrt{2/3} \approx 55^o$, $\varphi_0 = \pi/4$; $\Phi(\theta_0, \varphi_0) = 1$. This direction makes the same angle with all three Cartesian axes (the plane intersections). Introduce the auxiliary spherical coordinate system (θ', φ') with the same origin and the polar axis directed into (θ_0, φ_0). In these coordinates the given Gaussian amplitude pattern is of the form

$$\Phi(\theta, \varphi) = e^{-A\sin^2(\theta'/2)}. \quad (5.35)$$

At any fixed (θ, φ) the pattern depends only on the angle θ' between the directions (θ_0, φ_0), (θ, φ):

$$\cos\theta' = \cos\theta_0 \cos\theta + \sin\theta_0 \sin\theta \cos(\varphi - \varphi_0). \quad (5.36)$$

The condition that the sum of seven terms in the series should be larger than the eighth one, is fulfilled for all the angles if it is fulfilled for the angle (θ_0, φ_0), at which the largest of these terms is maximal, and the ratio of the sum of the remaining seven terms to the maximal one, is the least. If this condition is broken at any point (e.g., at $(\theta, \varphi) = (\theta_0, \varphi_0)$), then the amplitude nonapproximability occurs.

The values $\sin^2(\theta'/2)$ for the angles being the arguments of V in (5.34) are found from (5.36) after substituting these angles instead of θ, φ. For three summands in the first square

Generally speaking, the analogous statement is not valid for the three-dimensional vector fields. By definition (see (5.8)), two tangential components of the field **E** must be equal to zero on the specific surface. This means that the vector **E** must be normal to the surface at every point. Such a surface does not exist for an arbitrary vector field **A**. Its existence is connected with the fulfilment of the condition

$$\mathbf{A}\mathrm{rot}\mathbf{A} = 0. \tag{5.41}$$

If the field **E**, together with **H**, satisfies the homogeneous Maxwell equations, then this condition has the form

$$\mathbf{E}\mathbf{H} = 0. \tag{5.42}$$

The correlation between conditions (5.42) and (5.8) depends on whether (5.42) is valid in a volume or on the surface C only. Consider these two cases separately.

1. If (5.42) is valid in a volume, then the field **E** can be presented in the form

$$\mathbf{E} = \alpha \mathrm{grad}\beta, \quad \alpha = \alpha(x, y, z), \quad \beta = \beta(x, y, z). \tag{5.43}$$

Vice versa, if **E** cannot be presented in the form (5.43), then the condition (5.42) is not fulfilled (see, e.g. [48, Sections 79, 122]). The fields having this property are investigated in detail in [49]; in particular, the equations are given and investigated there, with which the functions α and β must be consistent for the field (5.43) to satisfy the Maxwell equations (more exactly, the Helmholtz equation and the equation $\mathrm{div}\mathbf{E} = 0$).

It follows from (5.42) that the surface $\beta(x, y, z) = D$ with an arbitrary constant D is a specific one, and the condition (5.8) is fulfilled on it. The same auxiliary field $(\hat{\mathbf{E}}, \hat{\mathbf{H}})$ (and, therefore, the same function of the orthogonal complement in the pattern space) corresponds to the family of specific surfaces continuously depending on the parameter D. The field $\hat{\mathbf{E}}$ can be complex. Such a situation has no analogy in the scalar case.

The simplest example is the field expressed in the form (5.43) with $\beta = x, \alpha = f(y)$. Then $\hat{E}_x = f(y)$, $E_y \equiv 0$, $E_z \equiv 0$. This field satisfies the Maxwell equations (see more precise formulation above) if $f''(y) + k^2 f = 0$. Any plane $x = D$ is a specific surface. The function $f(y)$ can represent a propagated wave, for instance, the plane one: $f(y) = \exp(iky)$. For this field equality (5.42) is satisfied in the whole space, because only one nonzero component $\hat{H}_z$ of the magnetic field exists, as does only one such component $\hat{E}_x$ of the electrical field.

A discrete set of the specific surfaces exists in the field if $f(y)$ is real. For instance, this set is $y = \pi n/k$, $(n = 0, \pm 1, \pm 2, ...)$ if $f(y) = sin(ky)$. It is analogous to the set of specific surfaces in the scalar case with $\hat{u} = \sin(ky)$.

2. If the condition (5.42) is fulfilled on some surface C, but not in a volume, then it is only necessary for the fulfilment of (5.8) on C. The necessity is verified by the fact that (5.8) causes the validity (5.42) on C. Indeed, by the Stokes theorem, the component of the vector $\mathrm{rot}\hat{\mathbf{E}}$, normal to C (which is proportional to $\hat{H}_N$) is the integral of $\hat{E}_t$. According to (5.8), it equals zero, so that **H** lies in the plane tangential to C. The vector **E** is normal to C, hence **E** and **H** are mutually perpendicular.

The insufficiency of the condition (5.42) for the fulfilment of two conditions (5.8) is clear from the following example, where (5.42) is fulfilled, but (5.8) are not. In this example $\hat{E}_x =$

$f(z)$, $\hat{E}_y = F(x,z)$, $\hat{E}_z = 0$, where the functions f and F satisfy the Helmholtz equation. Calculating $\mathbf{H}$ from the second Maxwell equation, multiplying it by $\mathbf{E}$ and equating to zero, according to (5.42), we obtain the differential equation

$$f\frac{\partial F}{\partial z} + F\frac{\partial f}{\partial z} = 0. \tag{5.44}$$

Integrating (5.44) gives an equation of type $\Phi(x,z) = 0$, which describes a cylindrical surface with the director parallel to the y-axis. The condition (5.42) is fulfilled on this surface, but the components of $\mathbf{E}$ tangential to it do not equal zero, so that conditions (5.8) are not fulfilled.

Thus, condition (5.42) is necessary for the existence of the specific surface in a field. If this condition is not fulfilled anywhere, then the specific surface is absent from the field. The example of the field $E_x = \cos(kz)$, $H_x = -i\cos(kz)$, $E_y = \sin(kz)$, $H_y = -i\sin(kz)$ illustrates this statement. The left-hand side of (5.42) is equal to $-i$, so (5.42) is not valid anywhere; it is easily seen that there are no surfaces perpendicular to the vector $\hat{\mathbf{E}}$ of this field.

2. Every closed specific surface $\hat{C}$ is resonant, so at the given frequency an eigenoscillation exists inside $\hat{C}$, satisfying the homogeneous Maxwell equations and the boundary conditions (5.8); it follows from this that condition (5.42) is fulfilled as well. However, not every resonant field can be *automatically* continued outside the surface: it is not necessary for this field to coincide with an auxiliary field $(\hat{\mathbf{E}}, \hat{\mathbf{H}})$ inside the volume. The problem can be formulated as an exterior Cauchy problem for the vector field satisfying the Maxwell equations. The continuation is possible only in the case when this problem is solvable. Namely, a field should exist which has no singularities in the whole space (or at least in some part of it), has zero tangential components of the electrical vector and the given (as in the interior resonant field) such components of the magnetic vector on $\hat{C}$. These tangential components are not independent: they satisfy the condition that the interior Cauchy problem with the same data should have a solution (the eigenoscillation).

If the formulas describing the field of the eigenoscillation have a sense outside the surface, then they describe the field $(\hat{\mathbf{E}}, \hat{\mathbf{H}})$ in the whole space and, in particular, the function $(\hat{F})$. Of course, the surface is a specific one in this case.

Remember that the analytical solution of the *interior* problem can be written for many surfaces, for which the *exterior* one can be solved only numerically. This fact relates not only to the vector problems, but to two-dimensional scalar ones as well.

For the rectangular parallelepiped, the interior field is expressed as a product of the trigonometrical functions and this expression is also valid outside the surface. The corresponding function of the orthogonal complement contains δ-functions and therefore cannot be normalized. The eigenoscillation inside any resonant surface of revolution can be expressed by two scalar functions (E_z and H_z in the obvious notation), and in the case when the director is smooth enough, the expression is kept outside $\hat{C}$. The fields independent of the azimuthal angle are expressed by one scalar function.

The field $(\hat{\mathbf{E}}, \hat{\mathbf{H}})$ can be considered as a result of "diffraction" of the incident field of a special type on the metallic surface $\hat{C}$. The currents do not arise on the surface by this "diffraction".

The formulation of the analytical continuation problem in terms of the Cauchy problem can be used for the proof of the fact that there exists a specific surface in the neighborhood of $\hat{C}$, associated with another orthogonal complement function $(\hat{F})$. However, the known results on both mentioned theories probably do not give the possibility (especially in the three-dimensional case) to formulate the necessary and sufficient conditions for a given surface C to be a specific one, i.e., the condition that there exists a field $(\hat{\mathbf{E}}, \hat{\mathbf{H}})$ satisfying equations (5.1) and conditions (5.8) on C. For the closed surfaces the necessary condition is known: the existence of the eigenoscillation at the given frequency.

3. In the scalar case, any function satisfying condition (2.18) can be considered as an orthogonal complement function $\hat{F}(\varphi)$ of some real field $\hat{u}(r, \varphi)$. This field can be found using the technique of Section 2.1 (or by any other method [32]). Its zero lines are the specific lines $\hat{C}$ corresponding to the given function $\hat{F}(\varphi)$. The analogous statement is not valid in the three-dimensional vector case. The field $(\hat{\mathbf{E}}, \hat{\mathbf{H}})$ satisfying the condition (5.8) on some surface C does not exist for any function pair $\{F_1(\theta, \varphi), F_2(\theta, \varphi)\}$, even if these functions satisfy the conditions (5.24) being analogous to (2.18).

Assume that the conditions (5.24) are valid for the functions F_1, F_2. They are necessary and sufficient for the fields $\hat{\mathbf{E}}$, $\hat{\mathbf{H}}$ reconstructed in the whole volume to be real and imaginary, respectively, although these conditions are not necessary for the existence of the specific surface. Such surfaces exist in any field, represented in the form (5.43), even when the functions α, β are complex. However, the class of functions which can be of the form (5.43) (i.e. for which the condition (5.42) is valid) is narrower than the class of fields in which a discrete set of specific surfaces exists.

We do not know any *explicit form of the connection between two functions* $F_1(\theta, \varphi)$, $F_2(\theta, \varphi)$ satisfying conditions (5.24), at which the reconstructed field $\hat{\mathbf{E}}$ satisfies (5.8) on some surface, that is, $(F) = \{F_1, F_2\}$ is an orthogonal complement function. We may only point out some procedure which can clear up whether (F) is such a function or not. At first, the real electrical field $\hat{\mathbf{E}}$ should be reconstructed by the given F_1, F_2 in its asymptotics (5.6), and the magnetic field $\hat{\mathbf{H}}$ should be calculated according to (5.3b). Then one should solve equation (5.42) with respect to the surface $\hat{C}$, that is, find the surfaces, on which this equation holds. Finally, one should verify that conditions (5.8) are fulfilled at least on a part of one of these surfaces.

Several methods exist for reconstruction of the real field by its asymptotics at infinity. They repeat the known methods for the reconstruction of the outgoing (complex) field by radiation pattern (see, e.g. [32, Subsection 3.5]). The sought real field is the real part of the found outgoing field. The determination of the surface, on which the scalar product $\mathbf{EH}$ is zero, is a simple computation problem. In the general case, such surfaces exist for any functions F_1, F_2, but it can be possible that there are no surfaces where the above product is constant (as in example in Subsection 1, above, where this product is constant).

However, conditions (5.8) can be fulfilled on these surfaces only in the case when the functions F_1, F_2 are not chosen independently. This does not mean that there are "less" specific surfaces than the specific lines in the two-dimensional case. It only means that one can choose arbitrarily *only one* of the functions F_1, F_2 satisfying condition (5.24).

Consider an obvious special case when the problem is reduced to the one-potential one, i.e., to the scalar problem, which is solved in an elementary way similar to that given in Section 2.1. Let $F_1 \equiv 0$, $F_2 = F_2(\theta)$, that is, the electrical field has only an azimuthal component, independent of φ. Put $\hat{U} \equiv 0$, $\hat{V} = \hat{V}(\theta)$; then, according to (5.18), $\hat{E}_\theta \equiv 0$, $\hat{E}_\rho \equiv 0$, $\hat{E}_\varphi = ik\partial\hat{V}/\partial\theta$. The potential $\hat{V}$ can be presented in the form of a series by the functions

(5.19) at $m = 0$, so that

$$\hat{E}_\varphi = ik \sum_{n=0}^{\infty} \frac{C_n}{\sqrt{k\rho}} J_{n+1/2}(k\rho) P_n'(\cos\theta). \tag{5.45}$$

Passing to the limit $k\rho \to \infty$ in (5.45) and using (5.6b) we have (a nonessential factor is omitted)

$$\hat{F}_2(\theta) = \sum_{n=0}^{\infty} C_n i^n P_n'(\cos\theta). \tag{5.46}$$

To obtain the coefficients C_n one should use the property of orthogonality of the Legendre polynomials $P_n(\cos\theta)$ and the known condition $(\sin\theta P_n')' = -n(n+1)P_n$. As a result one obtains

$$C_n = \frac{2n+1}{(n+1)!} \int P_n(\cos\theta) \frac{d}{d\theta}[\sin\theta \hat{F}(\theta)] d\theta. \tag{5.47}$$

The specific surface $\hat{C}$ can be found as a surface $\rho = \rho(\theta)$, where $\hat{E}_\varphi$ calculated by (5.45) is equal to zero. This surface is a surface of revolution.

The approximability condition $(F, \hat{F}) = 0$ is in this case the restriction only to the patterns (F) generated by the azimuthal current on $\hat{C}$. The restriction is imposed only on the zero term in the expansion of this current by $\cos(m\varphi)$. For the patterns generated by any other currents, this condition is fulfilled automatically at such a function $(\hat{F})$.

4. The difference between the cases reducible to the scalar problems and to essentially vector ones, is probably displayed, best of all, in the problem of shape reconstruction of the scattering body by scattered patterns. In Section 2.4 we pointed out some ways for solving this problem. The first one consists in comparison of the measured pattern f with the orthogonal complement function $\hat{F}$ associated with a given surface $\hat{C}$. The scattering surface is not like $\hat{C}$ if these functions are close to each other, that is, if their scalar product is not small. This method can be transferred to the vector problems without any essential alteration, only with the obvious corrections in the formulas. For the simple special surfaces, the function $(\hat{F})$ is approximately constructed in the same way as in the scalar case.

A quite different situation arises while transferring the methods described in Subsections 2.4.4, 2.4.5, to the vector case. In these methods the orthogonal complement function $(\hat{F})$ should be found by the given scattered patterns (f). This function should be close to (f) or, in the other case, be far from (f). Then the surface $\hat{C}$ should be constructed by this function $(\hat{F})$. However, the absence of the simple formulation of the conditions, at which a function pair can be considered as the orthogonal complement function, makes these two methods very laborious. The calculations are more complicated than in the scalar case.

At the end of Subsection 1.2.3 we noted that some physical questions considered in the book lead to mathematical problems, which can be easily formulated but hard to solve. One of these problems was formulated at the end of Subsection 2 above; the other one we have met in this subsection. The solution of these (and several other) problems mentioned in this book could give some completeness to the electromagnetic field approximation theory.

6 Long Narrow Beam of Electromagnetic Waves

6.1 Two-dimensional Fourier Transformation

1. In this section the properties of fields having the form of a long beam with small divergence are studied; the beam width is much smaller than its length. The interest in fields of this type is caused by the fact that such beams can serve as a guide for *energy transmission* for long distances. The problem of very long beams has arisen in connection with the development of the *aerial solar power stations*. These stations will look like large platforms with sizes of hundreds of meters, covered with solar batteries; they are supposed to move around the Earth in high orbits, for instance, in geostationary ones and therefore should hang over the same point of the Earth. The generators located on these platforms transform the electrical current produced by the batteries into microwaves of the decimeter band. An antenna composed of many synchronized radiators forms the electromagnetic wave beam directed onto the receiving antenna (rectenna) located on the Earth, where the beam is again transformed into the electric current. The beam should have the length of about dozens of thousands kilometers and the width (on the Earth) of about several kilometers. The linear sizes of rectenna should be the same as the beam width. The designed system is alternative to the existing energy sources and it is supposed to be ecologically pure.

The long narrow beam can only be created by the antenna of a size much larger than the wavelength. In this subsection, the key equation describing such a beam is obtained. The mathematical technique of the theory is the *two-dimensional Fourier transformation.* In the next sections of the chapter, the maximally attainable parameters of the beam and the optimal field distribution on the antenna are found, and the question of the optimal antenna and rectenna shapes is investigated.

Denote the linear antenna size by a, the rectenna size by α, and the distance between them by d. All these sizes are large in comparison with the wavelength: $ka >> 1$, $k\alpha >> 1$, $kd >> 1$. In the problem of orbital stations, the distance d is much larger than that from the antenna to the Fresnel zone $d_F = ka^2$.

The guide losses, caused by the fact that the entire beam is not incoming onto the rectenna, should be small (about several percent). For this condition to be satisfied, the field on the antenna should not be in-phase. The equi-phase surface should be a part of the sphere of the radius equal to the distance d to the rectenna. In geometrical optics terms the field is focused onto the rectenna. Here the rays (normal to the equi-phase surfaces) intersect. At very long distances from the antenna the diffraction extension of the beam turns out to be principal in spite of the fact that the wavelength is much smaller than all other linear sizes. Such an extension causes the field not to be concentrated in the geometrical focus. It creates a focal spot covering the whole rectenna. The losses can be made small if the dimensionless parameter c

$$c = \frac{ka\alpha}{d} \tag{6.1}$$

is not small in comparison with 2π. This number can be treated as a ratio between two large parameters $k\sqrt{a\alpha}$ and $d/\sqrt{a\alpha}$. Under the condition $d >> d_F$ this demand means that the rectenna should be much larger than the antenna, $\alpha >> a$. Dependence of the minimally possible losses on c is determined in the next section. The optimal field on the antenna is also found there. The larger the beam length d, the wider the beam in the rectenna plane, and the larger the rectenna should be. Enlarging d leads to a corresponding alteration (weakening) of the focusing. For the coefficient c not to decrease as d increases, α should decrease proportionally to d:

$$\alpha = c\frac{1}{ka}d \tag{6.2}$$

with the constant value of c (c has the order 2π). The beam widens "proportionally" to d with a small widening coefficient of the order $2\pi/ka$. The quotation marks mean that the proportional growth of α with d occurs only at simultaneous alteration of the focusing, that is, of the phase distribution of the field on the antenna. With constant focusing, the beam width varies along its axis by an expression more complicated than (6.2) [38].

Of course, the beam has no strict boundaries in its cross-sections. Its conditional boundary which determines the notion of "beam width" is some surface outside which the field intensity quickly decreases.

Note that in the *far zone* the antenna, large in comparison with the wavelength, can create the radiation pattern, the angular width of which is small; it has the order $2\pi/ka$. The linear size of the domain covered by the field at a distance d from the antenna is equal to the angular size multiplied by d, that is, it is given by the same expression (6.2) as the beam width at the same distance. For large antennas, the beam focused at a very large distance (at $d >> d_F$ and "weak focusing") is practically identical to the main lobe of the pattern of the sharp-directed antenna. The field diminishing in the central part of the beam is approximately inversely proportional to the beam width increase. At $d \sim d_F$ the field depends on d weakly, at $d >> d_F$ it is approximately proportional to $1/d$, as is expected in the far zone.

Introduce the Cartesian coordinate system (X, Y, Z) originating on the antenna. The Z-axis is directed perpendicularly to the antenna and rectenna planes. The field structure in the beam is close to that of the plane wave propagating in the Z-direction towards the rectenna, but its amplitude in the planes, perpendicular to the axis, decreases towards the periphery. This decrease is slow: the amplitude altering at the distance comparable with the wavelength is small. The longitudinal components E_Z and H_Z are not zero, but they are approximately $k\alpha$ times smaller than the cross ones. However, their existence is enough for the equations $div\mathbf{E} = 0$, $div\mathbf{H} = 0$ not to impose any restrictions on the cross components. Depolarization in the beam, that is, the connection between the components pairs E_X, H_Y and E_Y, H_X is weak or missing completely. Under these conditions it is sufficient to investigate only one scalar function $U(X, Y, Z)$ which can be treated as any of four cross components. The

Helmholtz equation

$$\frac{\partial^2 U}{\partial X^2} + \frac{\partial^2 U}{\partial Y^2} + \frac{\partial^2 U}{\partial Z^2} + k^2 U = 0 \tag{6.3}$$

which it satisfies, is equivalent to the Maxwell equations under these conditions. As the calculations show, taking into account the vector nature of the fields leads only to nonessential corrections.

2. Express the field in the beam, that is, the function $U(X,Y,Z)$ at $Z > 0$ by the field on the antenna. We do not introduce the currents on the radiators nor do we express the fields by these currents. This would requires to specify what kind of radiators are used, where they are located etc. These details define the field only in the neighborhood of the antenna at $Z \lesssim a$. Just at a distance several times longer than a, the field is averaged. We call this averaged field the field on the antenna, denote it by $U(X,Y,0)$ and express the field in the beam by means of it. Below, this field is determined from requirements on $U(X,Y,d)$, that is, from requirements on the field in the rectenna plane (at $Z = d$). Then one could determine in which way the radiators should be allocated and which currents should be excited in them. We will state that the function $|U(X,Y,0)|$ should have a maximum in the center, that is, at $X = 0$, $Y = 0$ and decrease toward the antenna edges. Hence the radiators should be allocated more densely in the central part of the antenna and/or the currents in them should be larger than in the periphery. The "averaged" current on the antenna should depend on the coordinates X, Y in the same way as the "averaged" field $U(X,Y,0)$ is determined from some requirements on the field $U(X,Y,d)$.

We find a function $U(X,Y,Z)$ which satisfies equation (6.2), equals $U(X,Y,0)$ at $Z = 0$ and satisfies the radiation condition at $X^2 + Y^2 + Z^2 \to \infty$. Introduce the Green function $p(X,Y,Z;X_0,Y_0,Z_0)$, where X_0, Y_0, Z_0 are coordinates of the point at which the function U should be determined. The function p, as a function of X, Y, Z, satisfies the nonhomogeneous equation

$$\frac{\partial^2 p}{\partial X^2} + \frac{\partial^2 p}{\partial Y^2} + \frac{\partial^2 p}{\partial Z^2} + k^2 p = \delta(X - X_0)\delta(Y - Y_0)\delta(Z - Z_0), \tag{6.4}$$

the boundary condition $p = 0$ at $Z = 0$, and the radiation condition. As is known, this function is

$$p = \frac{1}{4\pi}\left\{\frac{e^{-ikR}}{R} - \frac{e^{-ikR'}}{R'}\right\}, \tag{6.5}$$

where

$$R^2 = (X - X_0)^2 + (Y - Y_0)^2 + (Z - Z_0)^2, \tag{6.6a}$$
$$R'^2 = (X - X_0)^2 + (Y - Y_0)^2 + (Z + Z_0)^2. \tag{6.6b}$$

Applying the Green formula to $U(X,Y,Z)$ and p, we obtain

$$U(X_0,Y_0,Z_0) = \iint_D U(X,Y,0) \left.\frac{\partial p}{\partial Z}\right|_{Z=0} dX dY. \tag{6.7}$$

Here D is a domain in the plane $Z = 0$ occupied by the antenna. At $Z = 0$ the equality $\partial R/\partial Z = -\partial R'/\partial Z$ should hold. Since we are interested in the field at $kZ_0 >> 1$, then while calculating $\partial p/\partial Z$ we should differentiate only the exponents with values $1/R$ and $1/R'$ being put equal to $1/Z_0$, and factor them out the integral sign. Then (6.7) obtains the form

$$U(X_0, Y_0, Z_0) = \frac{ik}{2\pi Z_0} \iint_D U(X, Y, 0) \exp\{-ik[Z_0^2 + (X - X_0)^2 + (Y - Y_0)^2]^{1/2}\} dX dY. \quad (6.8)$$

Expand the square root in the exponent into a series by the small value $[(X - X_0)^2 + (Y - Y_0)^2]/Z_0^2$, which has the order $(\alpha/d)^2$. Keep the zeroth and first terms of the expansion, leaving out all further terms. This is permissible if $k\alpha^4 << d^3$; this condition always holds in our problem. Substituting Z_0 with d we have, for the field in the rectenna plane,

$$U(X_0, Y_0, Z_0) = \frac{ik}{2\pi d} e^{-ikd} e^{-\frac{ik}{2d}[X_0^2 + Y_0^2]} \times \iint_D U(X, Y, 0) \exp\left[-\frac{ik}{2d}(X^2 + Y^2)\right] \exp\left[-\frac{ik}{d}(XX_0 + YY_0)\right] dX dY. \quad (6.9)$$

Introduce the dimensionless coordinates

$$x = \frac{X}{a}, \quad y = \frac{Y}{a}, \quad \xi = \frac{X_0}{\alpha}, \quad \eta = \frac{Y_0}{\alpha}. \quad (6.10)$$

In these variables the coordinates of the points at the domain D (occupied by the antenna) boundary, have the order of unity. The points of the domain occupied by the beam have the coordinates of the order of unity as well.

Introduce the function $u(x, y)$ and $v(\xi, \eta)$ which differs from the field $U(X, Y, Z)$ on the antenna and in the rectenna plane, respectively, by the phase factors:

$$U(X, Y, 0) = u(x, y) e^{-\frac{ika^2}{2d}(x^2 + y^2)}, \quad (6.11)$$

$$U(X_0, Y_0, d) = v(\xi, \eta) e^{\frac{ik\alpha^2}{2d}(\xi^2 + \eta^2)}. \quad (6.12)$$

The function $v(\xi, \eta)$ is connected with $u(x, y)$ by the formula

$$v(\xi, \eta) = \frac{i}{2\pi} \frac{ka^2}{d} e^{-ikd} \iint_D u(x, y) e^{-ic(x\xi + y\eta)} dx dy \quad (6.13)$$

following from (6.7). The parameter c was introduced in (6.1). The pre-exponential factor in (6.13) can be written in the form $(a/\alpha)(c/2\pi)$. Further, we also omit the factor $i \exp(-ikd)$.

The formula (6.13) is a key one in beam theory. The function $v(\xi, \eta)$ depends on the dimensionless parameters c and a/α as well as on the domain D and function $u(x, y)$.

As will be shown, one should create such a field $U(X, Y, Z)$ on the antenna, that the function $u(x, y)$ introduced by (6.11) is real. Then the equation of the constant phase surface is $-Z + (X^2 + Y^2)/2d = const$; the normals to it intersect at $Z = d$, that is, on the rectenna. Under some symmetry conditions, which usually hold, the function $v(\xi, \eta)$ is real as well. This means that the normals to the surface of the constant phase of the field on the rectenna intersect on the antenna. According to the sign of the exponents in (6.11), (6.12), the field on the antenna is a convergent spherical wave, the field on the rectenna being a divergent spherical one with the same curvature radius. For brevity, we refer to the fields on the antenna and rectenna as $u(x, y)$ and $v(\xi, \eta)$, respectively.

The *approximation* based on substituting the square root in the exponent of (6.8) by two terms of expansion is sometimes referred to as *parabolic*. This notion is based on the fact that for the fields of quasi-plane structure the elliptical equation (6.3) can be substituted by an approximate parabolic equation for the function $V(X, V, Z)$ introduced instead of $U(X, Y, Z)$ by the formula $U = V\exp(-ikZ)$. Such an approximation is commonly used in quasi-optics theory.

The Green function of the parabolic equation is

$$\frac{1}{Z - Z_0}\exp\left\{\frac{-ik}{2(Z - Z_0)}\left[(X - X_0)^2 + (Y - Y_0)^2\right]\right\} \tag{6.14}$$

Using this technique one can obtain formula (6.9) avoiding the use of the Green function (6.5) for the elliptic equation (6.3).

3. The field energy density is equal, with accuracy to a nonessential factor, to $|U|^2$. The energy stream density, that is, Z-component of the Poynting vector is equal to $\mathrm{Im}(U\partial U^*/\partial Z)$ in the scalar problem. For the plane wave, in which $U \sim \exp(-ikZ)$, this value is $k\,|U|^2$. Below in Subsection 6.2.4 we use this simple expression, while comparing the energy radiated by the antenna with that incident onto the rectenna.

Note, however, that this expression holds only under the condition that the field is constant or very slowly varying in the direction perpendicular to the Z-axis. This condition holds more precisely on the rectenna; the field is appreciably changed only at a distance of the order a which is ka times larger than the wavelength. This change can always be neglected. However, nearby the antenna the field can vary faster, for instance, while passing from one radiator to another. Although the distance between radiators is much larger than the wavelength, the ratio of these values is essentially smaller than the analogical ratio ka near to the rectenna.

We find the more precise expression for the energy stream density near the antenna. For simplicity consider a two-dimensional model in which U depends only on X and Z and satisfies the corresponding Helmholtz equation. Express the field at $Z = 0$ as the Fourier integral

$$U(X, 0) = \int_{-\infty}^{\infty} F(\gamma)e^{-i\gamma X}\,d\gamma. \tag{6.15}$$

The extension of this field into the domain $Z > 0$, which satisfies the Helmholtz equation and the radiation condition, is

$$U(X, Z) = \int_{-\infty}^{\infty} F(\gamma)e^{-i\gamma X - i\sqrt{k^2-\gamma^2}Z}\,d\gamma, \tag{6.16}$$

where $\sqrt{k^2 - \gamma^2} = -i\sqrt{\gamma^2 - k^2}$ at $k^2 - \gamma^2 < 0$. The derivative of $U(X, Z)$ with respect to Z at $Z = 0$ is equal to

$$\left.\frac{\partial U(X,Z)}{\partial Z}\right|_{Z=0} = -i\int_{-\infty}^{\infty} F(\gamma)e^{-i\gamma X}\sqrt{k^2-\gamma^2}d\gamma. \tag{6.17}$$

The condition, that $U(X,0)$ alters at a distance of the order of a wavelength, only by a little, means that the function $F(\gamma)$ is large at $|\gamma| << k$ and small at $|\gamma| \approx k$. Therefore, although formally the variable γ in (6.17) alters in the infinite limits, the square root in (6.17) can be replaced by two terms of its expansion by the value γ^2/k^2

$$\sqrt{k^2-\gamma^2} = k - \frac{\gamma^2}{2k}. \tag{6.18}$$

Substituting (6.18) into (6.17) we obtain

$$\left.\frac{\partial U(X,Z)}{\partial Z}\right|_{Z=0} = -ik\int_{-\infty}^{\infty} F(\gamma)e^{-i\gamma X}d\gamma + \frac{i}{2k}\int_{-\infty}^{\infty}\gamma^2 F(\gamma)e^{-i\gamma X}d\gamma. \tag{6.19}$$

The first integral is equal to $U(X,0)$. Because from (6.14) it follows that

$$\frac{\partial^2 U(X,0)}{\partial X^2} = -\int_{-\infty}^{\infty}\gamma^2 F(\gamma)e^{-i\gamma X}d\gamma, \tag{6.20}$$

the second one can be expressed by the second order derivative of this function. Hence

$$\left.\frac{\partial U(X,Z)}{\partial Z}\right|_{Z=0} = -ik\left\{U(X,0) + \frac{1}{2k^2}\frac{\partial^2 U(X,0)}{\partial X^2}\right\}. \tag{6.21}$$

Compose the expression for the stream of energy radiated by the antenna

$$\int \mathrm{Im}[U(X,0)\left.\frac{\partial U^*(X,Z)}{\partial Z}\right|_{Z=0}]dX = k\left\{\int |U(X,0)|^2\,dX + \frac{1}{2k^2}\int U(X,0)\frac{\partial^2 U^*(X,0)}{\partial X^2}\right\}dX, \tag{6.22}$$

where integration is made over the antenna. While integrating the second summand by parts we leave out (for simplicity) the integrated terms, considering the field on the edge of the antenna to be small or equal to zero. In this way we obtain the sought formula for the energy stream. It equals

$$k\left\{\int |U(X,0)|^2\,dx - \frac{1}{2k^2}\int\left|\frac{\partial U(X,0)}{\partial X}\right|^2 dX\right\}. \tag{6.23}$$

This formula can be treated as the first two terms of the energy stream expansion by the degrees of the small parameter equal to the squared ratio of the wavelength to the distance along the antenna at which the field is significantly varied. Further, we confine ourselves to the first summand in (6.23). The more precise formula (6.23) is needed while calculating concrete antennas in which the position of their radiators is known.

6.2 Transmission of Field by Wave Beam: Possibilities and Restrictions

1. Section 6.1 shows that the magnitude $u(x,y)$ of the field on the antenna, that is, inside the domain D, creates a field with magnitude $v(\xi,\eta)$ in the rectenna plane, which is connected with $u(x,y)$ as

$$v(\xi,\eta) = \frac{a}{\alpha}\frac{c}{2\pi}\iint_D u(x,y)\exp[-ic(x\xi+y\eta)]dxdy \tag{6.24}$$

This is the main formula in all derivations within this and the next sections. It gives the magnitude of the field on the rectenna for the points inside the domain Δ occupied by the rectenna, and the field in the rectenna plane for the points lying outside the rectenna. Coordinates x, y and ξ, η are dimensionless. Both domains D and Δ are of unit size in these coordinates. Further, we omit the word "magnitude" which implies the presence of the factors with square phase and refer to the functions $u(x,y)$ and $v(\xi,\eta)$ as the fields.

According to (6.24), the transformation of the field on the antenna into the field in the rectenna plane lying in its Fresnel zone parallel to the antenna, which is made by the long beam, is described by the *double* Fourier transformation in the *finite domain*. This transformation is determined by the domains D, Δ and the parameter c.

Formula (6.24) is analogous to (4.26) describing the transformation of the current, given on the contour, into the pattern. However, the kernel and the domain of integration are different in these two problems. This is in accordance with the distinction of the processes of propagation of the electrodynamic wave in the long beam (or spherical wave with a very sharp pattern) and of the spherical (or cylindrical) wave with a pattern of finite width. The possibility to supply the antenna surface to the closed resonant surface, which is essential in the problem of the field of currents, does not make any relevant restrictions in the beam problem. The transformation (6.24) does not cause the problem of nonradiating currents. The formula (6.24) describes the transformation of one field into another. At $u(x,y)$ not equal to zero identically, the field $v(\xi,\eta)$ also cannot be the identical to zero.

The boundedness of the domain D in (6.24) leads to some problems in creating the field $v(\xi,\eta)$ with given properties. The key question considered in this section is: by what field $v(\xi,\eta)$ can the given field $u(x,y)$ be created in an "optimal way"? It is connected with the boundedness. Different optimization criteria lead to different fields $u(x,y)$.

2. Write (6.24) in the operator form:

$$Pu = v, \tag{6.25}$$

where the operator P maps the function $u(x,y)$ of the point (x,y) in D into the function $v(\xi,\eta)$ of the point in the plane ξ,η. The operator is integral:

$$Pu = \frac{a}{\alpha}\iint_D \mathcal{P}(x,y;\xi,\eta)u(x,y)dxdy \tag{6.26}$$

with the kernel

$$\mathcal{P}(x,y;\xi,\eta) = \frac{c}{2\pi}e^{-ic(x\xi+y\eta)}. \tag{6.27}$$

Define the inner product of the functions defined in D:

$$(u_1, u_2) = a^2 \iint_D u_1(x,y) u_2^*(x,y) dx dy. \tag{6.28}$$

We introduce two different inner products for the functions of variables ξ, η:

$$(v_1, v_2)_\infty = \alpha^2 \iint_{-\infty}^{\infty} v_1(\xi,\eta) v_2^*(\xi,\eta) d\xi d\eta, \tag{6.29}$$

$$(v_1, v_2)_\Delta = \alpha^2 \iint_\Delta v_1(\xi,\eta) v_2^*(\xi,\eta) d\xi d\eta. \tag{6.30}$$

Two adjoint operators P_∞^{ad} and P_Δ^{ad}, corresponding to the latter definitions, are introduced by the conditions:

$$(Pu, v)_\infty = (u, P_\infty^{\mathrm{ad}} v), \tag{6.31}$$

$$(Pu, v)_\Delta = (u, P_\Delta^{\mathrm{ad}} v). \tag{6.32}$$

The operator P_∞^{ad} maps the functions $v(\xi,\eta)$, defined in the whole plane ξ, η, into the functions $u(x,y)$, defined in the domain D. The operator P_Δ^{ad} maps the functions $v(\xi,\eta)$, defined in the domain Δ of the plane ξ, η into the functions $u(x,y)$, defined in the domain D. The functions $u(x,y)$ and $v(\xi,\eta)$ have the finite norms:

$$N(u) \equiv (u,u)^{1/2} < \infty, \tag{6.33a}$$

$$N_\infty(v) \equiv (v,v)_\infty^{1/2} < \infty \tag{6.33b}$$

The finiteness of the norm

$$N_\Delta(v) \equiv (v,v)_\Delta^{1/2} < \infty \tag{6.33c}$$

follows from (6.33b).

The operators P_∞^{ad} and P_Δ^{ad} are integral and analogous to P:

$$P_\infty^{\mathrm{ad}} v = \frac{\alpha}{a} \iint_{-\infty}^{\infty} \mathcal{P}^*(x,y;\xi,\eta) v(\xi,\eta) d\xi d\eta, \tag{6.34}$$

$$P_\Delta^{\mathrm{ad}} v = \frac{\alpha}{a} \iint_\Delta \mathcal{P}^*(x,y;\xi,\eta) v(\xi,\eta) d\xi d\eta. \tag{6.35}$$

We need the formula:

$$\iint_{-\infty}^{\infty} \mathcal{P}(x,y;\xi,\eta) \mathcal{P}^*(x',y';\xi,\eta) d\xi d\eta = \delta(x-x')\delta(y-y'), \tag{6.36}$$

which follows from the explicit form (6.37) of the kernel and the tabular integral

$$\int_{-\infty}^{\infty} e^{i\alpha\xi} d\xi = 2\pi\delta(\alpha). \tag{6.37}$$

Sequentially applying the operators P and P_{∞}^{ad} is equivalent to applying the unit operator, that is, the identity

$$P_{\infty}^{\text{ad}} P u = u \tag{6.38}$$

holds for any function $u(x, y)$. This follows from the fact that the action of the operator $P_{\infty}^{\text{ad}} P$ consists, according to (6.34), (6.26), in the calculation of the integral

$$\iint_{-\infty}^{\infty} \mathcal{P}^*(x', y'; \xi, \eta) \iint_D \mathcal{P}(x, y; \xi, \eta) u(x, y) dx dy d\xi d\eta \tag{6.39}$$

as a function of x', y'. Exchanging the integrating order and using (6.36), we obtain (6.38). The operators P_{∞}^{ad} and P are not commutative. However, the operator PP_{∞}^{ad} is not unit, it maps any function of ξ, η into another one. It does not follow from (6.38) that P_{∞}^{ad} is inverse to P. If such an operator existed then any field $v(\xi, \eta)$ could be created by the antenna with a finite area and the problem considered in this section would not exist. This remark also relates to the operator (4.28), as was already noted in Section 4.2.

The first two norms given in (6.33) are equal to each other (the Parseval identity). Indeed, according to (6.31) and (6.38)

$$(v, v)_{\infty} = (Pu, Pu)_{\infty} = (u, P_{\infty}^{\text{ad}} P u) = (u, u) \tag{6.40}$$

The energy implication of this equality will be noted later.

We emphasize that equalities (6.38) and (6.40) hold only for P_{∞}^{ad} and $(v, v)_{\infty}$ and they do not hold, in general, for P_{Δ}^{ad} and $(v, v)_{\Delta}$.

3. The function $v_0(\xi, \eta)$ which we want to obtain in the rectenna plane, is given. For some purposes the "ideal" function $v_0(\xi, \eta)$ should be constant on the rectenna and equal to zero outside it. We take it in this form in the numerical examples.

Denote by $u_\sigma(x, y)$ the pseudo-solution to the equation[1]

$$Pu = v_0, \tag{6.41}$$

which gives (after substituting into (6.24)) the function $v_\sigma(\xi, \eta)$

$$v_\sigma = P u_\sigma \tag{6.42}$$

the closest to the given $v_0(\xi, \eta)$ in the L_2 metric, that is u_σ should minimize the functional of u

$$\sigma(u) = \frac{N_{\infty}^2 (Pu - v_0)}{N_{\infty}^2 (v_0)}. \tag{6.43}$$

[1] A more general problem with only the amplitude of the field v_0 given, is described in the Appendix.

If equation (6.41) with respect to $u(x,y)$ has a solution, then this solution is the sought function $u_\sigma(x,y)$, and the norm (6.43) is zero for it. However, (6.41) has a solution only for a small class of functions $v_0(\xi,\eta)$. This class is described in the Paley–Wiener theorem for the one-dimensional operator P and it is reasonable to name it as the PW-class. It includes the functions describing the fields in the rectenna plane, which can be created by the antenna of finite area. The PW-class widens when the parameter c grows. As was already mentioned, the parameter c is about several units in most practical problems. Generally speaking, equation (6.41) has no solution.

The known Lagrange–Euler equation

$$P_\infty^{\mathrm{ad}} P u_\sigma - P_\infty^{\mathrm{ad}} v_0 = 0 \tag{6.44}$$

for the functional (6.43) can be obtained in the usual way. Unlike (6.41), this equation has a solution for any function $v_0(\xi,\eta)$. According to (6.38), this solution is

$$u_\sigma = P_\infty^{\mathrm{ad}} v_0 \tag{6.45}$$

This is a known expression for the pseudo-solution to equation (6.41). In our problem (6.45) is the field on the antenna which creates, according to (6.42), the same field in the rectenna plane

$$v_\sigma = P P_\infty^{\mathrm{ad}} v_0, \tag{6.46}$$

which is the closest to the given function $v_0(\xi,\eta)$, that is, to the function providing a minimum to (6.43). The function $v_\sigma(\xi,\eta)$ (6.46) is a realizable function, that is, it describes a field created by the antenna occupying the domain D. Of course, if the given function $v_0(\xi,\eta)$ belongs to the PW-class, then the right-hand side of (6.46) is equal to $v_0(\xi,\eta)$.

In the general case, $v_\sigma \neq v_0$; however, the products of both these functions with v_σ are equal to (u_σ, u_σ), so that

$$(v_\sigma, v_\sigma)_\infty = (v_\sigma, v_0)_\infty = (u_\sigma, u_\sigma). \tag{6.47}$$

To prove these equalities, one should use (6.42), (6.31), (6.45) and (6.38).

The minimal normalized distance between the given function $v_0(\xi,\eta)$ and any realizable one $v(\xi,\eta)$, that is, the minimum of $\sigma(u)$ (6.43) can be expressed in the form

$$\sigma(u_\sigma) = 1 - \frac{(v_0, v_\sigma)_\infty}{(v_0, v_0)_\infty}. \tag{6.48}$$

This expression is easily obtained after opening the brackets in the expression $\sigma = (v_\sigma - v_0, v_\sigma - v_0)_\infty / (v_0, v_0)_\infty$ and using equality (6.47).

A more general approach to the problem on the best approximation of $v_0(\xi,\eta)$ by $v(\xi,\eta)$ consists in finding the function $u(x,y)$ which minimizes the functional

$$\sigma_w(u) = \iint\limits_{-\infty}^{\infty} \left[Pu - v_0(\xi,\eta)\right]^2 w(\xi,\eta)\, d\xi d\eta, \tag{6.49}$$

where the "*weight*" $w(\xi, \eta)$ is some nonnegative function. At $w = const$ the functional (6.49) transforms to (6.43) and the problem has the explicit solution (6.45). If the function $w(\xi, \eta)$ is large in some domain, then the corresponding function $v = Pu$ is, in this domain, closer to $v_0(\xi, \eta)$. For instance, if $v_0(\xi, \eta)$ is the "ideal" function mentioned above (see (6.66) below) and the values of $w(\xi, \eta)$ on the rectenna are dominant, then we obtain the function $v(\xi, \eta)$, closer to a constant on the rectenna than $v_\sigma(\xi, \eta)$. If one wants to obtain an essentially small field in some domain without rectenna, then one should choose the function $w(\xi, \eta)$ with large values in this domain. Of course, in this case, in the rest of the plane, the field $v(\xi, \eta)$ will differ from $v_0(\xi, \eta)$ by more than $v_\sigma(\xi, \eta)$. The minimization of (6.49) at nonconstant w can be carried out by numerical methods [50], [44].

In the less general formulation, arbitrarily giving not the wave function, as in (6.49), but only a few parameters, one can calculate numerically the fields $u(x, y)$ providing the compromise between different demands to the field in the rectenna plane. The functional to be minimized can be composed of several summands describing different properties of the field $v(\xi, \eta)$, multiplied by the coefficients which regulate contributions of these summands into the whole functional. For instance, minimization of one of these summands should provide the uniformity of the field on the rectenna, and minimization of the second one should provide the minimal field energy in some domains without the rectenna [51]. Other considerations can also be used while constructing such functionals [52].

4. The functions $|u(x, y)|^2$ and $|v(\xi, \eta)|^2$ describe (with accuracy to a common nonessential factor) the energy density on the antenna and rectenna respectively. If the fields $u(x, y)$ and $v(\xi, \eta)$ are slightly varied at the distance of the wavelength order, then the energy stream is proportional to the energy density. The expression "slightly varied" was specified in the preceding section. We assume that this condition is fulfilled. Then the squared norm (6.33a) is the energy stream radiated by the antenna, and the norm (6.33b) is the energy stream incident onto the plane where the antenna lies. The equality (6.40) describes the energy conservation law.

Only a part of the energy falls onto the rectenna, that is onto the domain Δ in the plane ξ, η. The ratio of this "efficient" energy to the entire radiated energy is called the *energy transmission coefficient*. It equals

$$l(u) = \frac{(v, v)_\Delta}{(u, u)}. \tag{6.50}$$

This ratio is a functional of the field $u(x, y)$ on the antenna.

Denote the function providing the maximum to $l(u)$ by $u_l(x, y)$. Similarly to Section 2.5, using the Lagrange multiplier method leads to the Lagrange–Euler equation

$$P_\Delta^{\mathrm{ad}} P u_n = \Lambda_n u_n \tag{6.51}$$

corresponding to the functional $l(u)$. It is a homogeneous equation with respect to Λ_n as eigenvalues and u_n as eigenfunctions. The numbers Λ_n are not negative, because they are equal to the values of nonnegative functional (6.50) on u_n; according to (6.40), $\Lambda_n \leq 1$. If Λ_1, u_1 are the largest eigenvalue of (6.51) and the eigenfunction corresponding to it, then $u_l = u_1$, $l(u_l) = \Lambda_1$, that is, the largest eigenvalue Λ_1 is the maximal value of the energy transmission coefficient l (6.49) and it is provided by the field, described by the function u_1.

Equation (6.51) is a two-dimensional generalization of (4.40a). The eigenfunctions $j_n(s)$ of equation (4.40a) are called, in Section 4.2, as the generalized functions of double orthogonality. The one-dimensional variant of equation (6.51), considered in Subsection 5 below, is equivalent to the equation for the function of double orthogonality considered in the antennas theory.

The functions $u_\sigma(x, y)$ and $u_l(x, y)$ solve two different optimization problems. The function $u_\sigma(x, y)$ minimizes the difference between the field in the rectenna plane and the given function $v_0(\xi, \eta)$; the function $u_l(x, y)$ maximizes the ratio of energy captured by the rectenna to the whole radiated energy. Setting $u(x, y)$ as equal to $u_\sigma(x, y)$ or $u_l(x, y)$ we provide the optimality to one of these criteria. The *disagreement* between two optimization criteria can be estimated by calculation of the "cross" characteristics $l(u_\sigma)$ and $\sigma(u_l)$. It is obvious, that

$$\sigma(u_l)/\sigma(u_\sigma) > 1 \tag{6.52}$$

$$l(u_\sigma)/l(u_l) < 1 \tag{6.53}$$

Both these ratios depend on the shapes of D and Δ, and the parameter c. Below we give numerical results for several cases in which the problem is reduced to the one-dimensional one.

5. Let the domains D and Δ be rectangles with sides $2a, 2b$ and $2\alpha, 2\beta$, respectively, and the sides of the rectangle Δ be parallel to those of the rectangle D. Direct the axes x, y and ξ, η parallel to the sides and place the coordinate origins of both systems into the centres of respective rectangles. Introduce dimensionless coordinates x, y and ξ, η so that the domains D and Δ are bounded by inequalities $|x| < 1, |y| < 1$, and $|\xi| < 1, |\eta| < 1$, respectively. Then the main formula (6.24) takes the form

$$v(\xi, \eta) = \sqrt{\frac{ab}{\alpha\beta}} \frac{\sqrt{c^x c^y}}{2\pi} \int_{-1}^{1} e^{-ic^x x\xi} dx \int_{-1}^{1} e^{-ic^y y\eta} u(x, y) dy, \tag{6.54}$$

where

$$c^x = \frac{ka\alpha}{d}, \quad c^y = \frac{kb\beta}{d}. \tag{6.55}$$

We confine ourselves to the case when the function $u(x, y)$ is of the form

$$u(x, y) = u^{(1)}(x) u^{(2)}(y). \tag{6.56}$$

In this case the function $v(\xi, \eta)$ has also the form of the product $v(\xi, \eta) = v^{(1)}(\xi) v^{(2)}(\eta)$. The operator P (6.25) is the product of two operators (for which we keep the same notation)

$$v^{(1)}(\xi) \equiv P u^{(1)} = \sqrt{\frac{a}{\alpha}} \int_{-1}^{1} \mathcal{P}(x; \xi) u^{(1)}(x) dx, \tag{6.57}$$

with the kernel

$$\mathcal{P}(x; \xi) = \sqrt{\frac{c^x}{2\pi}} e^{-ie^x x\xi}. \tag{6.58}$$

Under the assumption (6.56) the finding of $v(\xi,\eta)$ is reduced to applying the "one-dimensional" operator (6.57). The main formulas (6.28)–(6.36), (6.38), written above for the two-dimensional operator (6.26), can be easily transferred to its one-dimensional analogue (6.57).

Assume that the given function $v_0(\xi,\eta)$ is also expressed in the similar form

$$v_0(\xi,\eta) = v_0^{(1)}(\xi)v_0^{(2)}(\eta). \tag{6.59}$$

Then, according to (6.46), the pseudo-solution $v_\sigma(\xi,\eta)$ has the separated form

$$v_\sigma(\xi,\eta) = v_\sigma^{(1)}(\xi)v_\sigma^{(2)}(\eta). \tag{6.60}$$

The functions $v_\sigma^{(1)}(\xi)$ and $v_\sigma^{(2)}(\eta)$ are determined from the $v_0^{(1)}(\xi)$ and $v_\sigma^{(2)}(\eta)$ by the formulas analogous to (6.46):

$$v_\sigma^{(1)} = PP_\infty^{\text{ad}}v^{(1)}, \tag{6.61a}$$

$$v_\sigma^{(2)} = PP_\infty^{\text{ad}}v^{(2)}, \tag{6.61b}$$

where P is the operator (6.57) and P_∞^{ad} is the one-dimensional analogue of (6.34).

According to (6.60), the pseudo-solution to the two-dimensional equation (6.41) is the product of pseudo-solutions to two one-dimensional equations into which equation (6.41) is split; the "one-dimensional" operators have the form (6.57). If $v_0(\xi,\eta)$ is of the form (6.59), then the closest to it realizable function $v_\sigma(\xi,\eta)$ has the form (6.60), where $v_\sigma^{(1)}(\xi)$ and $v_\sigma^{(2)}(\eta)$ are the functions closest to $v_0^{(1)}(\xi)$ and $v_0^{(2)}(\eta)$ correspondingly. The squared "distance" $\sigma = N_\infty^2(v_\sigma(\xi,\eta)-v_0(\xi,\eta))$ in the initial (three-dimensional) problem is expressed by the squared "distances" $\sigma_1 = N_\infty^2(v_\sigma^{(1)}(\xi)-v_0^{(1)}(\xi))$ and $\sigma_2 = N_\infty^2(v_\sigma^{(2)}(\eta)-v_0^{(2)}(\eta))$ in two-dimensional problems by the formula

$$\sigma = \sigma_1 + \sigma_2 - \sigma_1\sigma_2. \tag{6.62}$$

This formula follows from (6.48), applied to σ, σ_1 and σ_2, after using the equality

$$(v_\sigma^{(1)}v_\sigma^{(2)}, v_0^{(1)}v_0^{(2)})_\infty = (v_\sigma^{(1)}, v_0^{(1)})_\infty(v_\sigma^{(2)}, v_0^{(2)})_\infty$$

(see (6.29)).

As already noted above, there exists an analogy between the formulas (6.57), (6.58) and the expression

$$F(\varphi) = \int_{-1}^{1} j(x)e^{-icx\cos\varphi}dx \tag{6.63}$$

for the *pattern* created by the current $j(x)$, flowing along the straight-line segment. In this expression $c = ka$, where a is the half-length of the segment, x is the dimensionless coordinate equal to ± 1 at the segment edges; a nonessential factor is omitted in front of the integral. The formula (6.63) is obtained from (4.26), (4.27) by putting $\theta = 0$, $r = ax$. If one denotes $\cos\varphi = \xi$, then the formal analogy between (6.63) and (6.57), (6.58) becomes complete. The condition for the point ξ in (6.57) to lie on the rectenna is $|\xi| \leq 1$. This means that in (6.63) the angle φ is real. The points $|\xi| > 1$ lying outside the rectenna correspond to the complex angles for which $|\cos\varphi| > 1$. The radiation resistance (4.125) of the antenna is analogous to the energy transmission coefficient (6.49). For instance, if one requires the pattern $F(\varphi)$ not to depend on φ in the whole range $0 \le \varphi < 2\pi$ then the current $j(x)$ should be the δ-function, $j(x) = \delta(x)$. Then the radiation resistance is equal to zero. The analogy is a point-antenna in the beam problem (in this limiting example it is not obligatory to take into account the demand of finiteness of the norm $N(u)$). Such an antenna creates a uniform field not only on the rectenna but also in its whole plane where the approximation (6.9) is valid. The energy transmission coefficient is equal to zero. The condition on the function $F(\varphi)$ for the angles φ, $|\cos\varphi| < \infty$, under which such a pattern can be created by a current on the finite segment $|x| < a$, given by the Paley–Wiener theory, is transformed to the condition on the function $v_0^{(1)}(\xi)$ in the limits $|\xi| < \infty$ under which such a field can be created by a finite antenna.

6. In the two-dimensional problem, where the operator P has the form (6.58), the expressions for pseudo-solution $u_\sigma(x)$ to equation (6.41) and for the function $v_\sigma(\xi)$ created by $u_\sigma(x)$ have the form

$$u_\sigma(x) = \sqrt{\frac{\alpha}{a}}\sqrt{\frac{c}{2\pi}}\int_{-1}^{1} v_0(\xi)e^{icx\xi}d\xi, \tag{6.64}$$

$$v_\sigma(\xi) = \frac{1}{\pi}\int_{-\infty}^{\infty} v_0(\xi')\frac{\sin[c(\xi-\xi')]}{\xi-\xi'}d\xi' \tag{6.65}$$

(the upper indices are omitted). Similar expressions for $u_\sigma(y)$ and $v_\sigma(\eta)$ are not given here.

The ideal field, which we want to obtain in the rectenna plane, should be constant on the rectenna and equal to zero outside it. Such a field is described by the function

$$v_0(\xi,\eta) = \begin{cases} 1/2, & \text{at } |\xi| \le 1, |\eta| \le 1 \\ 0 & \text{at } |\xi| > 1 \text{ or } |\eta| > 1 \end{cases} \tag{6.66}$$

This function is normalized by the condition $\iint_{-\infty}^{\infty} v_0^2(\xi,\eta)d\xi d\eta = 1$. Of course, this field does not belong to the PW-class and it cannot be created.

The one-dimensional functions $v_0(\xi)$ and $v_0(\eta)$ should be

$$v_0(\xi) = \begin{cases} 1/\sqrt{2}, & \text{at } |\xi| \le 1 \\ 0 & \text{at } |\xi| > 1 \end{cases} \tag{6.67}$$

Substituting this function into (6.64), (6.65) we have

$$u_\sigma(x) = \sqrt{\frac{\alpha}{a}}\frac{1}{\sqrt{\pi c}}\frac{\sin cx}{x} \tag{6.68}$$

$$v_\sigma(\xi) = \frac{1}{\pi\sqrt{2}}\int_{-1}^{1}\frac{\sin[c(\xi-\xi')]}{\xi-\xi'}d\xi'. \tag{6.69}$$

This integral can be expressed as

$$v_\sigma(\xi) = \frac{1}{\pi\sqrt{2}}\left\{\mathrm{Si}[c(1+\xi)] + \mathrm{Si}[c(1-\xi)]\right\}, \tag{6.70}$$

where

$$\mathrm{Si}(\alpha) = \int_0^\alpha \frac{\sin x}{x} dx \tag{6.71}$$

is the integral sine.

At $c < \pi$ the field (6.68) preserves its sign on the whole antenna, at larger c this "optimal" field on the antenna changes its sign. The field on the antenna edges is zero at $c = \pi, 2\pi, ...$

The field $v_\sigma(\xi)$ tends to $v_0(\xi)$ (6.67) when c grows. At the rectenna center $v_\sigma(0) = \sqrt{2}\,\mathrm{Si}(c)/\pi$, and, since $\mathrm{Si}(\infty) = \pi/2$, $v_\sigma(0)$ is close to $v_0(0)$ at large c. At the rectenna edges $v_\sigma(1) = \mathrm{Si}(2c)/\sqrt{2}\pi$ and at large c we have $v_\sigma(1) \approx 1/2\sqrt{2}$, that is, it is twice as small as at the centre. At $|\xi| \to \infty$ the function $|v_\sigma(\xi)|$ decreases inversely to $|\xi|$:

$$v_\sigma(\xi) = \frac{\sqrt{2}}{\pi} \frac{\sin c}{c} \frac{\sin c\xi}{\xi} + O\left((c\xi)^{-2}\right). \tag{6.72}$$

This decrease is nonmonotonic. If $c = \pi, 2\pi, ...$ then the decrease is faster, the function $|v_\sigma(\xi)|$ diminishes as $1/\xi^2$; this is in accordance with the fact that $u_\sigma(\pm 1) = 0$ at these values of c, that is, the field has no jump at the antenna edges.

The curves in Fig. 6.1, calculated by (6.70) for several values of c, describe the field $v_\sigma(\xi)$ on the rectenna ($|\xi| \leq 1$) and outside it ($|\xi| > 1$). The dashed curve shows the field $v_0(\xi)$ given by (6.67). The larger c, the closer $v_\sigma(\xi)$ to $v_0(\xi)$ in the mean-square sense. This fact is also illustrated by curve 1 in Fig. 6.2 for $\sqrt{\sigma_1(u_\sigma(x))}$, where

$$\sigma_1(u_\sigma(x)) = \int_{-\infty}^{\infty} [v_\sigma(\xi) - v_0(\xi)]^2 d\xi. \tag{6.73}$$

Curve 2 in this figure represents the value

$$\sigma(u_\sigma(x, y)) = \iint_{-\infty}^{\infty} [v_\sigma(\xi, \eta) - v_0(\xi, \eta)]^2 d\xi d\eta = 2\sigma_1 - \sigma_1^2 \tag{6.74}$$

in the case $c_x = c_y = c$ (i.e. $a\alpha = b\beta$). The function $v_\sigma(\xi, \eta)$ is obtained by (6.60) and both its factors are the same, given in (6.70); $v_0(\xi, \eta)$ is given in (6.66). The parameter c for this curve is $c = (c_x c_y)^{1/2}$. This curve illustrates the degree of approximation to the ideal field (6.65) in the rectenna plane maximally reachable in the three-dimensional problem for this form of the antenna and rectenna. Remember that values (6.73), (6.74) are calculated using the condition $\iint_{-\infty}^{\infty} v_0^2(\xi, \eta) d\xi d\eta = 1$.

7. In this subsection we calculate the maximally reached energy coefficient (6.49) for the same example, that is, for the case when the antenna and rectenna have rectangular forms, and the fields on them have the forms of products (6.56), (6.60). The optimal field $u_l(x, y)$ is the product of the fields $u_l(x)u_l(y)$, where, for instance, $u_l(x)$ equals the first eigenfunction of the homogeneous equation

$$\frac{c^x}{2\pi} \int_{-1}^{1} e^{ic^x x\xi} \int_{-1}^{1} e^{-ic^x x'\xi} u_n(x') dx' d\xi = \lambda_n^{(x)} u_n(x) \tag{6.75}$$

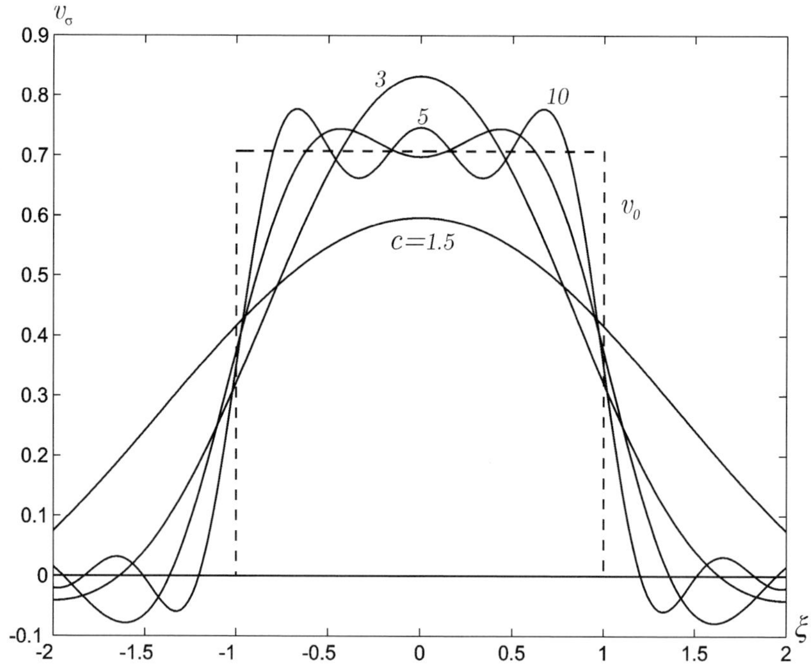

Figure 6.1:

which is the one-dimensional analogue of equation (6.51). The function $u_l(y)$ is the first eigenfunction of a similar equation, after replacing c^x with c^y and $\lambda_n^{(x)}$ with $\lambda_n^{(y)}$. The eigenvalues of equation (6.51) are

$$\Lambda_n = \lambda_n^{(x)}\lambda_n^{(y)}. \tag{6.76}$$

After integrating with respect to ξ, equation (6.75) has the form

$$\frac{1}{\pi}\int_{-1}^{1} u_n(x')\frac{\sin[c(x-x')]}{x-x'}dx' = \lambda_n u_n(x) \tag{6.77}$$

(the upper index (x) is omitted). The eigenfunctions of this equation are the prolate spheroidal functions (the so-called functions of double orthogonality); they are well studied. The first of them $u_1(x)$ is close to the Gaussian bell-shaped function. The curve describing it is sharper for larger c. The function $v_l(\xi) = Pu_1(x)$ has a similar form.

The coefficient $l(u)$ does not depend on the function $u(x, y)$ normalization. However, we will need this normalization again. The energy coefficient for the optimal field, that is, $l(u_1) = \Lambda_1 = \lambda_1^2$, where λ_1 is the maximal eigenvalue of (6.77), is given in Fig. 6.3 (curve 1) for the same case $c^x = c^y$, denoted by c.

The functions $u_\sigma(x, y)$ and $u_l(x, y)$ optimize two different characteristics of the function $v(\xi, \eta)$. The first one minimizes the norm of the difference between the function $v(\xi, \eta)$ and the given $v_0(\xi, \eta)$, the second one maximizes the energy transmission coefficient.

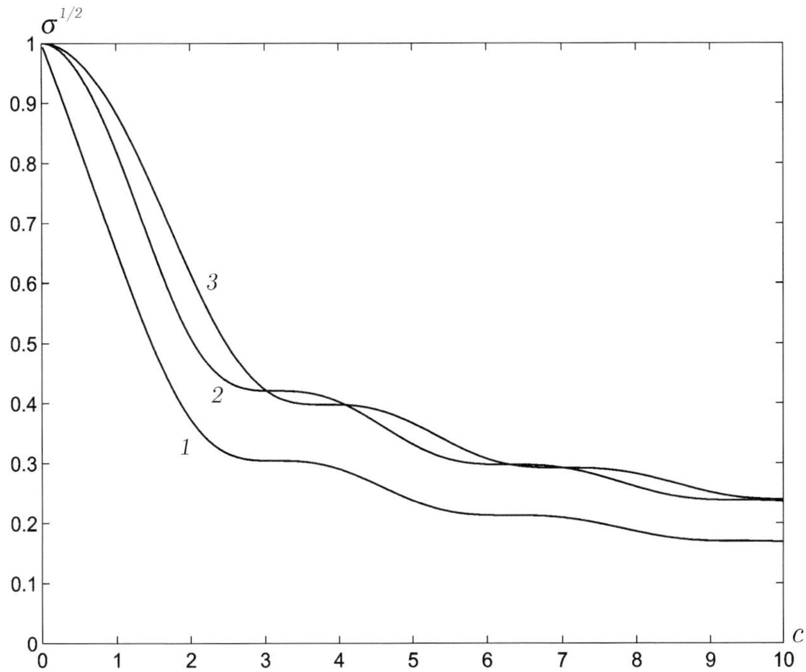

Figure 6.2:

In this subsection we give numerical results concerning the ratios (6.52), (6.53) introduced in Subsection 4 for the same square antenna and rectenna. These ratios characterize the degree of two optimization criteria mismatches. Begin by calculating $l(u_\sigma)$ from (6.50); the norms contained in this formula have the form of products (6.56) and (6.60), the squared norms for our case are

$$(u_\sigma(x,y), u_\sigma(x,y)) = \left\{ a \int_{-1}^{1} |u_\sigma(x)|^2 \, dx \right\}^2, \tag{6.78a}$$

$$(v_\sigma(\xi,\eta), v_\sigma(\xi,\eta))_\Delta = \left\{ \alpha \int_{-1}^{1} |v_\sigma(\xi)|^2 \, d\xi \right\}^2, \tag{6.78b}$$

$$(u_\sigma, u_\sigma) = \alpha^2 \left[\frac{2\sin^2 c}{\pi c} - \frac{1}{\pi} \operatorname{Si}(2c) \right]^2. \tag{6.79}$$

The ratio (6.53) of two energy coefficients (6.50), obtained using (6.68), (6.70), (6.78), (6.79) are presented in Fig.6.4 as a function of c. It is seen that the field u_σ providing the optimal field distribution v_σ in the rectenna plane in the sense of criterion σ reduces the optimal energy transmission coefficient but not significantly.

Proceed to calculating the “cross” parameter (6.52). Calculate the normalized difference

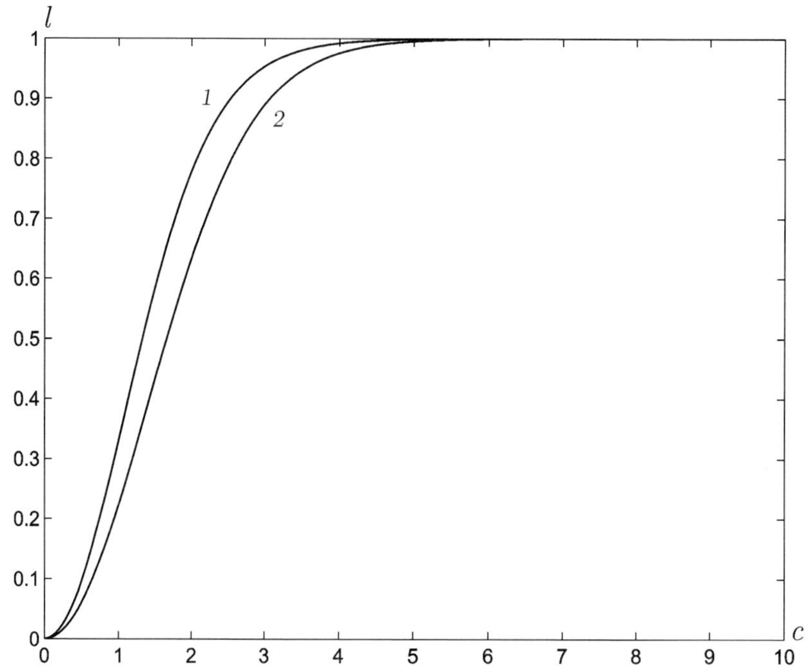

Figure 6.3:

between $v_0(\xi,\eta)$ and $v_l(\xi,\eta)$. The square of this distance is:

$$\sigma(u_l) = \frac{(v_l, v_l)_\infty + (v_0, v_0) - (v_0, v_l) - (v_l, v_0)}{(v_0, v_0)}. \tag{6.80}$$

Assume that $u_l(x,y) = Au_1(x.y)$, where $u_1(x,y)$ is the product of two eigenfunctions of the one-dimensional equation (6.77). The functional (6.80) depends on the coefficient A; choose this coefficient such that $\sigma(u_l)$ calculated by (6.80) is minimal. In contrast to Subsection 4, where the whole function $u(x,y)$ was varied, here only the numerical factor in front of it is varied.

Owing to the appropriate normalization of $u_1(x,y)$ the first term in (6.80) equals A^2, the second one equals unity. The sum of the two last terms is proportional to A. Denote it by $-2AI$, where, according to (6.58),

$$I = \frac{c}{\pi}\left[\int_{-\infty}^{\infty} v_0(\xi)\int_{-1}^{1} u_1(x)\cos cx\xi dx d\xi\right]^2. \tag{6.81}$$

It is easy to check that $I < 1$. The functional (6.80) treated as a function of A has the minimum at $A = I$, equal to $1 - I^2$. Substituting $v_0(\xi)$ given by (6.67) into (6.81), we obtain the sought minimum of the value (6.80):

$$\sigma(u_l) = 1 - \frac{4}{\pi^2 c^2}\left[\int_{-1}^{1} u_1(x)\frac{\sin cx}{x} dx\right]^4. \tag{6.82}$$

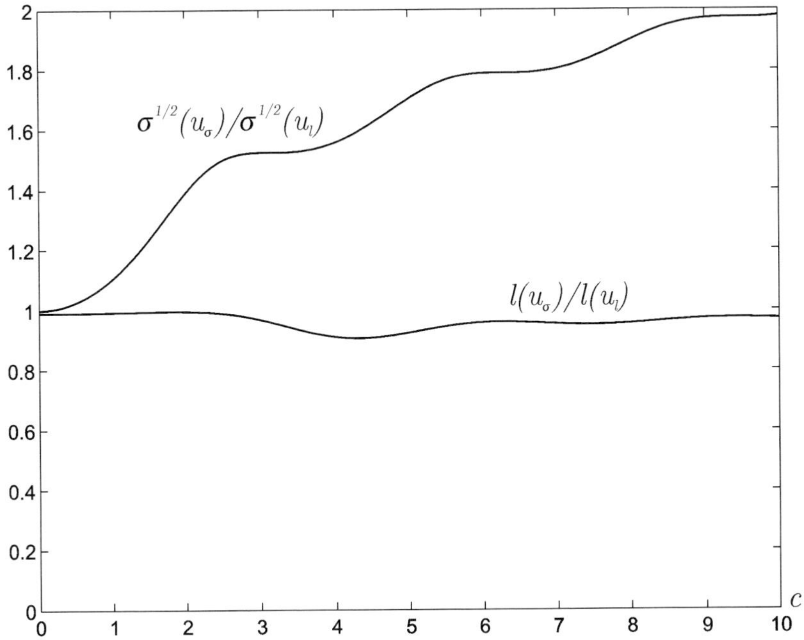

Figure 6.4:

The ratio (6.53) characterizing the mismatch of two criteria of the field optimality is presented in the same Fig.6.4.

It is seen that the field u_σ providing the optimal field distribution v_σ in the rectenna plane in the sense of criterion σ, reduces the optimal energy transmission coefficient l but not significantly at all values of c. This fact can be explained by the following arguments.

It follows from (6.46), (6.27), (6.34), (6.37) that $v_\sigma(\xi,\eta) \to v_0(\xi,\eta)$ at $c \to \infty$ for any v_0, which has an obvious physical sense. For the function (6.66) it holds $(v_0, v_0)_\Delta = (v_0, v_0)_\infty$, therefore, according to (6.40),

$$(v_\sigma, v_\sigma)_\Delta \to (u, u) \tag{6.83}$$

and

$$\lim_{c\to\infty} l(u_\sigma) = 1, \tag{6.84}$$

which explains the asymptotical behavior of the ratio $l(u_\sigma)/l(u_l)$ at large c.

According to (6.46), the function v_σ is the first iteration of the initial function v_0 in the power method for calculating the maximal eigenvalue Λ_1 of equation

$$PP_\Delta^{\mathrm{ad}} v_n = \Lambda_n v_n \tag{6.85}$$

equivalent to (6.51) (see [53]). The eigenfunctions of this equation are defined only in the domain Δ. Expand them into the whole plane ξ, η by zero. For these functions, it holds

$P^{\text{ad}}_{\Delta} v_n = P^{\text{ad}}_{\infty} v_n$. The function v_0 is equal to zero outside the domain Δ as well, and it can be represented in the form

$$v_0 = \sum_{n=1}^{\infty} C_n v_n. \tag{6.86}$$

Then

$$v_\sigma = \Lambda_1 \left(C_1 v_1 + \sum_{n=2}^{\infty} \frac{\Lambda_n}{\Lambda_1} C_n v_n \right). \tag{6.87}$$

At small c we have $\Lambda_2 << \Lambda_1$, thus v_σ is proportional to v_l and $l(u_\sigma)$ tends to $l(u_l)$ at $c \to 0$.

At intermediate c, as it follows from properties of the prolate spheroidal functions, only a few eigenvalues Λ_n are notably smaller than 1 and larger than 0. Only the eigenfunctions v_n corresponding to these intermediate eigenvalues in the expansion (6.86) makes the value $l(u_\sigma)$ smaller than $l(u_l)$. The range of values c, at which their relative contribution is noticeable, is not too wide.

9. In Subsections 5–8 the general theory has been applied to the case when the antenna and rectenna have a rectangular form. In this subsection another case is considered, where the problem is also reduced to a two-dimensional one.

Let the antenna and rectenna have the form of *circles*. Denote the antenna radius by a and the rectenna radius by α. Introduce the cylindrical coordinates R, θ on the antenna and ρ, φ in the rectenna plane. Both coordinates R and ρ are dimensionless. They equal unity at the edges of the antenna and rectenna, respectively. In these coordinates the exponent in (6.24) has the form

$$cR\rho\cos(\theta - \varphi), \tag{6.88}$$

where

$$c = \frac{ka\alpha}{d}. \tag{6.89}$$

The field on the antenna can be represented as the Fourier series with coefficients dependent on R. For simplicity we assume that this field does not depend on θ, that is, it contains only one term of this series, so that $u(x, y) = u(R)$. Then the field in the rectenna plane does not depend on φ either, $v(\xi, \eta) = v(\rho)$. If we assumed that $u(x, y)$ is proportional to $\cos(n\theta)$, then $v(\xi, \eta)$ would be proportional to $\cos(n\varphi)$ and the Bessel functions J_n would appear in the further formulas instead of J_0.

If $u(x, y)$ does not depend on θ, then one can make the integration with respect to θ in (6.24), after that the main formula (6.24) takes the form

$$v(\rho) = \frac{ac}{\alpha} \int_0^1 u(R) J_0(cR\rho) R dR. \tag{6.90}$$

In formula (6.25) one should take $u = u(R)$, $v = v(\rho)$; the operator P means

$$Pu = \frac{a}{\alpha} \int_0^1 \mathcal{P}(R, \rho) u(R) R dR, \tag{6.91}$$

where

$$\mathcal{P}(R,\rho) = cJ_0(cR\rho). \tag{6.92}$$

Similarly to (6.28)–(6.30), we determine the scalar products in the form

$$(u_1,u_2) = a^2 \int_0^1 u_1(R)u_2^*(R)RdR, \tag{6.93a}$$

$$(v_1,v_2)_\infty = \alpha^2 \int_0^\infty v_1(\rho)v_2^*(\rho)\rho d\rho, \tag{6.93b}$$

$$(v_1,v_2)_\Delta = \alpha^2 \int_0^1 v_1(\rho)v_2^*(\rho)\rho d\rho. \tag{6.93c}$$

Taking into account the reality of the Bessel function, the adjoint operators (6.34), (6.35) are

$$P_\infty^{\mathrm{ad}} v = \frac{\alpha c}{a} \int_0^\infty J_0(cR\rho)v(\rho)\rho d\rho, \tag{6.94a}$$

$$P_\Delta^{\mathrm{ad}} v = \frac{\alpha c}{a} \int_0^1 J_0(cR\rho)v(\rho)\rho d\rho. \tag{6.94b}$$

The Parseval identity (the energy conservation law) (6.40) for the norms $(v,v)_\infty$ and (u,u) is valid for the inner products (6.93) as well. Its immediate derivation is based on the formula

$$\int_0^\infty J_0(\alpha\rho)J_0(\beta\rho)\rho d\rho = \frac{1}{\alpha}\delta(\alpha-\beta) \tag{6.95}$$

analogous to (6.36). The formulas (6.45), (6.46) for the pseudo-solution to (6.41) are

$$u_\sigma(R) = \frac{\alpha c}{a} \int_0^\infty J_0(cR\rho)v_0(\rho)\rho d\rho, \tag{6.96}$$

$$v_\sigma(\rho) = c \int_0^\infty v_0(\rho') \frac{\rho J_0(c\rho')J_1(c\rho) - \rho' J_0(c\rho)J_1(c\rho')}{\rho^2 - \rho'^2} \rho' d\rho'. \tag{6.97}$$

These formulas are analogous to (6.64), (6.65) for the two-dimensional problem considered in Subsection 6.

Take the "ideal" function in the form

$$v_0(\rho) = \begin{cases} \sqrt{2}, & \text{at } \rho \le 1, \\ 0, & \text{at } \rho > 1. \end{cases} \tag{6.98}$$

Substituting (6.98) into (6.96) we obtain the formula

$$u_\sigma(R) = \sqrt{\frac{2\alpha}{a}} \frac{J_1(cR)}{R} \tag{6.99}$$

for the field on the antenna creating, in the rectenna plane, a field maximally close to the "ideal" one (6.98). At $c < 3.83$ this function is positive on the whole antenna. At $c > 3.83$

it changes its sign on a ring. If c is equal to the zeros of the Bessel function J_1 then the field (6.99) vanishes at the antenna edge. The field in the rectenna plane, maximally close to (6.98), is

$$v_\sigma(\rho) = c\sqrt{2}\int_0^1 \frac{\rho J_0(c\rho')J_1(c\rho) - \rho' J_0(c\rho)J_1(c\rho')}{\rho^2 - \rho'^2}\rho' d\rho'. \tag{6.100}$$

At the rectenna center ($\rho = 0$) it equals $\sqrt{2}[1 - J_0(c)]$, and at large c this value is close to $\sqrt{2}$. At $\rho >> 1$

$$v_\sigma(\rho) \approx \frac{J_1(c)}{\sqrt{2}}\frac{J_1(c\rho)}{\rho}, \tag{6.101}$$

that is, it decreases as $\rho^{-3/2}$, and even faster if $J_1(c) = 0$.

The formulas (6.99), (6.100) are analogous to (6.68), (6.69). In Fig. 6.5, analogous to Fig. 6.1, the function $v_\sigma(\rho)$ calculated by (6.100) is given for several values of c. The value $\sigma^{1/2}(u_\sigma)$ is represented in Fig. 6.2 by curve 3. This curve is similar to curve 2 describing the same value in the case of square antenna with sides $2a$.

While comparing two lines with circular and rectangular elements which have *equal areas* one should keep in mind that the parameter c is different for both lines (see formulas (6.127), (6.129)).

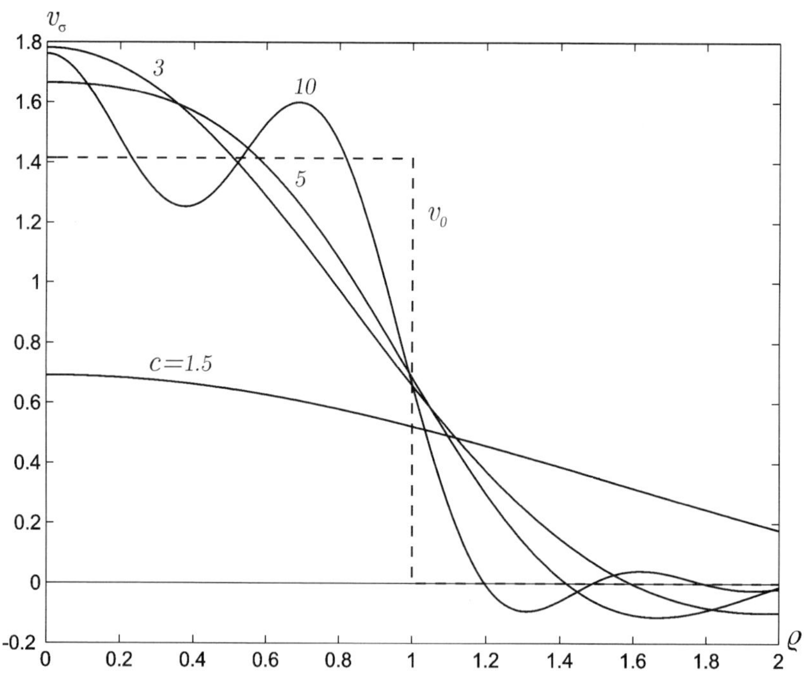

Figure 6.5:

Consider the function $u_l(R)$ providing a maximum of the energy transmission coefficient. It is the first eigenvalue of equation (6.51) which in our case is of the form

$$c \int_0^1 u_l(R') \frac{RJ_0(cR')J_1(cR) - R'J_0(cR)J_1(cR')}{R^2 - R'^2} R'dR' = \Lambda u_l(R). \tag{6.102}$$

This equation is analogous to (6.77). In Fig. 6.6 the first eigenfunction $u_1(R)$ of this equation is shown for several values of c. The function $u_1(R)$ is normalized to unity by the condition $\int_0^1 u_1^2(R)RdR = 1$. The curves have a bell-shaped form similar to that for equation (6.77).

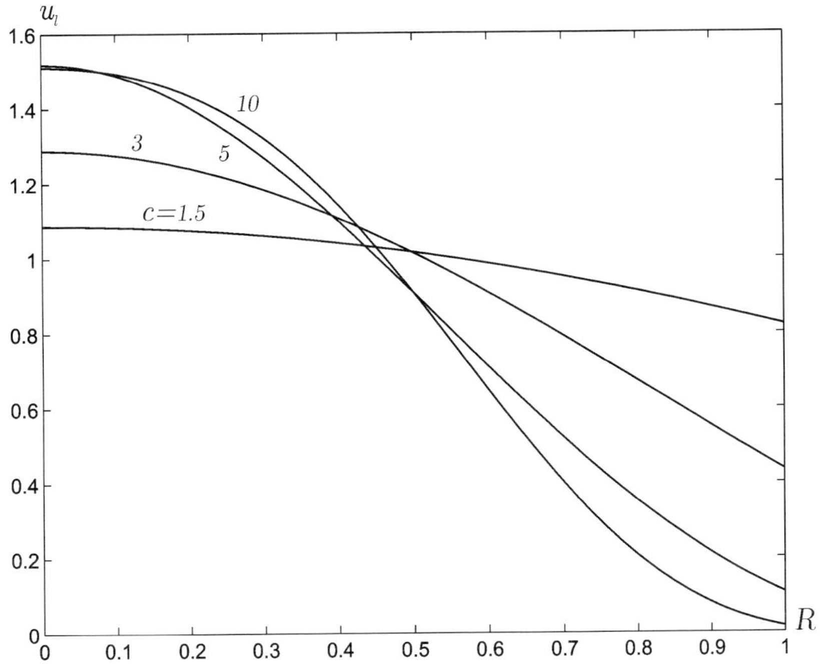

Figure 6.6:

The first eigenvalue Λ_1 of equation (6.100), namely, the maximally reachable energy transmission coefficient, is shown in Fig. 6.3 by curve 2. Just at $c \sim 3 \div 4$ the energy losses, equal to $1 - \Lambda_1$, become small, of the order $0.05 \div 0.03$.

Consequently, finding the pseudo-solution and the field, which provides the maximal energy transmission coefficient, is reduced to solving a one-dimensional equation (6.100) or (6.102) for the circle antenna and rectenna, respectively. Remember that in the case of rectangular antenna and rectenna these problems were reduced to two similar one-dimensional equations.

The numerical results obtained in this section give a quantitative estimation of the possibility of creating an "ideal" field on the plane located in the Fresnel zone ($d \sim d_F$) or in the far zone ($d >> d_F$) of the plane antenna with finite size which is very much larger than the wavelength. The larger the parameter c (e.g. the smaller the wavelength), the weaker the restrictions on creating a field close to the "ideal" one.

6.3 Shape of Antenna and Rectenna

1. In this subsection two lines are being compared; the antennas and rectennas corresponding to them are connected by some relationship. Constructions in this subsection are based on the fact that the kernel $\mathcal{P}$ of the integral operator P (6.26), contains the coordinates x, y and ξ, η only as the combination

$$x\xi + y\eta \tag{6.103}$$

This expression is invariant with respect to the coordinate transformation

$$\dot{x} = \mu x, \quad \dot{y} = \frac{1}{\mu} y, \tag{6.104a}$$

$$\dot{\xi} = \frac{1}{\mu}\xi, \quad \dot{\eta} = \mu\eta, \tag{6.104b}$$

that is,

$$x\xi + y\eta = \dot{x}\dot{\xi} + \dot{y}\dot{\eta}. \tag{6.105}$$

The Jacobians of these two transformations are equal to unity, so that

$$dxdy = d\dot{x}d\dot{y}, \quad d\xi d\eta = d\dot{\xi}d\dot{\eta}. \tag{6.106}$$

The transformations (6.104) mean that the domain D covered by the antenna expands μ times (at $\mu > 1$) in the x-axis direction and condenses μ times in the y-axis direction while moving from the x, y coordinate system to the $\dot{x}, \dot{y}$ one. The domain Δ occupied by the rectenna condenses in the ξ-axis direction and expands in the η-axis direction, respectively. Remember that the ξ-axis is parallel to the x-axis and the η-axis is parallel to the y-axis. According to (6.27), the areas of these domains do not change.

For instance, if D and $\dot{\Delta}$ are circles in the coordinate systems x, y and ξ, η (Fig. 6.7), then, in new coordinates $\dot{x}, \dot{y}$ and $\dot{\xi}, \dot{\eta}$, they are ellipses with the same ratio (equal to μ) of their large and small half-axes. (Fig. 6.8). The ellipses are oriented in two mutually perpendicular directions.

The invariance of expression (6.103) causes the operator P to remain the same after the simultaneous transformation of both coordinate systems x, y and ξ, η by (6.104). This means that, if the function $u(x, y)$ is transformed by (6.24) into $v(\xi, \eta)$, then $u(\dot{x}, \dot{y})$ is transformed into $v(\dot{\xi}, \dot{\eta})$. The value of $u(x, y)$ at the point A (Fig. 6.7(a)) coincides with the value of $u(\dot{x}, \dot{y})$ at the point $\dot{A}$ (Fig. 6.8(a)) if $\dot{x}, \dot{y}$ are connected with x, y by (6.104a). The value of $v(\xi, \eta)$ at the point B Fig. 6.7(b) coincides with the value of $v(\dot{\xi}, \dot{\eta})$ at the point $\dot{B}$ Fig. 6.8(b), if $\dot{\xi}, \dot{\eta}$ are expressed via ξ, η by (6.104b).

Thereby, it follows from the equality

$$u(\dot{x}, \dot{y}) = u(x, y) \tag{6.107a}$$

that

$$v(\dot{\xi}, \dot{\eta}) = v(\xi, \eta). \tag{6.107b}$$

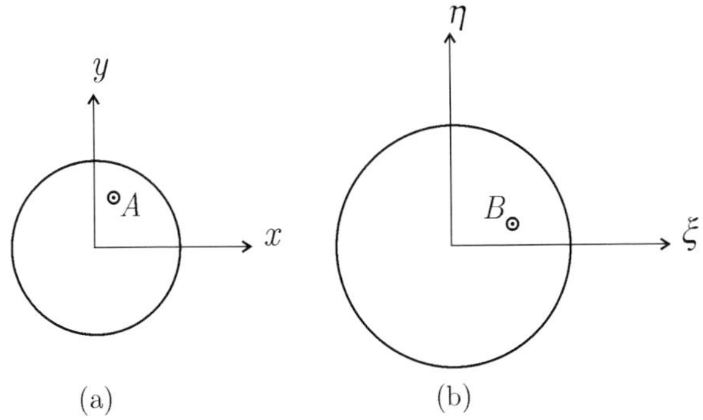

Figure 6.7:

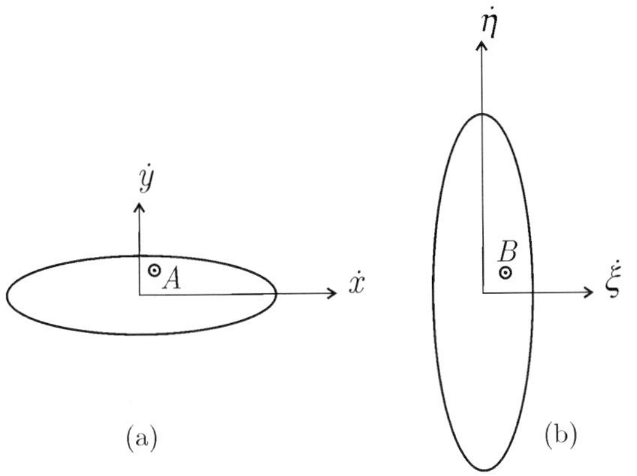

Figure 6.8:

From (6.107), (6.106) one can obtain the equality of the corresponding norms

$$(\dot{u}, \dot{u}) = (u, u), \tag{6.108a}$$
$$(\dot{v}, \dot{v})_\Delta = (v, v)_\Delta. \tag{6.108b}$$

Therefore, the energy transmission coefficient l defined by (6.50) is not changed when passing from one line to the other, that is, at simultaneous affine transformations of the domains occupied by the antenna and rectenna. Remember that the area of the antenna and rectenna remains the same at such transformations.

In the above example, two lines with circular and ellipse-shaped antennas and rectennas, have the same energy transmission coefficient if the fields on the antenna satisfy (6.107a), where the coordinates x, y and $\dot{x}, \dot{y}$ are connected by (6.104b).

2. It was shown in Subsection 1, that if the shape of both antenna and rectenna is simultaneously changed according to (6.104), then the energy transmission coefficient l (6.50) does not change. In this and the next subsections we consider the process which is opposite (in some sense) to the above one. We find a rule for how these shapes should be changed to obtain a larger coefficient l. We will also describe an iterative process of *sequential antenna and rectenna shape altering* but with preservation of their area, in every step of which the coefficient l increases (more specifically, l does not decrease; we omit this specification further on). Since the coefficient l is limited from above ($l \leq 1$), the process is convergent.

First, we give an elementary example of such an alteration. Let, as in Section 6.2.5, the antenna and rectenna be similar rectangles parallel to each other. The antenna is of size $2a \times 2b$, and the rectenna is $2\alpha \times 2\beta$. According to (6.76), the coefficient Λ, that is, the value of l, reached at the optimal field distribution on the antenna, is the product of maximal eigenvalues λ of two equations of the type (6.75) which differ from each other by the parameter c depending on the antenna and rectenna sizes. To emphasize this dependence we rewrite equality (6.76) in the form

$$\Lambda = \lambda\left(\frac{ka\alpha}{d}\right)\lambda\left(\frac{kb\beta}{d}\right) \tag{6.109}$$

(the indices in λ and Λ are omitted). In this example we consider only such alterations of antenna and rectenna shape which keep them as rectangles, similar and parallel to each other, preserving their area so that the four numbers a, b, α, and β are connected by the conditions

$$\frac{a}{\alpha} = \frac{b}{\beta}, \quad ab = const, \quad \alpha\beta = const. \tag{6.110}$$

In this case Λ depends only on the ratio a/b which we denote by m. The formula (6.109) can be rewritten as

$$\Lambda(m) = \lambda(qm)\lambda\left(\frac{q}{m}\right), \tag{6.111}$$

where $q^2 = (k^2/d^2)ab\alpha\beta$ is proportional to the product of the antenna and rectenna areas; it is constant. The maximum of (6.111) is reached at $m = 1$. Consequently, if both the antenna and rectenna fulfil the conditions (6.110), then Λ is the largest when they are square.

3. Consider, in the general case, the problem of choosing the antenna and rectenna shapes in such a way that the value of the transmission coefficient l, *reached at the optimal field distribution on the antenna*, is as large as possible at the fixed antenna and rectenna areas. We construct an iterative process which raises the value of the transmission coefficient

$$l(D, \Delta) = \max_u \frac{(v, v)_\Delta}{(u, u)_D}, \tag{6.112}$$

depending on two domains: D in the plane x, y and Δ in the plane ξ, η. We indicate the dependence of the squared norm of $u(x, y)$ on D by the lower index. The function $v(\xi, \eta)$ in the whole plane is obtained after acting by the operator P onto $u(x, y)$ according to (6.25).

Each step of the algorithm consists of two half-steps. At the fixed domains D and Δ, the maximal value on the right-hand side of (6.112) is reached if $u(x, y)$ is the first eigenfunction of the homogeneous equation (6.51)

$$P_\Delta^{\text{ad}} P u = \Lambda u. \tag{6.113}$$

Denote this function by $u_{\text{opt}}^{(0)}(x,y)$ (the upper index indicates the number of the iteration). It corresponds to the largest eigenvalue $\Lambda^{(0)}$ of this equation. The value of l is equal to $\Lambda^{(0)}$. Denote the corresponding function $v(\xi,\eta)$ by $v_{\text{opt}}^{(0)}(\xi,\eta)$: $v_{\text{opt}}^{(0)} = Pu_{\text{opt}}^{(0)}$. The squared norm $(v,v)_\Delta$ depends on the domain Δ according to (6.30). The function $v_{\text{opt}}^{(0)}(\xi,\eta)$ is defined in the whole plane ξ,η at a given $u_{\text{opt}}^{(0)}(x,y)$.

In the general case, the function $\left|v_{\text{opt}}^{(0)}(\xi,\eta)\right|^2$ is not constant on the boundary of Δ. We *replace* Δ with another domain, in which the integral of this function has the largest value. To this end, we construct the contours on which $\left|v_{\text{opt}}^{(0)}(\xi,\eta)\right|^2$ has constant values, and choose that of these contours, which bounds the domain, having the same area as Δ. Denote this domain by Δ_1 and replace the domain Δ with it. Then $(v_{\text{opt}}^{(0)}, v_{\text{opt}}^{(0)})_{\Delta_1} > (v_{\text{opt}}^{(0)}, v_{\text{opt}}^{(0)})_\Delta$. The ratio on the right-hand side of (6.112) grows after replacing Δ with Δ_1:

$$\frac{(v_{\text{opt}}^{(0)}, v_{\text{opt}}^{(0)})_{\Delta_1}}{(u_{\text{opt}}^{(0)}, u_{\text{opt}}^{(0)})_D} > \frac{(v_{\text{opt}}^{(0)}, v_{\text{opt}}^{(0)})_{\Delta}}{(u_{\text{opt}}^{(0)}, u_{\text{opt}}^{(0)})_D} = l(D,\Delta). \tag{6.114}$$

Now we find a new function $u_{\text{opt}}^{(1)}(x,y)$, optimal for the domain pair $\{D,\Delta_1\}$ (instead of $\{D,\Delta\}$), and the function $v_{\text{opt}}^{(1)}(\xi,\eta)$, corresponding to it. The value of the functional (6.112), calculated on these new functions, is larger than the left-hand side of inequality (6.114). It is equal to the largest eigenvalue $\Lambda^{(1)}$of the equation

$$P_{\Delta_1}^{\text{ad}} Pu = \Lambda u \tag{6.115}$$

and, consequently

$$\Lambda^{(1)} > \Lambda^{(0)}. \tag{6.116}$$

The replacement $\Delta \to \Delta_1$ described above is the first half-step of the iterative process. It results in increasing the maximal value of the transmission coefficient.

The operator P describes the illumination of the plane ξ,η by the field defined in the domain D of the plane x,y. The physical sense of the replacement $\Delta \to \Delta_1$ consists in the fact that the rectenna should be chosen of such a form and location that the greater part of the total energy falling onto the plane ξ,η is concentrated on the rectenna.

An analogous description of the second half-step of the iterative process, namely, replacement of the domain D, covered by the antenna, with another domain D_1 of the same area, requires the introduction of a virtual "reciprocal" line. The actions of antenna and rectenna are mutually exchanged in this line. The energy is propagated from the rectenna towards antenna. The field $v(\xi,\eta)$ defined in the domain Δ of the plane ξ,η illuminates the plane x,y.

Introduce the coefficient l^{rec} as the ratio of the energy incoming onto the domain D to the energy radiated by the domain Δ_1

$$l^{\text{rec}} = \max_v \frac{(u,u)_D}{(v.v)_{\Delta_1}}, \tag{6.117}$$

where

$$u = P^{\text{ad}}_{\Delta_1} v. \tag{6.118}$$

The coefficient l^{rec} is a functional of $v(\xi,\eta)$. At fixed domains D and Δ_1, the maximum of l^{rec} is reached at $v(\xi,\eta) = v^{\text{rec}(1)}_{\text{opt}}(\xi,\eta)$ which is the first eigenfunction of the equation

$$PP^{\text{ad}}_{\Delta_1} v^{\text{rec}} = \Lambda^{\text{rec}} v^{\text{rec}} \tag{6.119}$$

corresponding to the largest eigenvalue $\Lambda^{\text{rec}(1)}$. This fact can be easily obtained in a similar way to (6.113) .

There exists a simple connection between the real and "reciprocal" lines, which is the base of the second half-step of the iterative process. To find this connection, we apply the operator $P^{\text{ad}}_{\Delta} P$ to the equality

$$u^{\text{rec}(1)}_{\text{opt}} = P^{\text{ad}}_{\Delta_1} v^{\text{rec}(1)}_{\text{opt}} \tag{6.120}$$

which is a particular case of (6.118). Taking into account (6.119), we obtain

$$P^{\text{ad}}_{\Delta_1} P u^{\text{rec}(1)}_{\text{opt}} = \Lambda^{\text{rec}(1)} u^{\text{rec}(1)}_{\text{opt}}. \tag{6.121}$$

Comparing it with (6.113) we find that if $v^{\text{rec}(1)}_{\text{opt}} = v^{(1)}_{\text{opt}}$, then $u^{\text{rec}(1)}_{\text{opt}}$ and $u^{(1)}_{\text{opt}}$ are proportional to each other, and that

$$\Lambda^{\text{rec}(1)} = \Lambda^{(1)}. \tag{6.122}$$

This is the sought connection between two compared lines. It consists in the fact that the *maximally reached* coefficient of the energy transmission from the antenna to rectenna (in the real line) is equal to that in the reciprocal line.

Repeat the derivations made while passing from Δ to Δ_1. Replace the domain D with some domain D_1 chosen so that the integral of $\left|u^{\text{rec}(1)}_{\text{opt}}\right|^2$ over D_1 is larger than that over D. This is the second half-step of the iterative process. The coefficient (6.38) grows

$$\frac{(u^{\text{rec}(1)}_{\text{opt}}, u^{\text{rec}(1)}_{\text{opt}})_{D_1}}{(v^{\text{rec}(1)}_{\text{opt}}, v^{\text{rec}(1)}_{\text{opt}})_{\Delta_1}} > \frac{(u^{\text{rec}(1)}_{\text{opt}}, u^{\text{rec}(1)}_{\text{opt}})_{D}}{(v^{\text{rec}(1)}_{\text{opt}}, v^{\text{rec}(1)}_{\text{opt}})_{\Delta_1}} = l^{\text{rec}}(\Delta_1, D) \tag{6.123}$$

after such a replacement.

Now we find a new function $v^{\text{rec}(2)}_{\text{opt}}$ and, according to (6.39), a new function $u^{\text{rec}(2)}_{\text{opt}}$ corresponding to the domains D_1, Δ_1. The new functions provide the maximal value of the functional l^{rec} reached on these domains. It is seen that $l^{\text{rec}}(\Delta_1, D_1) > l^{\text{rec}}(\Delta_1, D)$. The new value of l^{rec} is equal to the eigenvalue of equation (6.43) for D_1, Δ_1; we denote it by $\Lambda^{\text{rec}(2)}_1$. Thus, $\Lambda^{\text{rec}(1)}$ is replaced with the larger value $\Lambda^{\text{rec}(2)}_1$ while passing from D to D_1. Consequently, according to (6.122), the maximally reached value of l is larger than that for the domains D, Δ_1, as well as for the real line with domains D, Δ.

The physical sense of the replacement $D \to D_1$ is the same as the replacement $\Delta \to \Delta_1$ made in the first half-step of the iteration process. However, it relates not to the real line, but to a "reciprocal" one. The operator P_{Δ}^{ad} describes the illumination of the plane x, y by a wave incoming from the domain Δ in the plane ξ, η. The replacement $D \to D_1$ has to shape and locate the antenna to receive the greatest possible energy.

Note that the operator $P_{\Delta}^{\text{ad}} P$ also describes a possible process. Replace the rectenna with a plane mirror. Being illuminated by the antenna with the field $u(x, y)$, it reflects the incoming field onto the plane x, y. This reflected field equals $P_{\Delta}^{\text{ad}} P u$ in the plane x, y. If the initial field is u_{opt}, then the reflected one equals Λu_{opt} in the domain D.

Both half-steps of the iterative process, that is, the alteration of Δ and of D, do not depend on each other. If one of the domains, for instance, the rectenna, is given not from the requirement of minimal losses but from some other considerations, then the entire process consists only in sequential alterations of domain D. The obtained value of coefficient l is maximal for the given rectenna and area of antenna.

4. Passing from Δ to Δ_1 as well as from D to D_1 should not necessarily consist in finding the contour on which $|v_{\text{opt}}|^2$ is constant or, correspondingly, the contour on which $\left|u_{\text{opt}}^{\text{rec}}\right|^2$ is constant and which bounds a domain of given area. Finding this contour can turn out to be a very cumbersome procedure. It can be replaced with the more simple one described below. In such a procedure both sides of (6.114) as well as (6.123) differ less, and each step of the iterative process results in a smaller increase in the transmission coefficient and therefore the number of steps is larger, but each step is simpler. The procedure, allowing to avoid the construction of contours with constant $|v_{\text{opt}}|^2$, is based on the comparison of its values on the boundary of Δ. The domain Δ_1 is constructed as follows: the contour is moved off and the domain is expanded in those parts of the domain Δ boundary where $|v_{\text{opt}}|^2$ is large. To preserve the domain D area, its boundary should be tightened and domain Δ condensed in the parts of small $|v_{\text{opt}}|^2$. The same procedure applies to the transformation from D to D_1. The physical explanation of this process is obvious. The domain Δ has to occupy the parts of the plane ξ, η, where the field is large and release those where it is small in order for the amount of energy captured by the antenna to grow.

We give without proof (see [54]) the formulas for alteration of the eigenvalue Λ of (6.113) caused by small alterations of domains D and Δ on which the operators P and P_{Δ}^{ad} depend. Let the transformation from Δ to Δ_1 consist in supplying the domain Δ with small domain $\delta\Delta$ prolate along the domain Δ boundary. We hold that $\delta\Delta > 0$ at the points, where the domain Δ is expanded , and $\delta\Delta < 0$, where it is constricted. In Fig. 6.9 the contours of Δ and Δ_1 are shown by the solid and dotted lines, respectively. The condition for the areas of domains Δ and Δ_1 to be equal gives

$$\int \delta\Delta d\sigma = 0, \tag{6.124}$$

where σ is the coordinate of a point on the boundary of Δ, and integration is made over the whole contour. The alteration of Λ at $\Delta \to \Delta_1$ is

$$\delta\Lambda = \int |v_{\text{opt}}(\sigma)|^2 \, \delta\Delta d\sigma, \tag{6.125}$$

where the notation $v_{\text{opt}}(\sigma)$ means the value of $v_{\text{opt}}(\xi, \eta)$ on the boundary of Δ. The fields are normalized by the condition $(u_{\text{opt}}, u_{\text{opt}})_D = 1$ here. It follows from (6.125) that for $\delta\Lambda$ to be positive, δD should be positive at the points σ for which $|v_{\text{opt}}(\sigma)|$ is large and negative where it is small.

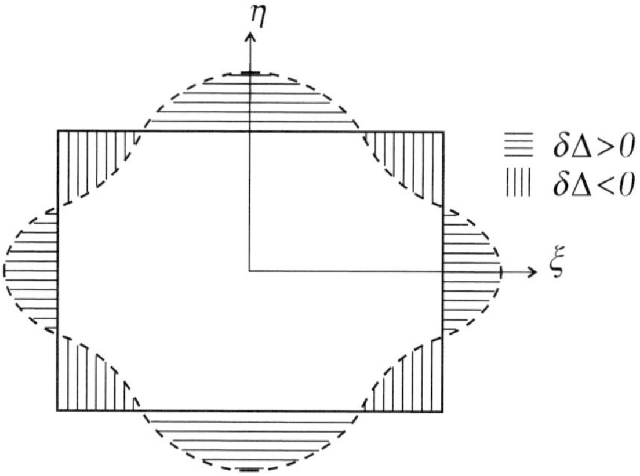

Figure 6.9:

Similarly, alteration of the domain D, covered by the antenna, by supplying it with small domain δD, prolate along the contour of D, leads to the alteration of Λ by

$$\delta\Lambda = \Lambda \int \left|u_{\text{opt}}^{\text{rec}}(s)\right|^2 \delta D ds, \tag{6.126}$$

where integration is made over the boundary of D. The same algorithm of transformation $D \to D_1$ follows from (6.126), which provides the growth for Λ, that is, positiveness of $\delta\Lambda$.

5. If in some step of the iterative process the obtained domain Δ possesses the property that $|v_{\text{opt}}(\sigma)|^2$ is constant on its boundary, then the process of changing Δ stops. The coefficient l, as a functional of shape and position of this domain, has the maximal value. If the domain D also possesses the same property, that is, $\left|u_{\text{opt}}^{\text{rec}}(s)\right|^2$ is constant on its boundary, then l cannot be increased by any small alterations of domains D and Δ. They can be altered in the way described in Subsection 1, that is, by simultaneous affine transformation. However, this does not change the value of l. Since the points of the contour remain the points of the contour at the affine transformation, the value of the field modulus on the contour remains constant. Further, we do not mention this possibility to change the domains D and Δ.

Of course, if the functions $u_{\text{opt}}(x,y)$ and $v_{\text{opt}}^{\text{rec}}(\xi,\eta)$ do not depend on the angles, then the line, for which both domains D and Δ, covered by the antenna and rectenna, respectively, are circular, possesses the property that "the energy stream at contour points of the real line as well as of the "reciprocal" one is constant". Small alteration of the antenna and rectenna shape does not increase the transmission coefficient for the line with circular antenna and rectenna. The transmission coefficient in such a line at the optimal (in particular, symmetrical) field on the antenna, that is, the eigenvalue Λ of (6.113) is maximal among those for lines with close shapes and the same antenna and rectenna areas.

We accept a *hypothesis:* "the maximum of l, reached on the circles, is global", that is, there are no other shapes of domains D and Δ with the same area, for which l is larger. There exist essential physical considerations supporting this hypothesis, however, a proof of its validity is unknown. Supposedly, a stronger hypothesis is valid saying that this maximum is unique.

This hypothesis allows to find the maximum of transmission coefficient l between the antenna of area S and rectenna of area Σ. The maximum depends only on S and Σ. Circles of those areas have the radii $a = (S/\pi)^{1/2}$ and $d = (\Sigma/\pi)^{1/2}$, respectively. The value Λ depends only on the coefficient c (6.89), which is expressed by S and Σ as

$$c = \frac{k}{\pi d}(S\Sigma)^{1/2}. \tag{6.127}$$

The value Λ is the first eigenvalue of (6.101)). It is presented in Fig.6.3 as a function of c calculated by (6.127) (curve 2).

Consequently, the energy transmission coefficient l, for any line and any field created on the antenna, is subject to

$$l \leq \Lambda(c), \tag{6.128}$$

where c is expressed by the area of the antenna and rectenna by (6.127), and function $\Lambda(c)$ is shown in Fig. 6.3.

Note that, for the line with rectangular elements, we have

$$c = \frac{k}{4d}(S\Sigma)^{1/2}. \tag{6.129}$$

If the areas of elements in both lines are equal, then the curve of l treated as a function of $k/d \cdot (S\Sigma)^{1/2}$ is higher for the line with circular elements, than for the line with rectangular ones.

Bibliography for Chapters 1–6

[1] B. Z. Katsenelenbaum. *The Approximability Problem of Electromagnetic Field.* Nauka, Fizmatlit, Moscow 1996 (Russian).

[2] H. Hönl, A. V. Maue, K. Wespfal. *Theory der Beugung.* Springer-Verlag, Berlin 1961.

[3] A. N. Kolmogorov, S. V. Fomin. *Reele Funktionen und Funktionalanalysis.* Deutscher Verlag der Wissenschaften, Berlin 1975 (Russian).

[4] M. Abramowitz and I. A. Stegun. *Handbook of Mathematical Functions.* Dover Publ., New York 1970.

[5] B. Z. Katsenelenbaum, I. M. Polishchuk. On the derectivity patterns of currents, distributed on closed and open curves or surfaces. *Soviet J. of Communications Technology and Electronics* **35** (11) (1990), 2363–2367.

[6] A. Colton, A. Kirsch. Dense sets and far field patterns in acoustic wave propagation. *SIAM J. Math. Anal.* **15** (1984), 996–1006.

[7] A. Colton, R. Kress. Dense sets and far field patterns in electromagnetic wave propagation. *SIAM J. Math. Anal.* **16** (1985), 1049–1060.

[8] A. Kirsch. Denseness of the far field patterns for the transmission problem. *IMA J. Appl. Math.* **37** (1986), 213–225.

[9] D. Colton, D. Kress. *Inverse Acoustic and Electromagnetic Scattering Theory.* Springer-Verlag, Berlin 1991.

[10] N. Inagaki, R. Garbacz. Eigenfunction of composite Hermitian operators with application to discrete and continuous radiating systems. *IEEE Trans.* **AP-30** (4) (1982), 571.

[11] D. Lin, R. Garbacz, and D. Pozar. Antenna synthesis and optimization using generalized characteristic modes. *IEEE Trans.* **AP-38** (6) (1990), 860–868.

[12] B. Z. Katsenelenbaum, M. Yu. Shalushin. Optimum current synthesis on a circle. *Soviet J. of Communications Technology and Electronics* **34** (5) (1989), 25–33.

[13] B. Z. Katsenelenbaum, M. Yu. Shalushin. Properties of fields of currents distributed on nonclosed surfaces. *Pisma v ZhTF* **14** (21) (1988), 2012–2015 (Russian).

[14] B. Z. Katsenelenbaum, M. Yu. Shalushin. On the impossibility of approximating an arbitrary field by the field of a current, flowing in an open surface or curve. *Soviet J. of Communications Technology and Electronics* **34** (12) (1989), 75–82.

[15] B. Z. Katsenelenbaum. The properties of horn patterns (nonapproximability). *Soviet J. of Communications Technology and Electronics* **35** (5) (1990), 65–70.

[16] B. Z. Katsenelenbaum. Nonapproximability of amplitude radiation patterns by the patterns of currents located on two intersected straight lines. *Pisma v ZhTF* **15** (11) (1988), 17 (Russian).

[17] M. Yu. Shalukhin. Optimum currents synthesis on the arc of a circle. *Soviet J. of Communications Technology and Electronics* **35** (14) (1990), 16–21.

[18] M. Yu. Shalukhin. Properties of the fields of currents, distributed on certain open curves. *Soviet J. of Communications Technology and Electronics* **35** (14) (1990), 72–76.

[19] B. Z. Katsenelenbaum, M. Yu. Shalukhin. On determination of body shape by scattered pattern. *Pisma v ZhTF* **16** (5) (1990), 60–63 (Russian).

[20] B. Z. Katsenelenbaum. Condition on the current density, localized in a given region, necessary in order that the patterns, produced by it, can be approximated by a given function. *Soviet J. of Communications Technology and Electronics* **36** (5) (1991), 97–101.

[21] B. Z. Katsenelenbaum, M. Yu. Shaluhin. Properties of surface currents and solvability of inverse electrodynamics problems (current synthesis). *IEICE Trans. (Japan)* **E74** (9) (1990), 2910–2914.

[22] B. Z. Katsenelenbaum, M. Yu. Shaluhin. Approximate current synthesis, its limitations and possibilities and solvability of inverse electrodynamics problems (current synthesis). In: *IV Int. Sem. on Mathematical Methods in Electromagnetic Theory,* Alushta 1991, 306.

[23] B. Z. Katsenelenbaum. On the norm of the current, producing a given pattern. *Soviet J. of Communications Technology and Electronics* **37** (8) (1992), 17–22.

[24] B. Z. Katsenelenbaum. Nonapproximable patterns and Nonradiating currents. *Soviet J. of Communications Technology and Electronics* **38** (6) (1993), 998–1005.

[25] B. Z. Katsenelenbaum. The approximability problem of electromagnetic field. *Physics–Uspekhi* **37** (9) (1994), 903–913.

[26] B. Z. Katsenelenbaum, G. M. Matviyiv, N. N. Voitovich. How to find the shape of the body when its several scattering patterns are known. In: *Proc. 1995 URSI Int. Symp. on Electromagn. Theory,* St. Peterburg 1995, 133–135.

[27] B. Z. Katsenelenbaum. On the recognition of the shape of a scatterer, located near other bodies, from its several scattering patterns. *Doklady Akad. Nauk SSSR* **368** (5) (1999), 609–611 (Russian).

[28] B. Z. Katsenelenbaum. Near-region properties of field, scattered by nonplanar screens. *J. of Communications Technology and Electronics* **41** (14) (1996), 1193–1198.

[29] B. Z. Katsenelenbaum. The field in the vicinity of a transmitting antenna. *J. of Communications Technology and Electronics,* **47** (8) (2002), 844–847.

[30] H. Hristov and A. Kumar. *Microwave Cavity Antennas.* Artech House, London 1989.

[31] P. L. Overfelt and D. J. White. The TE- and TM-modes in some triangular cross-section waveguides using superposition of plane waves. *IEEE Trans.* **MTT-34** (1) (1964), 161–167.

[32] N. N. Voitovich, B. Z. Katsenelenbaum, E. N. Korshunova, L. I. Pangonis, M. L. Pereyaslavets, A. N. Sivov, A. D. Shatrov. *Electrodynamics of Antennas with Semitransparent Surfaces. Methods of Constructive Synthesis.* Eds. B. Z. Katsenelenbaum and A. N. Sivov. Nauka, Fizmatlit, Moscow 1989 (Russian).

[33] M. S. Agranovich, B. Z. Katsenelenbaum, A. N. Sivov, N. N. Voitovich. *Generalized Method of Eigenoscillations in Diffraction Theory.* WILEY-VCH, Berlin 1999.

[34] N. N. Voitovich, O. F. Zamorska, R. I. Germanyuk. Optimization of plane antennas with semitransparent aperture. *Electromagnetics* **18** (5) (1998), 481–495.

[35] I. S. Gradshtein and I. M. Ryzhik. *Tables of Integrals, Sums, Series, and Products.* Academic Press, New York 1965.

[36] N. N. Voitovich, O. V. Kusyi. Body shape reconstruction by its measured scattering patterns. *Matematychni Metody i Fizyko-Mekhanichni Pola* **46** (2) (2003) (Ukrainian).

[37] *Linear Equations of Mathematical Physics.* Ed. S. G. Mikhlin. Nauka, Moscow 1964 (Russian).

[38] R. B. Vaganov, B. Z. Katsenelenbaum. *Bases of Diffraction Theory.* Nauka, Fizmatlit, Moscow 1982 (Russian).

[39] V. F. Apeltsin, A. G. Kiurkchan. *Analytical Properties of the Wave Fields.* Izdatelstvo Moskovskogo Universiteta, Moscow 1990 (Russian).

[40] A. G. Kiurkchan, B. Yu. Sternin, V. E. Shatalov. Singularities of solutions to the Maxwell equations. *Radiotekhnika i Elektronika* **37** (5) (1992), 777–796 (Russian).

[41] T. B. Savina, B. Yu. Sternin, V. E. Shatalov. On reflection formula for the Helmholtz equation. *Radiotekhnika i Elektronika* **38** (2) (1993), 229–240 (Russian).

[42] R. S. Popovidi-Zaridze, D. D. Karkashadze, V. Z. Akhvlediani, D. Sh. Khatiashvili. Investigation of possibilities of auxiliary sources method in two-dimensional electrodynamics problems. *Radiotekhnika i Elektronika* **26** (2) (1981), 254–262 (Russian).

[43] R. Zaridze, G. Bit-Babik, K. Tavzarashvili, N. K. Uzunoglu, D. Economou. The Method of Auxiliary Sources (MAS)–Solution of Propagation, Diffraction and Inverse Problems Using MAS. In: *Applied Computational Electromagnetics. State of the Art and Future Trends. Eds. N. K. Uzunoglu, K. S. Nikita, D. I. Kaklamani. NATO Advanced Science Institutes Series, Series F: Computer and Systems Sciences, vol. 171*, Springer 2000, 33–45.

[44] M. I. Andriychuk, N. N. Voitovich, P. A. Savenko, V. P. Tkachuk. *Antenna Synthesis by Amplitude Radiation Pattern. Numerical Methods and Algorithms.* Naukova Dumka, Kiev 1993 (Russian).

[45] L. D. Bakhrakh, S. D. Kremenecky. *Synthesis of Radiating Systems. Theory and Calculation Methods.* Sovetskoe Radio, Moscow 1974 (Russian).

[46] M. I. Andriychuk, N. N. Voitovich. Synthesis of closed plane antenna by the given amplitude pattern. *Radiotekhnika i Elektronika* **30** (2) (1985), 276–283 (Russian).

[47] D. Pozar. Antenna synthesis and optimization using weighed Inagaki modes. *IEEE Trans.* **AP-32** (2) (1984), 159–165.

[48] V. I. Smirnov. *Course of Higher Mathematics, Part II.* Nauka, Moscow 1974 (Russian).

[49] Yu. N. Khudak. On local structure of a class of solutions to the Maxwell equations. *Radiotekhnika i Elektronika* **32** (2) (1987), 225 (Russian).

[50] P. A. Savenko, N. N. Voitovich. Generalized criterion of the pattern closeness in antenna synthesis problem by V.V.Semenov's method. *Radiotekhnika i Elektronika* **18** (9) (1973), 1794–1798 (Russian).

[51] O. V. Kusyi, S. S. Shaposhnikov, R. B. Vaganov, N. N. Voitovich. Bicriterion optimization problems for power transmitting line. In: *Direct and Inverse Problems of Electromagnetic and Acoustic Wave Theory (DIPED–2002)*, Lviv–Tbilisi 2002, 127–130.

[52] V. G. Fadeev, V. A. Vanke. Optimization of the transmitting antenna of solar space power system. *J. of Communications Technology and Electronics* **44** (7) (1999), 775–779 (Russian).

[53] A. G. Fox and T. Li. Resonant modes in a laser interferometer. *Bell Sys. Tech. J.* **40** (1961), 453.

[54] B. Z. Katsenelenbaum. The antenna shape in the line for microwave power transmission by a long beam. In: *Direct and Inverse Problems of Electromagnetic and Acoustic Wave Theory (DIPED–2000)*, Lviv–Tbilisi 2000, 12–20.

Appendix

Antenna Synthesis by Amplitude Radiation Pattern and Modified Phase Problem *(by N. N. Voitovich)*

A.1 Synthesis of Antennas by Amplitude Radiation Pattern

A.1.1 Introduction

One of the topics discussed in the book is related to problems in which only the amplitude pattern is given, but the phase pattern can be chosen at will. This question was a subject of consideration in Section 2.3 wholly devoted to the amplitude nonapproximability, as well as in Subsections 4.1.5 and 4.3.3, while discussing the optimal current synthesis problem. In both of these cases two aspects of the topic were emphasized, namely, the great practical significance of the problems and their complication both from the point of view of theoretical investigation and their numerical solution. The first aspect is connected with the fact that the majority of applications impose requirements only on the power distribution of the far field, which is described by the amplitude radiation pattern. In such applications the free choice of the phase pattern can by used for better approximation of the given amplitude pattern or for fulfilment of some additional demands on the radiating system. The complication of the problems with only given amplitude pattern is caused by the nonlinearity of the problems, nonuniqueness of their solutions, and a poor choice of efficient numerical methods for their solution. The appendix is exclusively devoted to such problems. It can also be used for generalization of the problems considered in Section 2.2.

The works [1]–[4] were probably the first ones on nonlinear statements of the antenna synthesis problems. The approach discussed here is based on the idea proposed in paper [5], developed in a series of publications, and summarized in the book [6]. Generalization of the approach to different antenna types is given in [7].

Here the results concerning the analytical solutions to some of these problems are briefly presented. They have been obtained by M. I. Andriychuk, O. O. Bulatsyk (Reshnyak), O. M. Gis, Yu. P. Topolyuk, and N. N. Voitovich. For a detailed exposition of the results we refer the reader to [8]– [20].

In this section we give the variational statement of the synthesis problem by the amplitude radiation pattern for curvilinear antennas and reduce it to the Lagrange–Euler equation. The equation is a nonlinear integral one. It belongs to the Hammerstein type. An iterative technique is proposed for solving it. Separately, the linear antenna case is considered. In this case the direct problem is described by the Fourier transformation. This extends the domain of application of the results, outside antenna theory. Some main properties of the solutions are briefly analyzed.

Section A.2 is devoted to a detailed investigation of the obtained equations. The connection of the problem considered here with the so-called phase one is established, which allows to consider it as a modified phase problem. Solutions to this problem are obtained in explicit form. A technique for investigating the bifurcation phenomenon is described and realized in numerical examples.

In Section A.3 the above results are transferred to the case of the linear antenna array, in which the problem is described by the discrete Fourier transformation.

Some notations are used in the appendix which are different from those used in the previous chapters. This relates mainly to the notation of complex conjugated values by dashed letters (not by the asterisked ones, as in the previous text), and adjoint operators denoted by the asterisk letters (not by the index "ad"). The bibliography is also given as a separate list.

A.1.2 Problem formulation for the curvilinear antenna

The problem of antenna synthesis by the given amplitude radiation pattern can be formulated as finding pseudo-solutions to the equation

$$|Au| = F. \tag{A.1}$$

Here A is an operator mapping the current u into the pattern

$$f = Au; \tag{A.2}$$

in general, we assume A to be a linear bounded operator from $H_1 = L_2(D_1)$ to $H_2 = L_2(D_2)$; D_1, D_2 are the definition domains of the currents and patterns, respectively; F is a given amplitude pattern: nonnegative function from H_2. If the antenna has a form of a line C (closed or not) described by the equation $r = r(\theta)$ in polar coordinates, then the formula (A.2) transforms to

$$f(\varphi) = \int_C u(s_\theta)K(\theta,\varphi)ds_\theta, \quad K(\theta,\varphi) = \frac{k}{4}\sqrt{\frac{2}{\pi}}e^{3\pi i/4}e^{ikr(\theta)\cos(\varphi-\theta)} \tag{A.3}$$

(see (1.1)). In this case the domains of definition of u and f are: $D_1 = \{s_\theta : s_\theta \in C\}$, $D_2 = \{\varphi : 0 \le \varphi < 2\pi\}$.

We look for the functions u in H_1 which minimize the functional

$$\sigma(u) = \| \, |Au| - F\|_2^2, \tag{A.4}$$

where $\|\cdot\|_2$ means the least square norm in H_2; similar notations will be used for the norm in H_1, as well as for the inner products in both spaces.

As it follows from the main text of the book, the minimum of $\sigma(u)$ may be not reached in H_1 (the amplitude pattern F cannot be supplied by any phase to be realizable), or the function

u (the current distribution), on which the minimum is reached, has a large norm (the antenna contour coincides with or is close to a specific line $\hat{C}$, or the amplitude pattern F is too sharp). For these reasons the functional (A.4) should be replaced with

$$\sigma_t(u) = \|\,|Au| - F\|_2^2 + t\|\,u\|_1^2. \tag{A.5}$$

The functional (A.5) can be treated in three different ways depending on the formulation of the initial problem. If the problem is on the optimal current, similar to that in Chapter 4 (the conditional minimization problem: to minimize the mean square discrepancy σ at bounded current norm: $\|\,u\|_1^2 \leq N^2$, or, conversely, to minimize the current norm at bounded discrepancy σ), then $\sigma_t(u)$ is the Lagrange unconditional minimization functional; t is the Lagrange multiplier which should be chosen for the condition to be fulfilled. If A is compact (as in (A.3)), then the problem on $\sigma(u)$ minimization is ill-posed, its solution can be unstable with respect to perturbations of the given function F. The Tikhonov regularization method [33] leads to the regularization functional (A.5) with t as a regularization parameter. It can be determined, for instance, by the so-called discrepancy principle. Finally, the problem can be formulated as a two-criterion one, consisting in simultaneous minimization of both σ and $\|\,u\|_1^2$. Then $\sigma_t(u)$ is the weighted combined functional to be minimized, t is a weight factor, determined from the balance demand on contributions of both terms into the functional.

A.1.3 Reducing to the Lagrange–Euler equation

The problem on minimization of the functional (A.5) can be reduced to the nonlinear Lagrange–Euler equation. To this end, we calculate the first variation of the functional: the linear part $\delta\sigma_t$ in the perturbation of the functional

$$\sigma_t(u + \delta u) = \sigma_t(u) + \delta\sigma_t(u, \delta u) + o(\delta u), \tag{A.6}$$

caused by an arbitrary small perturbation $\delta u \in H_1$ of the function u. After the usual derivations we have

$$\delta\sigma_t(u, \delta u) = 2\,\mathrm{Re}(A^*[Au - Fe^{i\arg Au}] + tu, \delta u)_1. \tag{A.7}$$

Here A^* is the operator from H_2 to H_1, adjoint of A. The Lagrange–Euler equation is obtained from the condition $\delta\sigma_t(u, \delta u) = 0$ at arbitrary δu. Putting, sequentially, δu to be real and imaginary, we obtain

$$tu + A^*Au = A^*[Fe^{i\arg Au}]. \tag{A.8}$$

Acting by A on both sides of (A.8) leads to an equivalent form of the Lagrange–Euler equation

$$tf + AA^*f = AA^*[Fe^{i\arg f}]. \tag{A.9}$$

This equation is simpler than (A.8) for both analytical investigation and numerical solution.

After (A.9) is solved, the function u can be calculated from (A.8) modified by replacing $\arg Au$ on its right-hand side with $\arg f$. Then (A.8) becomes linear. We emphasize that the Lagrange–Euler equation is solved not only by the functions which are the minimum points of the appropriate functional, but also by all its stationary points, that is, by all the points for which $\delta\sigma_t(u, \delta u) = 0$ identically with respect to δu (this condition is only necessary for u to be the minimum point of σ_t).

Note that the only specific property of L_2 used while deriving (A.7) is the identity

$$(f_1 e^{i\psi}, f_2)_2 = (f_1, f_2 e^{-i\psi})_2 \tag{A.10}$$

which is not inherent for arbitrary Hilbert spaces H_1, H_2. In this case the operator adjoint of the multiplication one can by used in the equality (A.10) (see [14]).

In the case of the curvilinear antenna C, the operators in (A.8), (A.9) are integral:

$$\begin{aligned} tu(s_\theta) + \int_C K_1(\theta, \theta')u(\theta')ds_{\theta'} \\ = \int_0^{2\pi} \bar{K}(\theta, \varphi)F(\varphi)\exp\left[i\arg\int_C K(\theta', \varphi)u(s_{\theta'})ds_{\theta'}\right]d\varphi, \end{aligned} \tag{A.11}$$

$$K_1(\theta, \theta') = \frac{k^2}{8\pi}\int_0^{2\pi} e^{ik\left[r(\theta)\cos(\varphi-\theta') - r(\theta')\cos(\varphi-\theta)\right]}d\varphi; \tag{A.12}$$

$$tf(\varphi) + \int_0^{2\pi} K_2(\varphi, \varphi')f(\varphi')d\varphi' = \int_0^{2\pi} K_2(\varphi, \varphi')F(\varphi)e^{i\arg f(\varphi')}d\varphi', \tag{A.13a}$$

$$K_2(\varphi, \varphi') = \frac{k^2}{8\pi}\int_C e^{ikr(\theta)[\cos(\varphi-\theta)-\cos(\varphi'-\theta)]}d\theta. \tag{A.13b}$$

Equation (A.13a) is a little simpler than (A.11) owing to the same kernel in both integrals. When $f(\varphi)$ is determined from (A.13a), then the linear equation

$$tu(s) + \int_C K_1(\theta, \theta')u(\theta')ds_{\theta'} = \int_0^{2\pi} \bar{K}(\theta, \varphi)F(\varphi)e^{i\arg f(\varphi)}d\varphi \tag{A.14}$$

should be solved to calculate $u(s)$.

The kernel $K_2(\varphi, \varphi')$ possesses the obvious property

$$K_2(\varphi, \varphi') = \bar{K}_2(\varphi', \varphi). \tag{A.15}$$

If the line C is symmetrical with respect to $\theta = 0$, that is, if $r(-\theta) = r(\theta)$, then

$$K_2(-\varphi, -\varphi') = K_2(\varphi, \varphi'). \tag{A.16}$$

Equations (A.11), (A.13a) belong to the class of nonlinear integral equations of the Hammerstein type. They have nonunique solutions, which can bifurcate as the frequency k (or the size of C) grows. We analyze a similar situation in the example of linear antenna in the next section.

An obvious iterative scheme can be used for solving equations (A.11). It consists in substituting $\arg f(\varphi)$ on the right-hand side, by its value from the previous step. So, to solve (A.13a) we should solve the linear equation

$$tf_{p+1}(\varphi)+\int_0^{2\pi} K_2(\varphi,\varphi')f_{p+1}(\varphi')d\varphi' = \int_0^{2\pi} K_2(\varphi,\varphi')F(\varphi)e^{i\arg f_p(\varphi')}d\varphi' \quad (A.17)$$

in each step. Comparing $\sigma_t(u)$ with an auxiliary functional

$$\sigma_t^{(p)}(u) = \|\, Au - Fe^{i\arg f_p}\|_2^2 + t\|\, u\|_1^2, \quad (A.18)$$

one can easily obtain the relaxation property of the algorithm:

$$\sigma_t(u_{p+1}) \le \sigma_t(u_p). \quad (A.19)$$

The sequence $\{u_p\}$ converges in L_2 metric everywhere except for the bifurcation points (see [15]). Of course, the convergence rate decreases near to these points.

The functional (A.5) can be minimized by the standard minimization methods as well.

A.1.4 Case of linear antenna. Main properties of solutions

The linear antenna can be considered as a particular case of the curvilinear one. Then the above formulation of the problem can be used with some modification. Here we give another formulation which is a generalization of the usual synthesis problem according to the complete (complex) radiation pattern (see, e.g.,[29]). To this end, we introduce the new angle variable instead of φ: $\xi = \sin\varphi/\sin\alpha$ (the value α, $\alpha \le \pi/2$, will be introduced below) and extend it onto the entire axis: $-\infty < \xi < \infty$ (for symmetry the antenna is assumed to be directed perpendicularly to the axis $\varphi = 0$). In this way we introduce the complex angle φ, in order to describe the so-called "reactive power"[1]. For the sake of single-valuedness and owing to symmetry, the "real" angles are bounded by $-\pi/2 \le \varphi \le \pi/2$ which corresponds to $-1/\sin\alpha \le \xi \le 1/\sin\alpha$.

It is natural to assume that the given amplitude pattern F differs from zero only in some interval of angles φ: $|\varphi| \le \alpha$. In new coordinates $F(\xi) = 0$ when $|\xi| > 1$. In these notations the usual formula for the radiation pattern has the form of Fourier transformation of the compactly supported function:

$$f(\xi) = [Au](\xi) = \frac{1}{\sqrt{2\pi}}\int_{-c}^{c} u(x)e^{i\xi x}dx. \quad (A.20)$$

Here $c = ka\sin\alpha$ is the joint physical parameter, $2a$ is the antenna length, $x = cs/a$, s is the linear coordinate of the antenna point. Operator A is isometric from $L_2(-c,c)$ to $L_2(-\infty,\infty)$: the Parseval identity

$$\|u\|_1^2 = \|Au\|_2^2 \quad (A.21)$$

[1] Similar formulation was used in Chapter 6 for optimization of the power transmitting lines.

holds in this case. Functions of the form (A.20) are said to have a bounded spectrum.

An isometric operator has a bounded inverse one in its range space. Therefore, we confine ourselves to the use of the functional (A.4) instead of (A.5). The Lagrange–Euler equations for the functional σ with isometric operator A, are of the form

$$u = A^*[Fe^{i\arg Au}], \tag{A.22}$$

or of the equivalent one

$$f = AA^*[Fe^{i\arg f}]. \tag{A.23}$$

An iterative scheme analogous to that described in the previous subsection can be applied to these equations.

In the case of linear antenna, the above equations are

$$u(x) = \frac{1}{\sqrt{2\pi}} \int_{-1}^{1} F(\xi) e^{i(\arg f(\xi) - \xi x)} d\xi \tag{A.24}$$

and

$$f(\xi') = \frac{1}{\pi} \int_{-1}^{1} \frac{\sin[c(\xi - \xi')]}{(\xi - \xi')} F(\xi) e^{i\arg f(\xi)} d\xi. \tag{A.25}$$

If one uses the Lagrange–Euler equation in the form (A.24), then $f(\xi)$ on its right-hand side is calculated by the formula (A.20). If equation (A.25) is used for determining $f(\xi)$, then the sought function $u(x)$ is calculated by (A.24). For our further consideration equation (A.25) is more convenient.

It is easy to check the following simplest properties of solutions to equation (A.25):

1. If the function

$$f_0(\xi) = \frac{1}{\pi} \int_{-1}^{1} \frac{\sin[c(\xi - \xi')]}{(\xi - \xi')} F(\xi') d\xi' \tag{A.26}$$

   is nonnegative at $|\xi| < 1$, then it solves (A.25). At sufficiently small c such a solution exists for arbitrary $F(\xi)$.

2. If $f(\xi)$ solves (A.25), then $\alpha f(\xi)$, $|\alpha| = 1$, also solves (A.25).

3. If $f(\xi)$ solves (A.25), then $\bar{f}(\xi)$ also solves (A.25).

We describe analytical solutions to (A.25) in the next section.

A.2 Modified Phase Problem. Continuous Case

A.2.1 Modified phase problem and related mathematical and physical problems

The topic discussed here is closely related to the so-called phase problem [21, 22, 23, 24, 25, 26, 27]. The latter consists in reconstructing the argument (phase) of a complex function $f(\xi)$, which is the Fourier transform of a compactly supported function $u(x)$, according to the given modulus (amplitude) $|f(\xi)|$, and then in reconstructing $u(x)$ itself. This problem arises in different applications such as the image, fields, and signal transformers, diagnostic systems and devices, X-ray crystallography, etc.

Our problem differs from the mentioned one in two principal aspects. First, the amplitude function $F(\xi)$, which we are going to approximate, may be not the modulus of a function with bounded spectrum (in the case of the antenna synthesis problem it is not an amplitude of a realizable pattern), so that we do not solve the equation exactly, but only seek its pseudo-solutions. This allows us to impose on this function all essential requirements, except for the requirement to belong to any class not defined constructively. Second, in contrast to the reconstruction (diagnostic) problems, the synthesis (construction) problems can have nonunique solutions: existence of several solutions permits to choose the best of them by additional criteria.

We call the problem of functional (A.4) minimization, with the operator A of the form (A.24), (as well as others connected with the Fourier transformation) *the modified phase problem.* Laying such emphasis on this problem (an important, but only a particular one in the antenna synthesis theory) is first of all caused by wide usage of the Fourier transformation in applications. On the other hand, the problem gives rise to nonlinear integral equations which belong to an interesting nonstudied class of equations of the Hammerstein type. Here we try to describe analytically all the solutions of one of them, namely, equation (A.25), as well as a similar one which arises in the analogous synthesis problem of the antenna array theory. The existence of analytical solutions to these special equations allows us to understand the specific properties of such equations in general.

A.2.2 Analytical solutions to the Lagrange–Euler equation for linear antenna. Theoretical results

In this subsection we construct analytical solutions to equation (A.25). These solutions depend on a finite number of complex parameters which can be calculated from a set of transcendental equations. The results are stated in the form of a lemma and four propositions.

We seek the solutions to equation (A.25) in the form

$$f(\xi) = \alpha \hat{f}(\xi) P_N(\xi), \tag{A.27}$$

where α is a complex constant with $|\alpha| = 1$, $\hat{f}(\xi)$ is a real function, positive[2] on $[-1;1]$ (all its zeros lie outside this segment),

[2] We confine ourselves to the case when solutions to eq. (A.25) do not vanish at any point of $[-1,1]$. The general case is considered in [10], [16].

$$P_N(\xi) = \prod_{n=1}^{N} (1 - \eta_n \xi) \tag{A.28}$$

is a polynomial of degree N with complex zeros η_n^{-1} nonconjugated pairwise:

$$\eta_n - \bar{\eta}_m \neq 0, \quad n, m = 1, \cdots, N. \tag{A.29}$$

Note that according to the Paley–Wiener theorem (see, e.g. [31]) functions of the form (A.20) are entire functions of exponential type and they can have, in general, an infinite number of complex zeros. Under the assumption (A.27), the phase factor on the right-hand side of (A.25) takes the form

$$e^{i \arg f(\xi)} = \alpha \frac{P_N(\xi)}{|P_N(\xi)|}. \tag{A.30}$$

Introduce the following polynomial in two real variables:

$$R_{N-1}(\xi, \xi') = \frac{2i \operatorname{Im}(P_N(\xi)\bar{P}_N(\xi'))}{(\xi - \xi')} = \sum_{n,m=1}^{N} d_{nm} \xi^{n-1} (\xi')^{m-1}, \tag{A.31}$$

with the matrix of coefficients

$$D = \{d_{nm}\} \tag{A.32}$$

(the finiteness of (A.31) at $\xi = \xi'$ follows from the obvious equality $\operatorname{Im} |P_N(\xi)|^2 = 0$). The property of this matrix to be nondegenerated is significant for the next considerations. This property was established in [34] and described in [35]. We give another proof of this property here.

Lemma 1 *The following formula holds:*

$$\det D = (-1)^{[\frac{N}{2}]} \prod_{n,m=1}^{N} (\bar{\eta}_m - \eta_n), \tag{A.33}$$

where the square brackets denote the integer part of a quantity. In particular, inequalities (A.29) imply $\det D \neq 0$.

Proof. Write $P_N(\xi)$ as

$$P_N(\xi) = a_0 + a_1 \xi + \cdots + a_N \xi^N, \quad a_0 = 1, \tag{A.34}$$

and denote

$$b_{nm} = a_n \bar{a}_m - \bar{a}_n a_m. \tag{A.35}$$

From (A.31) one gets the recurrence

$$d_{n,m+1} = d_{n+1,m} + b_{nm}, \quad n, m = 1, 2, \cdots, N, \tag{A.36}$$

with initial data $d_{n1} = b_{n0}$, $n = 1, 2, \cdots, N$. The solution of (A.36) is given by

$$d_{nm} = \sum_{k=0}^{\min(N-n,m-1)} b_{n+k,m-k-1}. \tag{A.37}$$

One can readily check the identity

$$D = (B_1 \bar{B}_2 - \bar{B}_1 B_2) E', \tag{A.38}$$

where

$$B_1 = \begin{pmatrix} a_N & a_{N-1} & \dots & a_1 \\ 0 & a_N & \dots & a_2 \\ \dots & \dots & \dots & \dots \\ 0 & 0 & \dots & a_N \end{pmatrix}, \qquad B_2 = \begin{pmatrix} 1 & 0 & \dots & 0 \\ a_1 & 1 & \dots & 0 \\ \dots & \dots & \dots & \dots \\ a_{N-1} & a_{N-2} & \dots & 1 \end{pmatrix}, \tag{A.39}$$

$$E' = \begin{pmatrix} 0 & \dots & 0 & 1 \\ 0 & \dots & 1 & 0 \\ \dots & \dots & \dots & \dots \\ 1 & \dots & 0 & 0 \end{pmatrix}. \tag{A.40}$$

Calculate the following block matrix product:

$$\begin{pmatrix} -\bar{B}_1 & B_1 \\ E & 0 \end{pmatrix} \begin{pmatrix} B_1 & B_2 \\ \bar{B}_1 & \bar{B}_2 \end{pmatrix} = \begin{pmatrix} B_1 \bar{B}_1 - \bar{B}_1 B_1 & B_1 \bar{B}_2 - \bar{B}_1 B_2 \\ B_1 & B_2 \end{pmatrix}. \tag{A.41}$$

The determinant of the second factor on the left-hand side of (A.41) is the resultant of polynomials $P_N(\xi)$ and $\bar{P}_N(\xi)$. Its value is (see, e.g.[28])

$$r(P_N, \bar{P}_N) = \prod_{n,m=1}^{N} (\bar{\eta}_m - \eta_n). \tag{A.42}$$

Matrix $B_1 \bar{B}_1$ is real due to the specific form of B_1, hence $B_1 \bar{B}_1 - \bar{B}_1 B_1 = 0$. Consequently, we can write

$$\det D = r(P_N, \bar{P}_N) \det E'. \tag{A.43}$$

Since $\det E' = (-1)^{[N/2]}$, we obtain (A.33). □

Conditions under which a function $f(\xi)$ of the form (A.27) solves equation (A.25) are established in:

Proposition 2 *Let a real $c > 0$ and positive $F(\xi) \in L_2(-1;1)$ be given. Suppose that a function $f(\xi)$ of the form (A.27) does not vanish on $[-1,1]$. Then $f(\xi)$ is a solution to equation (A.25) if and only if the parameters $\eta_n, n = 1, 2, \cdots, N$, satisfy the following set of transcendental equations:*

$$\int_{-1}^{1} \frac{\xi^{n-1} \sin(c\xi) F(\xi)}{|P_N(\xi)|} d\xi = 0, \tag{A.44a}$$

$$\int_{-1}^{1} \frac{\xi^{n-1} \cos(c\xi) F(\xi)}{|P_N(\xi)|} d\xi = 0, \tag{A.44b}$$

$$n = 1, 2, \ldots, N.$$

Before proving the proposition , we put $\alpha = 1$ in (A.27) without loss of generality.

Proof. Necessity. Let (A.27) be a solution of (A.25). Substituting (A.27) into (A.25), multiplying by $\bar{P}_N$ and separating imaginary parts, yields:

$$\int_{-1}^{1} \frac{\sin[c(\xi - \xi')] F(\xi) R_{N-1}(\xi, \xi')}{|P_N(\xi)|} d\xi = 0. \tag{A.45}$$

Using notation (A.31) and linear independence of the functions $\xi^n \sin c\xi$, $\xi^n \cos c\xi$, $n = 0, 1, \cdots, N-1$, one gets from (A.45) the set of equations

$$\sum_{n=1}^{N} d_{nm} \int_{-1}^{1} \frac{\xi^{n-1} \sin(c\xi) F(\xi)}{|P_N(\xi)|} d\xi = 0, \tag{A.46a}$$

$$\sum_{n=1}^{N} d_{nm} \int_{-1}^{1} \frac{\xi^{n-1} \cos(c\xi) F(\xi)}{|P_N(\xi)|} d\xi = 0, \tag{A.46b}$$

$$n = 1, 2, ..., N.$$

Since, according to Lemma 1, the determinant of the matrix D is not equal to zero, equations (A.46), treated as sets of linear algebraic equations with respect to the values of integrals, have only zero solutions. This means that the system (A.44) is satisfied.

Sufficiency. Let now (A.44) be satisfied. Then (A.46) and, going backwards, (A.45) are satisfied as well. Equation (A.45) together with (A.30), (A.31) results in

$$\operatorname{Im}(\bar{P}_N \cdot AA^*(Fe^{i \arg f})) = 0. \tag{A.47}$$

Taking into account (A.27), we obtain from (A.47) that

$$\arg f = \arg AA^*(Fe^{i\arg f}), \tag{A.48}$$

which leads to (A.25), or, alternatively, to

$$\arg f = \arg AA^*(Fe^{i\arg f}) + \pi. \tag{A.49}$$

The latter is equivalent to the equation

$$f = -AA^*(Fe^{i\arg f}). \tag{A.50}$$

Equation (A.50) cannot have a solution. Indeed, if f were such a solution, then

$$(Fe^{i\arg f}, f) = -(Fe^{i\arg f}, AA^*(Fe^{i\arg f})) = -||A^*(Fe^{i\arg f})||^2 \leq 0. \tag{A.51}$$

On the other hand,

$$(Fe^{i\arg f}, f) = (F, |f|) > 0. \tag{A.52}$$

Consequently, equation (A.25) must be satisfied. □

After equations (A.44) are solved, the solution to (A.25) can be calculated by the formula

$$f(\xi) = \frac{1}{\pi}\int_{-1}^{1} \frac{\sin[c(\xi - \xi')]}{(\xi - \xi')} F(\xi') \frac{P_N(\xi')}{|P_N(\xi')|} d\xi' \tag{A.53}$$

The next proposition gives an upper bound for the polynomial P_N degree.

Proposition 3 *Equations (A.44) can have solutions only with*

$$N < \frac{2c}{\pi}. \tag{A.54}$$

Proof. Equation (A.44a) means that the function

$$\Phi_s(\xi) = \frac{F(\xi)}{|P_N(\xi)|} \sin(c\xi) \tag{A.55}$$

is orthogonal to any polynomial Q_m of degree less than N. This function must change its sign on the interval (-1,1) at least N times. Because such an alternation can belong only to the factor sin $c\xi$, inequality

$$c > \left[\frac{N}{2}\right]\pi \tag{A.56a}$$

has to hold. Similarly, from equation (A.44b) one gets the inequality

$$c > \left(\left[\frac{N+1}{2}\right] - \frac{1}{2}\right)\pi. \tag{A.56b}$$

These two inequalities together give (A.54).

Note that the inequality (A.54) is universal: it does not contain the given function F. For this reason it can be redundant for a concrete F. □

Proposition 4 *If $f(\xi) = \hat{f}(\xi)P_N(\xi)$ is a solution to equation (A.25), with $\hat{f}(\xi)$ and $P_N(\xi)$ having the properties described above, then*

$$f_n(\xi) = \hat{f}(\xi)P_N(\xi)\frac{1-\bar{\eta}_n\xi}{1-\eta_n\xi} \tag{A.57}$$

is also a solution to (A.25) for any $n = 1, 2, \cdots, N$.

Proof. Let $f(\xi) = \hat{f}(\xi)P_N(\xi)$ solve equation (A.25). Then, according to Proposition 1, the equation set (A.44) is satisfied. Simultaneously, this set is satisfied by substituting P_N by any polynomial of degree N with the same modulus, in particular, of $p_n(\xi)(1-\bar{\eta}_n\xi)$ where $p_n(\xi) = P_N(\xi)/(1-\eta_n\xi)$. This means that the following two identities are satisfied:

$$\hat{f}(\xi')p_n(\xi')(1-\eta_n\xi') = \frac{1}{\pi}\int_{-1}^{1}\frac{\sin[c(\xi-\xi')]F(\xi)p_n(\xi)(1-\eta_n\xi)}{|P_N(\xi)|(\xi-\xi')}d\xi, \tag{A.58a}$$

$$\hat{f}_n(\xi')p_n(\xi')(1-\bar{\eta}_n\xi') = \frac{1}{\pi}\int_{-1}^{1}\frac{\sin[c(\xi-\xi')]F(\xi)p_n(\xi)(1-\bar{\eta}_n\xi)}{|P_N(\xi)|(\xi-\xi')}d\xi \tag{A.58b}$$

with some unknown function $\hat{f}_n(\xi')$. Multiplying identity (A.58a) by $(1-\bar{\eta}_n\xi')$ and (A.58b) by $(1-\eta_n\xi')$, and subtracting the results, yields:

$$(\hat{f}(\xi')-\hat{f}_n(\xi'))p_n(\xi')|(1-\eta_n\xi')|^2 = \frac{\bar{\eta}_n-\eta_n}{\pi}\left(\cos(c\xi')\int_{-1}^{1}\frac{\sin(c\xi)F(\xi)p_n(\xi)}{|P_N(\xi)|}d\xi - \sin(c\xi')\int_{-1}^{1}\frac{\cos(c\xi)F(\xi)p_n(\xi)}{|P_N(\xi)|}d\xi\right). \tag{A.59}$$

Since $p_n(\xi)$ is a polynomial of degree $N-1$, the integrals in (A.59) are equal to zero owing to (A.44), which gives

$$\hat{f}_n(\xi') = \hat{f}(\xi'). \tag{A.60}$$

Together with (A.58b) this means that f_n solves (A.25). □

Corollary. Any solution of equation set (A.44) generates an equivalent group of solutions which correspond to the polynomials P_N of the type

$$P_N^{(s)} = \prod_{m=1}^{s} (1 - \eta_{n_m}\xi) \prod_{m=s+1}^{N} (1 - \bar{\eta}_{n_m}\xi),$$

$$s = 1, 2, \cdots, N-1, \; n_{m_1} \neq n_{m_2} \text{ if } m_1 \neq m_2. \quad \text{(A.61)}$$

All these solutions give the same value of $|f|$ and, consequently, the same value of the functional σ.

All the above proofs are based on the assumption that N is finite. The question, whether or not solutions of the type (A.27) with $N = \infty$ exist, is still open. However, the following important proposition is proved.

Proposition 5 *Let the function $f_*(\xi)$ of the form (A.20) be a global-minimum point of the functional σ and have no zeros in $[-1, 1]$. Then this function has the form (A.27) with finite N.*

Proof. It is known that any $f(\xi)$ of the form (A.20), not vanishing at $\xi = 0$, can be written as

$$f(\xi) = f(0)e^{i\beta\xi} \prod_{n=1}^{\infty} (1 - \eta_n \xi), \quad \operatorname{Im}\beta = 0, \tag{A.62}$$

with some complex (in general) η_n. Suppose that such a function is the global-minimum point of the functional σ and has an infinite number of complex zeros η_n^{-1}, nonconjugated pairwise. Recall that according to the Paley–Wiener theorem, the function $f(\xi) \in L_2(-\infty; \infty)$ can be presented in the form (A.20) with $u(x) \in L_2(-c; c)$ if and only if its analytical extension into the complex plane is an entire function satisfying the condition

$$|f(\xi)| \leq Be^{c|\xi|}, \quad B < \infty. \tag{A.63}$$

The function $f_*(\xi)$ belongs to $L_2(-\infty; \infty)$ and it has the form (A.20); hence it satisfies the condition (A.63). As an extremum point of the functional (A.4), it satisfies equation (A.25).

Let us consider the function

$$f_m(\xi) = f_*(\xi)\frac{1 - \bar{\eta}_m\xi}{1 - \eta_m\xi}, \tag{A.64}$$

where η_m is one of the parameters η_n satisfying (A.29). This function is entire, because it is expressed in the form (A.62) with the mth factor $1 - \eta_m\xi$ in the product replaced with $1 - \bar{\eta}_m\xi$. It satisfies the condition (A.63), because

$$\lim_{|\xi|\to\infty}\left|\frac{1-\bar{\eta}_m\xi}{1-\eta_m\xi}\right| = 1, \tag{A.65}$$

and therefore it is expressible in the form (A.20). At real ξ, $|f_m(\xi)| = |f_*(\xi)|$. This means that $|f_m(\xi)|$ gives the same (global-minimal) value of σ as $f_*(\xi)$ does and that is why $f_m(\xi)$ also satisfies equation (A.25). It is easily seen that any function $f(\xi)$ satisfying equation (A.25) and having the same modulus as $|f_*(\xi)|$, solves the linear homogeneous integral equation

$$\nu f(\xi) = \frac{1}{\pi}\int_{-1}^{1}\frac{F(\xi)}{|f_*(\xi)|}\frac{\sin[c(\xi-\xi')]}{(\xi-\xi')}f(\xi)d\xi \tag{A.66}$$

with $\nu = 1$. According to our assumption, the number of such functions is infinite.

Since $F(\xi)$ is integrable and $|f_*(\xi)|$ is bounded from below, the kernel

$$\sqrt{\frac{F(\xi)}{|f_*(\xi)|}\frac{F(\xi')}{|f_*(\xi')|}\frac{\sin[c(\xi-\xi')]}{(\xi-\xi')}} \tag{A.67}$$

of the symmetrized equation belongs to $L_2((-1;1)\times(-1;1))$. Consequently, equation (A.66) cannot have the eigenvalue $\nu = 1$ of infinite multiplicity. Therefore, the number of parameters η_n in expression (A.62) for $f_*(\xi)$, satisfying (A.29), is finite and one can write

$$f_*(\xi) = \alpha e^{i\beta\xi}\hat{f}(\xi)P_N(\xi), \tag{A.68}$$

where $\alpha = e^{i\arg f(0)}$.

It remains to prove that $\beta = 0$. Substitute (A.68) into (A.25) and calculate the asymptotic behavior of $f_*(\xi)$ as $\xi = \pm iz$, $\operatorname{Im} z = 0$, $z \to \infty$:

$$|f_*(\pm iz)| \simeq |A_{\pm}^{(m)}||z|^{-(m+1)}e^{c|z|}, \tag{A.69}$$

where $A_{\pm}^{(m)}$ is the first non-zero coefficient of

$$A_{\pm}^{(n)} = \frac{1}{2\pi}\int_{-1}^{1}\frac{F(\xi)P_N(\xi)\xi^n e^{i(a\pm c)\xi}}{|P_N|}d\xi,\ n = 1, 2, \cdots \tag{A.70}$$

It is obvious that all of these coefficients do not equal zero simultaneously, except for the case $F(\xi) \equiv 0$. On the other hand, since $\hat{f}(-iz) \simeq \bar{\hat{f}}(iz)$, we have from (A.68)

$$|f_*(\pm iz)| \simeq |\hat{f}(iz)a_N z^N|e^{\mp\beta z}. \tag{A.71}$$

This means that f_* has different exponential behavior on the upper and lower imaginary half-axes, which does not agree with (A.69). Consequently, $\beta = 0$, and the proposition is proved.
□

A.2.3 Solution branching

The above propositions allow to find all the solutions to (A.25) if one can find all complex roots of the finite set of transcendental equations (A.44) at fixed c and N. The following questions arise: a) how many of such solutions exist at fixed c; b) in which way does the number of them change as c grows; c) which of them minimizes the functional (A.4)? These questions are briefly considered in this subsection.

It follows from Proposition 3 that the solution (A.26) corresponding to $N = 0$ is unique at least for $c \leq \pi/2$ (we do not draw a distinction between solutions with different constant factors α, $|\alpha| = 1$). At some value of c new solutions with $N = 1$ appear. They exist together with the above solution f_0. The value of N, permitted by the inequality (A.54), grows as c grows, and the number of existing solutions grows in the manner of an avalanche. The values of c at which the number of solutions changes are said to be *bifurcation points*. These points can be of three types:

a) the *branching points* (solutions continuously appear from the existing ones);

b) the isolated *appearance points* (in the neighborhood of the solution there exist no solutions at smaller c);

c) the isolated *disappearance points* (in the neighborhood of the solution there exist no solutions at larger c).

The first case is typical, the last two are rather exotic. We analyze their nature in an example, which permits their investigation using the same methods as are used in the first case. Note that, owing to analyticity of the left-hand side of equations (A.44) treated as functions of $c, \eta_n, n = 1, ...N$, (recall that P_n has no zeros on the real axis), the curves of solutions in the coordinate space $\{c, \eta_n, n = 1, ...N\}$ are continuous.

We start by determining the bifurcation points of the first type. Assume a solution $f_j(\xi) = f(\xi; c_j)$ to equation (A.25) to be known at some $c = c_j$. Following the nonlinear equation branching theory [32], apply the perturbation method for investigation of the solution in the neighborhood of c_j. Perturb c_j with ε: $c = c_j + \varepsilon$, and calculate the perturbation δf of the solution:

$$f(\xi) = f_j(\xi) + \delta f(\xi). \tag{A.72}$$

In the first approximation the equation for $\delta f(\xi)$ is of the form

$$\delta f(\xi) - i\int_{-1}^{1} \frac{\sin c_j(\xi - \xi')}{\xi - \xi'} F(\xi') e^{i \arg f_j(\xi')} \operatorname{Im} \frac{\delta f(\xi')}{f_j(\xi')} d\xi'$$
$$= \varepsilon \int_{-1}^{1} \cos c_j(\xi - \xi') F(\xi') e^{i \arg f_j(\xi')} d\xi'. \tag{A.73}$$

(Of course, this equation can be obtained by straightforward differentiation of (A.25) with respect to c and notation $\delta f = \varepsilon df/dc$.) This equation is linear with respect to the function vector $\{w, v\} \equiv \{\mathrm{Re}\,(\delta f/f_j)\,,\mathrm{Im}\,(\delta f/f_j)\}$. It has a unique solution only when the corresponding homogeneous equation

$$f_j(\xi)\left[w(\xi)+iv(\xi)\right]-i\int_{-1}^{1}\frac{\sin c_j(\xi-\xi')}{\xi-\xi'}F(\xi')e^{i\arg f_j(\xi')}v(\xi')d\xi'=0 \qquad \text{(A.74)}$$

has only the trivial one. The existence of nontrivial solutions to (A.74) at some c_j means that at this point the function $f(\xi;c_j)$ is not differentiable with respect to c, that is, the function f cannot be uniquely smoothly continued from the point $c=c_j$ (this point can be a branching one).

Since the integrand in (A.74) does not depend on $w(\xi)$, the equation can be transformed to the equation for unknown $v(\xi)$ only:

$$|f_j(\xi)|^2 v(\xi)-\int_{-1}^{1}\frac{\sin c_j(\xi-\xi')}{\xi-\xi'}F(\xi')\,\mathrm{Re}\left[e^{i\arg f_j(\xi')}f_j(\xi)\right]v(\xi')d\xi'=0. \qquad \text{(A.75)}$$

Observing that

$$f_j(\xi)=\frac{1}{\pi}\int_{-1}^{1}\frac{\sin c_j(\xi-\xi')}{\xi-\xi'}F(\xi')\frac{P_N(\xi')}{|P_N(\xi')|}d\xi', \qquad \text{(A.76)}$$

after simple derivations we obtain

$$\int_{-1}^{1}\frac{\sin c_j(\xi-\xi')}{\xi-\xi'}\frac{F(\xi')}{|P_N(\xi')|}\,\mathrm{Re}\left[P_N(\xi')\bar{P}_N(\xi)\right]\left[v(\xi')-v(\xi)\right]d\xi'=0. \qquad \text{(A.77)}$$

Thus we have the homogeneous equation nonlinear with respect to the spectral parameter c_j. Instead of solving this nonlinear eigenvalue problem, we can seek the values $c=c_j$ at which the linear problem

$$\lambda_n v_n(\xi)\int_{-1}^{1}\frac{\sin c(\xi-\xi')}{\xi-\xi'}\frac{F(\xi')}{|P_N(\xi')|}\,\mathrm{Re}\left[P_N(\xi')\bar{P}_N(\xi)\right]d\xi'$$
$$=\int_{-1}^{1}\frac{\sin c(\xi-\xi')}{\xi-\xi'}\frac{F(\xi')}{|P_N(\xi')|}\,\mathrm{Re}\left[P_N(\xi')\bar{P}_N(\xi)\right]v_n(\xi')d\xi' \qquad \text{(A.78)}$$

has eigenvalues $\lambda_n=1$.

It is easily seen that equation (A.77) is solved by $v(\xi)\equiv 1$ at all c_j. Correspondingly, the eigenpair $v_0(\xi)\equiv 1$, $\lambda_0=1$ solves equation (A.78) at all c. This fact is connected with property 2 of solutions to (A.25) (see end of the previous section). It means, that the arbitrary factor α, $|\alpha|=1$, makes any solution ununiquely continuable at all c. Of course, we are not interested in such a "bifurcation". Therefore, only the points $c=c_j$ at which (A.78) has multiple eigenvalues $\lambda_n=1$ can be considered as the branching points.

After the polynomial $P_N(\xi)$, at some N, is found from the transcendental equation set (A.44), one can solve the homogeneous equation (A.78) and determine (at least, approximately) all the points c_j at which the eigenvalue $\lambda_n=1$ has a multiplicity larger than one. The following cases can occur at these points:

a) a new polynomial $P_N(\xi)$ of the same degree N, branches off from the old one;

b) the polynomial $P_{N+s}(\xi)$ of degree $N+s$, $s \neq 0$, branches off from $P_N(\xi)$;

c) no solution branching takes place at $c = c_j$.

Consider different cases occurring at these points separately. To recognize the solutions corresponding to different polynomials, introduce the following notations: $f_N(\xi)$ instead of $f(\xi)$ in (A.27), and η_{Nn} instead of η_n in (A.28).

a) Branching off the polynomial of the same degree N. For simplicity introduce the temporary notations: $\eta'_{Nn} = \operatorname{Re}\eta_{Nn}$, $\eta''_{Nn} = \operatorname{Im}\eta_{Nn}$. The equation set (A.44) can be treated as an implicit form of the functions $\eta'_{Nn}(c)$, $\eta''_{Nn}(c)$:

$$\Phi_n(\eta'_{N1}, ..., \eta'_{NN}; \eta''_{N1}, ..., \eta''_{NN}) = 0, \tag{A.79a}$$

$$\Psi_n(\eta'_{N1}, ..., \eta'_{NN}; \eta''_{N1}, ..., \eta''_{NN}) = 0, \tag{A.79b}$$

$$n = 1, 2, \ldots, N.$$

According to implicit function theory (see, i.e.[36]), the Jacobian

$$J(c) = \frac{D(\Phi_1, ..., \Phi_N; \Psi_1, ..., \Psi_N)}{D(\eta'_{N1}, ..., \eta'_{NN}; \eta''_{N1}, ..., \eta''_{NN})} \tag{A.80}$$

of these functions equals zero at the points $c = c_j$ where the uniqueness of the solutions can be violated. The equation

$$J(c) = 0 \tag{A.81}$$

should be added to the system (A.44) for finding these points.

Equation (A.81) holds also at the points where one of $\eta_{Nn}(c)$ has an infinite derivative $d\eta_n/dc$. These points are the isolated bifurcation points from which continuation of the mentioned function $\eta_{Nn}(c)$, to the right and left, is nonunique or impossible. We call them the *appearance* and *disappearance points*, respectively.

b) Branching off the polynomial of degree $N+1$. Since both $f_N(\xi)$ and $f_{N+1}(\xi)$ are solutions to (A.25) and at the point $c = c_j$ they must coincide, then, according to (A.30),

$$\frac{P_N(\xi)}{|P_N(\xi)|} = \frac{P_{N+1}(\xi)}{|P_{N+1}(\xi)|} \tag{A.82}$$

or, observing (A.28),

$$\prod_{n=1}^{N} \frac{1 - \eta_{Nn}\xi}{|1 - \eta_{Nn}\xi|} = \prod_{n=1}^{N+1} \frac{1 - \eta_{N+1,n}\xi}{|1 - \eta_{N+1,n}\xi|}. \tag{A.83}$$

Equating the factors in both sides of (A.83) term by term, we obtain

$$\eta_{Nn} = \eta_{N+1,n}, \quad n = 1, 2 ... N; \tag{A.84a}$$

$$\operatorname{Im}\eta_{N+1,N+1} = 0. \tag{A.84b}$$

A.2.4 Numerical example. Problem with symmetrical data

In the following numerical examples we illustrate the technique of the problem solving, including investigation of branching points. First, consider the case when the given function $F(\xi)$ is even. Take the simplest (but very important) case $F(\xi) = 1/\sqrt{2}$ (the constant is chosen from the normalizing condition $||F||^2 = 1$). Numerical results related to this case are presented in Figs.A.1–A.6. As a rule, we show only one representative (with $\mathrm{Im}\,\eta_{Nn} \geq 0$) of each equivalent group of solutions. Labels at curves correspond to the indices Nn; the curves describing different solutions corresponding to polynomials with the same N are indicated as $N'n$, $N''n$, etc.

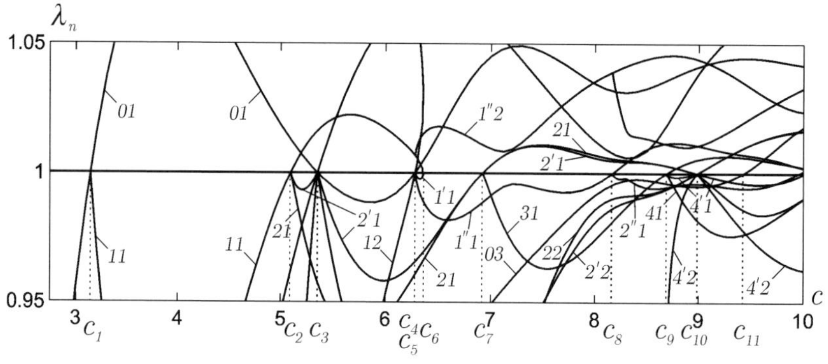

Figure A.1:

We start by analyzing the real solution (A.26) to equation (A.25), which corresponds to $N = 0$. Recall that we consider only solutions not vanishing at $\xi \in (-1, 1)$; the solution (A.26) of such a type exists for any c. For the case $F(\xi) = 1/\sqrt{2}$ this solution was considered in Section 6.2 (see (6.70)).

Several eigenvalues λ_{0n} of equation (A.78) for the polynomial $P_0 = 1$ are shown in Fig.A.1 by the curves indicated as $0n$. Eigenvalue $\lambda_0 = 1$ is common for all N; it is described by curve 0. Curve 01 intersects the line $\lambda = 1$ for the first time at the point $c = c_1$. Two solutions corresponding to $N = 1$ arise at this point. The values c_1, η_{11} are calculated from the equation set (A.88). For even $F(\xi)$ this system has solutions with $\eta_{11} = 0$, and we have the equation

$$\int_{-1}^{1} \cos(c\xi)F(\xi)d\xi = 0 \tag{A.95}$$

for c_j. At $F(\xi) = 1/\sqrt{2}$ this equation is transformed into $\sin c = 0$ which gives $c_1 = \pi$. Two complex conjugate solutions branch off from $f_0(\xi)$ at $c = c_1$; they correspond to the polynomial $P_1(\xi)$ with imaginary $\pm\eta_{11}$ (curve 11 in Fig. A.2(b)). These solutions form the first equivalent group.

The next branching of the solution $f_0(\xi)$ is observed at $c = c_3$, where curve 01 in Fig. A.1 intersects the line $\lambda = 1$ for the second time. This point is specific: the second eigenvalue λ_{02} equals one here. Solutions with complex parameters η_{21}, η_{22}, corresponding to $N = 2$, branch off from $f_0(\xi)$ there. At $c = c_3$ these parameters are imaginary, $\eta_{22} = -\eta_{21}$; they are calculated from the equation set (A.94) transformed at $N = 0$ into

$$\int_{-1}^{1} \frac{\xi \sin(c\xi) F(\xi)}{1 - \eta_{21}^2 \xi^2} d\xi = 0, \tag{A.96a}$$

$$\int_{-1}^{1} \frac{\cos(c\xi) F(\xi)}{1 - \eta_{21}^2 \xi^2} d\xi = 0. \tag{A.96b}$$

At $c > c_3$ the values η_{21}, η_{22} become complex with the same property $\eta_{22} = -\eta_{21}$(curves 21, 22 in Fig. A.2(a) and the curve symmetrical to 22 in Fig. A.2(b). We will return to these solutions while analyzing those corresponding to $N = 2$.

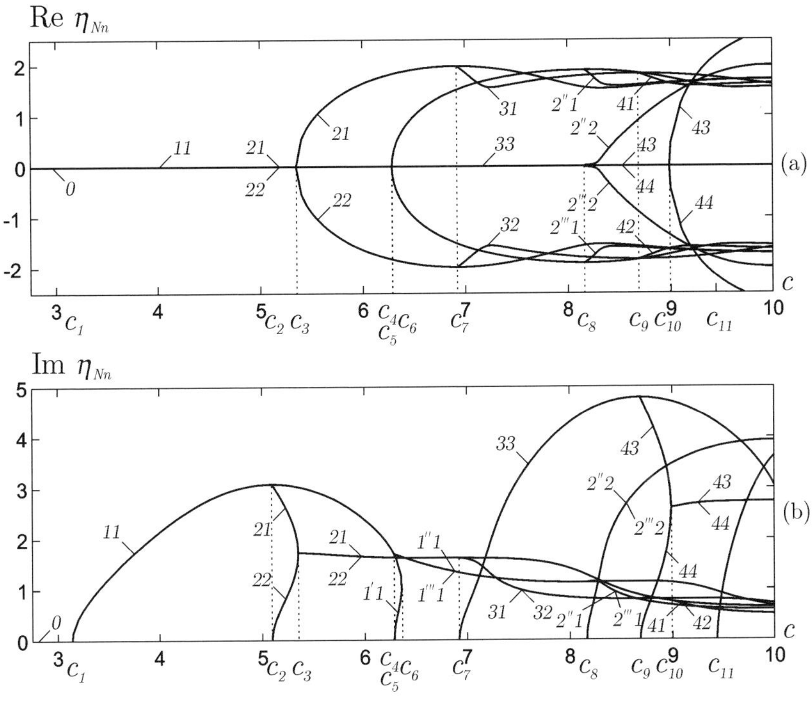

Figure A.2:

Curve 01 in Fig. A.1 intersects the level $\lambda = 1$ for the third time at the point $c = c_4$. This is the second root of equation (A.95). For $F(\xi) = 1/\sqrt{2}$ it equals 2π. At this point the second solution corresponding to $N = 1$ (we denote it by $f_{1'}(\xi)$) with the imaginary $\eta_{1'1}$, branches off from $f_0(\xi)$ (curves $1'1$ in Fig. A.2).

The next branching point of solution $f_0(\xi)$ is the third root $c = c_{11}$ of (A.95) at which $\lambda_{03} = 1$. For $F(\xi) = 1/\sqrt{2}$ this root is $c_{11} = 3\pi$. This point is qualitatively similar to c_1 and the whole branching process described above is repeated with a shift along the c-axis.

Return to the solution corresponding to $N = 1$, which has branched off from $f_0(\xi)$ at $c = c_1$. Denote it by $f_1(\xi)$. The first eigenvalue λ_{11} of equation (A.78) corresponding to

$P_1(\xi)$ starts at $c = c_1$ from the value $\lambda = 1$. This means that this point can be treated as a branching one, at which the solution $f_0(\xi)$ branches off from $f_1(\xi)$.

At $c = c_2$, where $\lambda_{11} = 1$ for the second time, two solutions corresponding to $N = 2$ arise; each of them has two imaginary parameters: η_{21}, $\pm\eta_{22}$ (curves 21, 22 in Fig. A.2). They start from $\eta_{21} = \eta_{11}$, $\eta_{22} = 0$. The values η_{21}, c_2 are obtained from the transcendental equations

$$\int_{-1}^{1} \frac{\cos(c\xi)F(\xi)}{\sqrt{1-\eta_{21}^2\xi^2}}d\xi = 0, \tag{A.97a}$$

$$\int_{-1}^{1} \frac{\xi\sin(c\xi)F(\xi)}{\sqrt{1-\eta_{21}^2\xi^2}}d\xi = 0, \tag{A.97b}$$

to which (A.44) together with (A.87) are reduced in our case. Two solutions of such a type (corresponding to different signs of η_{22}) branch off from each of $f_1(\xi)$ and $\bar{f}_1(\xi)$; they form the equivalent group consisting of four solutions. (We do not further point out the contents of these groups in such simple cases).

At $c = c_5$ (coinciding for $F(\xi) = const$ with c_4) the eigenvalue λ_{12} equals 1. At this point two new solutions $f_{1''}(\xi)$, $f_{1'''}(\xi)$ of the same equivalent group corresponding to $N = 1$ branch off from $f_1(\xi)$. These solutions have complex parameters: $\eta_{1''1} =- \bar{\eta}_{1'''1}$(curves $1''1$, $1'''1$ in Fig. A.2) coinciding with η_{11} at the branching point. The equation set for η_{11}, c_5 obtained after supplying (A.44) with (A.81) is

$$\int_{-1}^{1} \frac{\cos(c\xi)F(\xi)}{\sqrt{1-\eta_{11}^2\xi^2}}d\xi = 0, \tag{A.98a}$$

$$\eta_{11}\int_{-1}^{1} \frac{\xi\sin(c\xi)F(\xi)}{(1-\eta_{11}^2\xi^2)^{3/2}}d\xi \int_{-1}^{1} \frac{\xi^2\cos(c\xi)F(\xi)}{(1-\eta_{11}^2\xi^2)^{3/2}}d\xi = 0. \tag{A.98b}$$

The point $c = c_5$ is the first root of the equation set (A.98), where the first integral in (A.98b) equals zero.

Note that the solutions $f_{1''}(\xi)$, $f_{1'''}(\xi)$ have the same nonsymmetrical amplitude: $|f_{1''}(-\xi)| \neq |f_{1''}(\xi)|$. Therefore they cannot be minimum points (they are only stationary ones) of the functional (A.4). One can easily check that

$$\sigma(u_{1''}(x)) > \sigma([u_{1''}(x) + u_{1''}(-x)]/2), \tag{A.99}$$

where $u_{1''}(x)$ is connected with $f_{1''}(\xi)$ by (A.20), (A.24).

The next specific point is $c = c_6$, where two eigenvalues λ_{11} and $\lambda_{1'1}$ corresponding to the solutions $f_1(\xi)$ and $f_{1'}(\xi)$, respectively, become equal to unity together. At this point these solutions coincide. The values η_{11} (equal to $\eta_{1'1}$) and c_6 are calculated from the above system (A.98). The point is the second root of the equation set (A.98): the second integral in (A.98b) equals zero at this point.

The bifurcation point $(\eta_{11}; c_6)$ can be qualified as the isolated disappearance point, because no solution exists in its neighborhood at $c > c_6$. We will return to this point while analyzing solutions to the problem with non-even $F(\xi)$.

We have analyzed one branch of solutions corresponding to $N = 1$ consisting of curves 11, 1′1. According to the remark made at the beginning of Subsection A.2.3, they make up a smooth continuous curve in the phase space (c, η_{11}).

With the aim of investigating all the solutions to equation (A.25) in the chosen range of c, we follow up the solution $f_{1''}(\xi)$ with c increasing. Similarly to the above cases, the eigenvalue $\lambda_{1''1}$ of (A.78) corresponding to this solution starts from the value $\lambda = 1$ at $c = c_5$ and intersects this value for the second time at $c = c_8$. At this point the solutions $f_{2''}(\xi)$ with two complex parameters $\eta_{2''1}$, $\eta_{2''2}$ branch off from it. At the branching point $c = c_8$ these parameters are: $\eta_{2''1} = \eta_{1''1}$, $\operatorname{Im} \eta_{2''2} = 0$. The equation set for complex $\eta_{1''1}$, and real $\eta_{2''2}$ and c_8 is obtained from (A.44) supplied with (A.87) which, in our case, gives

$$\int_{-1}^{1} \frac{\sin(c\xi)F(\xi)}{|1 - \eta_{11}\xi|} d\xi = 0, \tag{A.100a}$$

$$\int_{-1}^{1} \frac{\cos(c\xi)F(\xi)}{|1 - \eta_{11}\xi|} d\xi = 0, \tag{A.100b}$$

$$\int_{-1}^{1} \frac{\xi \sin(c\xi)F(\xi)}{|1 - \eta_{11}\xi|\,(1 - \eta_{22}\xi)} d\xi = 0, \tag{A.100c}$$

$$\int_{-1}^{1} \frac{\xi \cos(c\xi)F(\xi)}{|1 - \eta_{11}\xi|\,(1 - \eta_{22}\xi)} d\xi = 0. \tag{A.100d}$$

Return to the solutions corresponding to $N = 2$, which have branched off from $f_1(\xi)$ at $c = c_2$. In this case we should separately consider two different representatives of the corresponding equivalent group of solutions, because they have different branching behavior. Denote such solutions by $f_2(\xi)$ (with the same signs of η_{21}, η_{22}) and by $f_{2'}(\xi)$ (with the opposite signs of η_{21}, η_{22}), respectively. The first eigenvalues λ_{21} and $\lambda_{2'1}$of (A.78), corresponding to $f_2(\xi)$ and $f_{2'}(\xi)$, respectively, start at $c = c_2$ from $\lambda = 1$, but their behavior is different: the first one smoothly varies up to $c = c_7$ where it intersects $\lambda = 1$ for the second time. However, $\lambda_{2'1}$ promptly returns to the line $\lambda = 1$ and touches it at $c = c_3$. At this point the imaginary parameters $\eta_{2'1}$, $\eta_{2'2}$ of the solution $f_{2'}(\xi)$ become complex-conjugated: $\eta_{2'1} = -\eta_{2'2}$, and the solution itself becomes real, it coincides with $f_0(\xi)$ here. Recall that this point has been already observed as a branching point of $f_0(\xi)$. The point can be treated as a branching one of $f_{2'}(\xi)$ by reducing N by 2.

Note that, owing to Proposition 4, the solutions $f_2(\xi)$ and $f_{2'}(\xi)$ (as well as the complex-conjugates to them) belong to the same equivalent group both before and after the branching point c_3.

Follow up the solution $f_2(\xi)$ as c grows. The eigenvalue λ_{21}corresponding to it (as well as $\lambda_{2'1}$corresponding to $f_{2'}(\xi)$ from the same equivalent group) intersects the value $\lambda = 1$ the next time at $c = c_7$. Here two solutions with $N = 3$ branch off from each solution of the mentioned equivalent group; denote one of them by $f_3(\xi)$. Each new solution has two complex η_{31}, $\eta_{32} = -\bar{\eta}_{31}$ and one imaginary $\pm\eta_{33}$ parameter. All these parameters form the polynomial $P_3(\xi)$ having the symmetry property of type (A.93). At $c = c_7$ we have $\eta_{31} = \eta_{21}$, $\eta_{33} = 0$; the values η_{21}, c_7 are calculated from the equation set (A.44) (consisting of two real equations owing to the symmetry of $F(\xi)$ and $|P_2(\xi)|$) supplied with the equation

$$\int_{-1}^{1} \frac{\xi^2 \cos(c\xi)F(\xi)}{\left|1-\eta_{21}^2\xi^2\right|} d\xi = 0 \tag{A.101}$$

following from (A.87).

The last branching of the solution $f_2(\xi)$ in the considered range of c is observed at $c = c_{10}$, where the eigenvalue λ_{21} becomes equal to unity the second time (curve 21 in Fig.A.1). This point is similar to $c = c_3$: the second eigenvalue λ_{22} equals unity here. Solutions with complex parameters η_{41}, η_{42},η_{43}, η_{44}, corresponding to $N = 4$, branch off from $f_2(\xi)$. At $c = c_{10}$ two of these parameters are complex: $\eta_{41} = \eta_{21}$, $\eta_{42} = \eta_{22}$, and two others are imaginary: $\eta_{44} = -\eta_{43}$. Both they are calculated from the equation set (A.94) for $N = 2$. At $c > c_{10}$ the values η_{43}, η_{44} become complex having the same property $\eta_{44} = -\eta_{43}$ (curves 43, 44 in Fig. A.2(a) and symmetrical to 43 in Fig. A.2(b)). We will return to these solutions while analyzing those corresponding to $N = 4$.

Consider the branching of the solution $f_3(\xi)$. The first eigenvalue λ_{31} corresponding to it starts at $c = c_7$ from $\lambda = 1$ and intersects this level the next time at $c = c_9$. Here two solutions with $N = 4$ branch off from it, each of them has two complex parameters η_{41}, $\eta_{42} = -\bar{\eta}_{41}$ and two imaginary ones η_{43}, $\pm\eta_{43}$. At the branching point $\eta_{41} = \eta_{31}$, $\eta_{43} = \eta_{33}$, $\eta_{44} = 0$. The additional equation supplied to (A.44) for calculating η_{31}, η_{33}, c_9 is

$$\int_{-1}^{1} \frac{\xi^3 \sin(c\xi)F(\xi)}{\left|1-\eta_{31}^2\xi^2\right|\sqrt{1-\eta_{33}^2\xi^2}} d\xi = 0. \tag{A.102}$$

We conclude the branching analysis by considering the solutions corresponding to $N = 4$, which have branched off from $f_3(\xi)$ at $c = c_9$. Qualitatively these solutions are similar to those corresponding to $N = 2$ which have branched off from $f_1(\xi)$ at $c = c_3$. We have to consider separately two different representatives of the corresponding equivalent group: $f_4(\xi)$ (with η_{43}, η_{44} having the same sign) and $f_{4'}(\xi)$ (with η_{43}, η_{44} having the opposite signs). The first eigenvalues λ_{41} and $\lambda_{4'1}$of (A.78), corresponding to $f_4(\xi)$ and $f_{4'}(\xi)$, respectively, start at $c = c_9$ from $\lambda = 1$. Their behavior is different: the first of them is smoothly varied, it intersects $\lambda = 1$ for the second time without the range $c < 10$, considered here. The second one $\lambda_{2'1}$ touches the line $\lambda = 1$ at $c = c_{10}$. At this point the imaginary parameters $\eta_{4'3}$, $\eta_{4'4}$ of the solution $f_{4'}(\xi)$ become complex-conjugated: $\eta_{4'3} = -\eta_{4'4}$, the value of N decreases by 2, and the solution coincides with $f_2(\xi)$. This point has been already analyzed as a branching point of $f_2(\xi)$ by growing N by 2.

Note that at $c = c_9$ the solution pairs like $f_4(\xi)$, $f_{4'}(\xi)$ have branched off from each of eight solutions corresponding to $N = 3$. All the solutions with $N = 4$ belong (in the considered range of c) to the same equivalent group both before and after the branching point c_{10}.

The whole branching process is shown in Fig. A.3 in graph form. The curve labels correspond to the polynomial degree N.

In Fig. A.4 the values of functional $\sigma(u)$ corresponding to the solution $f_0(\xi)$ and to the optimal ones $f_N(\xi)$ (which provide the global minimum for σ) are compared. Numerical results show that the optimal solutions at fixed c correspond to the largest possible value

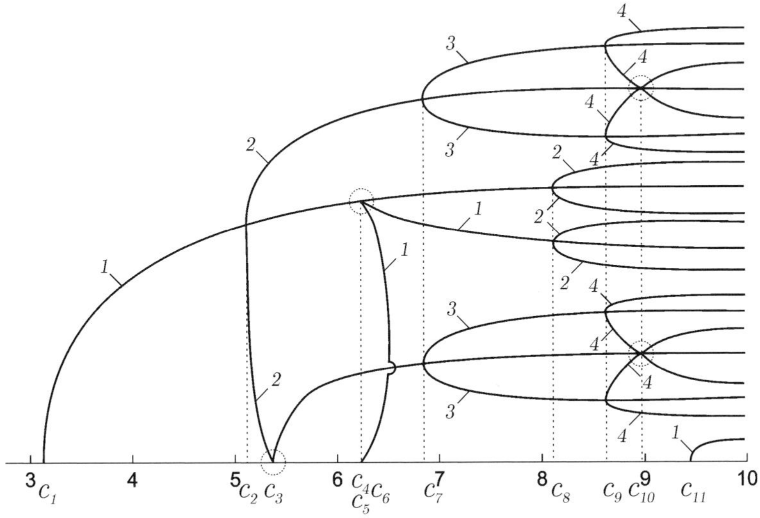

Figure A.3:

$N = N_{\max}$ at this c. However, at $N < N_{\max}$ the dependence σ on N is not monotonic. This is clearly observed, for instance, at the point c_{10}: since $f_4(\xi) = f_2(\xi)$ here, $f_2(\xi)$ provides the same (global minimum) value to σ as $f_4(\xi)$ does, and this value is smaller than that for $f_3(\xi)$. Note that the opposite conclusion was made in [13], because the value of N considered there was limited by $N \leq 3$; that is why the nonmonotonic dependence σ_{opt} on N was observed at $c > c_{10}$.

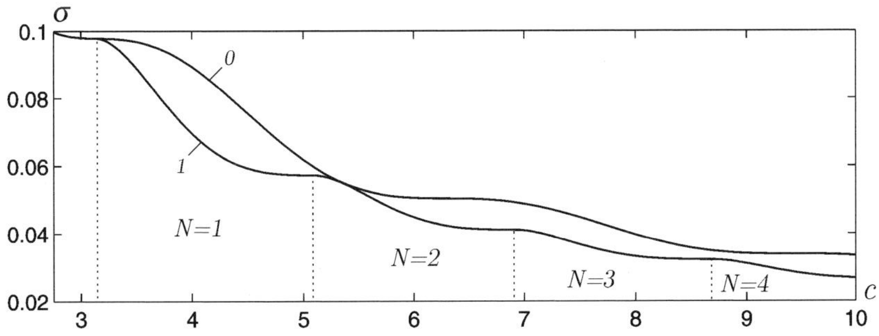

Figure A.4:

Several optimal functions $u(x)$ corresponding to different solutions $f_4(\xi)$ of the same equivalent group at $c = 9.5$ are shown in Fig. A.5. The functions of three different types are shown: complex symmetrical ($u(x) \neq \bar{u}(x)$, $u(-x) = u(x)$), real nonsymmetrical ($\operatorname{Im} u(x) \equiv 0$, $u(-x) \neq u(x)$), and complex nonsymmetrical ($u(x) \neq \bar{u}(x)$, $u(-x) \neq u(x)$). Recall that $|f_4(\xi)|$, calculated by applying (A.20) to all these functions, is the same and hence the value of $\sigma(u)$ is the same. For comparison, the functions $|f_N(\xi)|$ for all $N \leq 4$ are shown in Fig. A.6 at the same c.

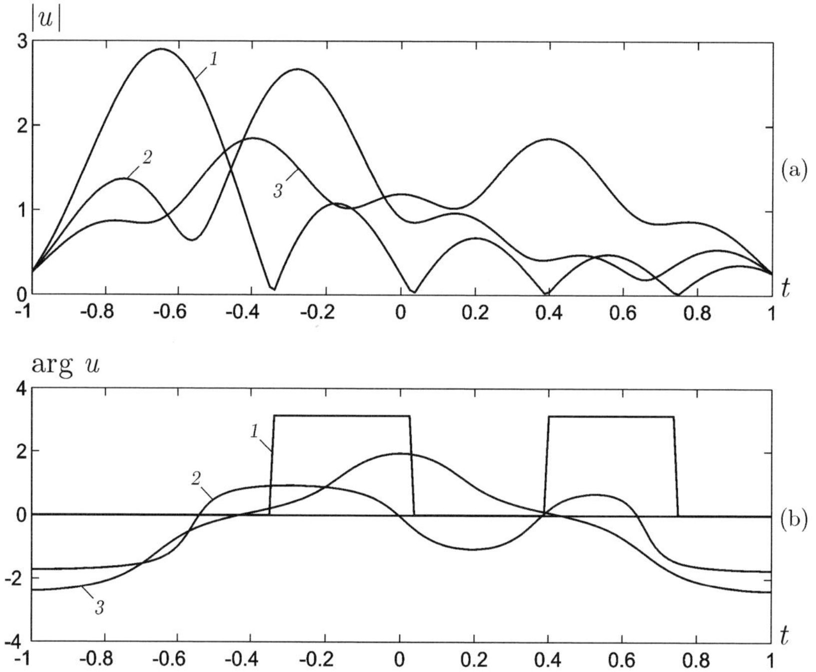

Figure A.5:

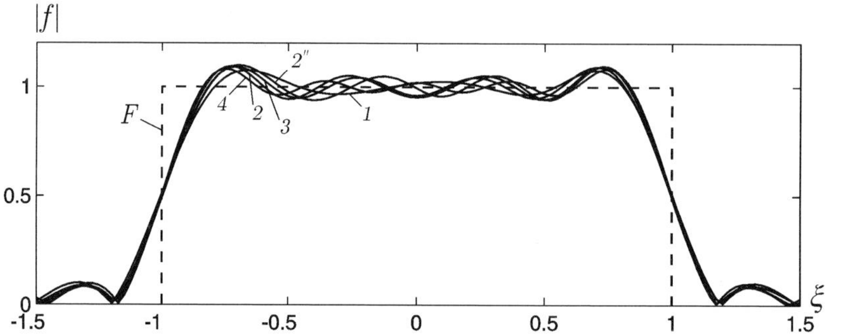

Figure A.6:

We conclude our analysis of the case $F(\xi) = 1/\sqrt{2}$ with the following remarks. In the chosen range of the parameter c we have almost all possible types of bifurcation. A rather surprising situation occurs at the points c_3, c_4, c_5, c_6, c_{10} (see dotted circles in Fig.A.3). To explain the nature of these points, in the next subsection we consider perturbations of $F(\xi)$ disturbing its symmetry.

A.2.5 Problems with nonsymmetrical data

The numerical results concerning the nonsymmetrical given function $F(\xi) = 1 - \beta\xi$ (the nonessential normalizing factor $1/\sqrt{2}$ is omitted here), analogous to those of the previous subsection, are given in Figs.A.7–A.9 for $\beta = 0.1$. They illustrate, in sequence, the eigenvalues of the homogeneous equation (A.78) (Fig.A.7), solutions to the transcendental equation set (A.44) (Fig. A.8), and the graph form of the branching process (Fig. A.9).

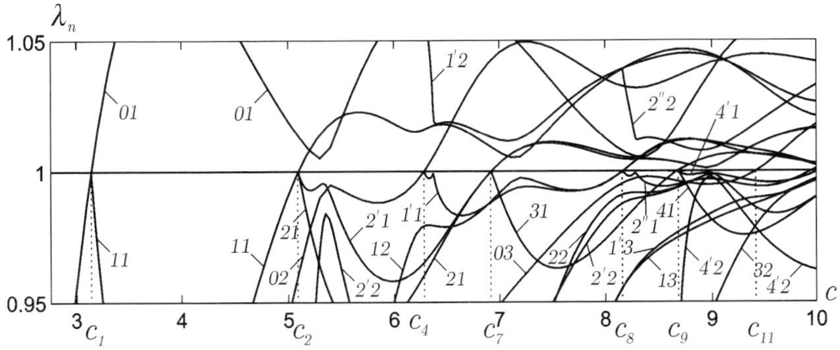

Figure A.7:

In contrast to the symmetrical case, for nonsymmetrical $F(\xi)$ all parameters η_{Nn} in solutions $f_N(\xi)$ (except for $N = 0$) are complex (with nonzero real parts). This relates also to the branching points: new parameters η_{Nn} in the engendered solutions start not from zero, but from some real values. This difference reveals itself most of all around the values of c, indicated at the end of the previous subsection.

At the point c_3 in the symmetrical case, two different solutions of the same equivalent group at $N = 2$, corresponding to imaginary parameters η_{2n}, have transformed into the complex ones (with nonzero real parts) in different ways. One of them had the coinciding parameters $\eta_{22} = \eta_{21}$, which obtained real parts with the opposite signs. The second one had the conjugated ones $\eta_{22} = -\eta_{21}$ which gave the real function $f_2(\xi)$ coinciding with $f_0(\xi)$. The point $c = c_3$ was the branching point with alteration of N by 2. In the nonsymmetrical case the curves of the eigenvalues for $N = 2$ (which intersected at $\beta = 0$ (see Fig.A.1)) break, they do not intersect the value $\lambda = 1$ (Fig.A.7). The branching point $c = c_3$ fails to exist. A similar situation occurs at $c = c_{10}$ where, in the symmetrical case, solutions with $N = 4$ meet those with $N = 2$. This effect is clearly illustrated by Fig.A.9, where the mentioned points are indicated by dotted circles.

A specific situation arises around the points c_4, c_5, c_6. Fig.A.10 shows the comparative behavior of the eigenvalues of equation (A.78) with $N = 0$ and $N = 1$ for the symmetrical ($\beta = 0$ – Fig.A.10(a)) and nonsymmetrical ($\beta \neq 0$ – Figs.A.10(b–d)) cases. Corresponding dependencies of the parameter η_{11} on c in different solutions are given in Figs.A.11(a–d). The values $\beta = 0$ (a), 0.01 (b), 0.0725 (c), 0.15 (d) are chosen such that all typical stages of solution transformation are shown. Since the topology of the eigenvalue curves is changed during this transformation, we must change the indication of some curves in comparison with Figs.A.1, A.7.

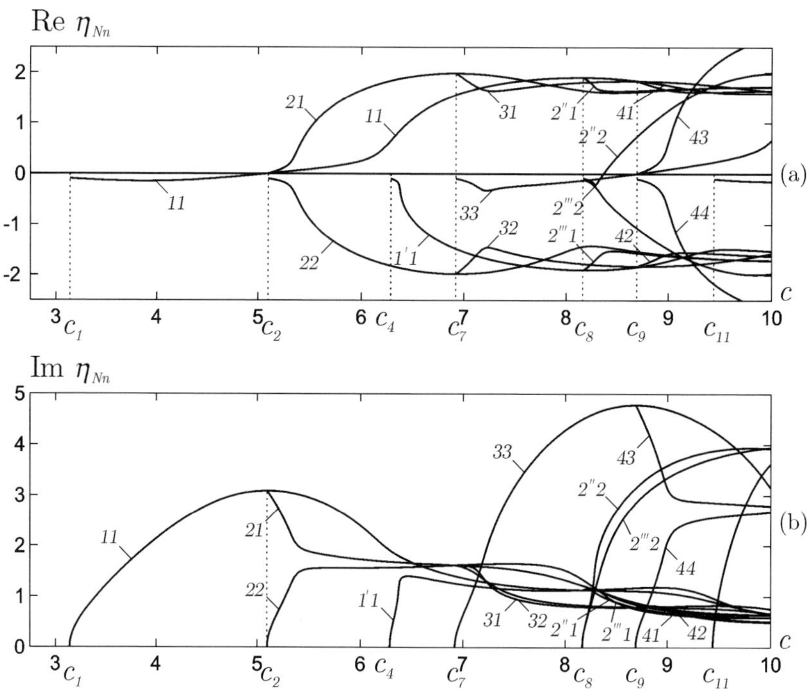

Figure A.8:

At $\beta = 0$ one observes two coinciding branching points: c_4, where the eigenvalue λ_{01} intersects the value $\lambda = 1$ (branching off a solution $f_{1'}$ with $N = 1$ from $f_0(\xi)$), and c_5, where the eigenvalue λ_{12} intersects the value $\lambda = 1$ (branching off two new solutions $f_{1''}(\xi)$, $f_{1'''}(\xi)$ with $N = 1$ from $f_1(\xi)$). The eigenvalues $\lambda_{1'1}$ and $\lambda_{1''1} = \lambda_{1'''1}$, corresponding to new solutions, start from the value $\lambda = 1$ at this point (recall that the first index in the eigenvalue coincides with the index of f_n to which it corresponds, and the second one indicates the eigenvalue number for the same solution). Simultaneously the eigenvalue $\lambda_{1'2}$ and two coinciding ones $\lambda_{1''2}$, $\lambda_{1'''2}$ start from λ_{02} and λ_{11}, respectively. Another point is significant in Fig.A.10(a), where curves 11 and 12 are mutually intersecting. At this point the homogeneous equation (A.78) for $f_1(\xi)$ has the double eigenvalue $\lambda_{11} = \lambda_{12}$. Three points, significant for further analysis, are indicated in Fig.A.10(a) by small dotted circles and indicated by capital letters A, B, C.

The third bifurcation point at $\beta = 0$ is $c = c_6$ where two solutions $f_1(\xi)$ and $f_{1'}(\xi)$ coincide and then disappear. The corresponding eigenvalues λ_{11}, $\lambda_{1'1}$ and λ_{12}, $\lambda_{1'2}$ coincide pairwise at this point and create two smooth continuous curves in the coordinates c, λ.

At a small nonsymmetrical perturbation of $F(\xi)$ the curves describing the eigenvalues λ_{1n}, for different solutions with the same N, break at the points of their multiplicity (see Fig.A.10(b) corresponding to $\beta = 0.01$). The point $c = c_4$ remains the branching point: the solution $f_{1'}(\xi)$ branches off from $f_0(\xi)$ here. Two eigenvalues λ_{01} and $\lambda_{1'1}$corresponding to these solutions equal unity at this point.

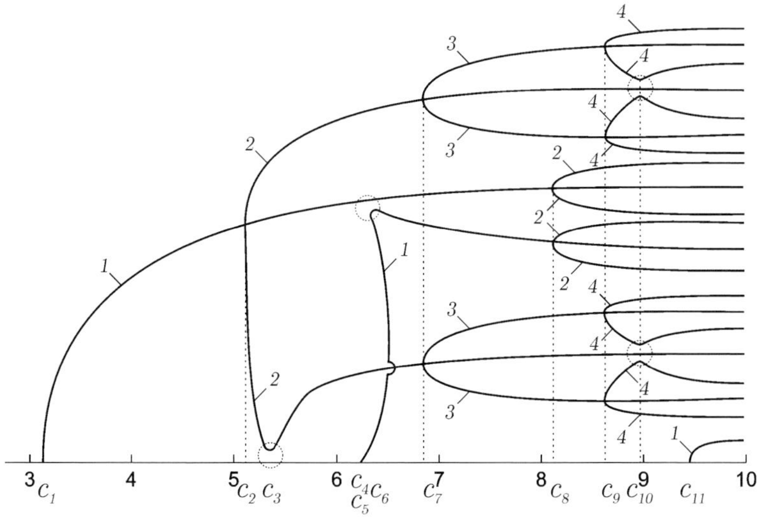

Figure A.9:

At the point $c = c_5$ curve 12 breaks. Its left part joins curve $1''1$ and goes down from the value $\lambda = 1$. The branching of the solution $f_1(\xi)$ into $f_{1''}(\xi)$, $f_{1'''}(\xi)$ fails to happen at this point. Instead of this, the solution $f_1(\xi)$ joins $f_{1''}(\xi)$ and together they form a continuous branch of solutions.

The right part of curve 12 breaks for the second time at point B (see Fig.A.10(a) where $\lambda_{11} = \lambda_{12}$ for $\beta = 0$). Its lower and upper pieces participate in constructing two loops of the eigenvalue curves at small β. The first of them is significant for investigating the bifurcation process. It is constructed with curve $1'1$, the right part of curve 11 (broken at point B as well) from $c = c_6$ to point B, the lower piece of the right part of curve 12 from point B to A, and curve $1'''1$. The loop intersects the value $\lambda = 1$ twice: at the point $c = c_6$ (which remains the disappearance point) and at $c = c_5$. The latter is transformed from the branching point into the appearance point. The second loop, formed by the upper piece of the right part of curve 12, the piece of curve 11 between points B and C, and curve $1'''2$, describes the second eigenvalues $\lambda_{1'2}$ corresponding to new solutions.

Two isolated solutions which appeared at $c = c_5$ are continuously (with respect to β) transformed from f_1 and $f_{1'''}$, having branched off from f_1 at $\beta = 0$. One of them, namely, $f_{1'''}$, exists at all larger c, the latter (denoted now by $f_{1''''}$) exists only up to $c = c_6$, when it coincides with $f_{1'}$ and disappears (Fig.A.11(b)). At $c_5 < c < c_6$ we have four solutions with $N = 1$.

The bifurcation points c_5, c_6 draw together (more specifically, c_5 tends to c_6) as β grows. At $\beta \approx 0.0725$ they coincide: $c_5 = c_6$ (Fig.A.10(c)). The loops on the eigenvalue curves shrink into the angle points. The eigenvalue $\lambda_{1'1} = 1$, but no solution appears and disappears at this point. The derivative $d\eta_{1'1}/dc = 0$, but that is only an inflection point on its graph (Fig.A.11(c)).

At $\beta > 0.0725$ the eigenvalue $\lambda_{1'1}$ becomes less than one, the points c_5, c_6 disappear, and

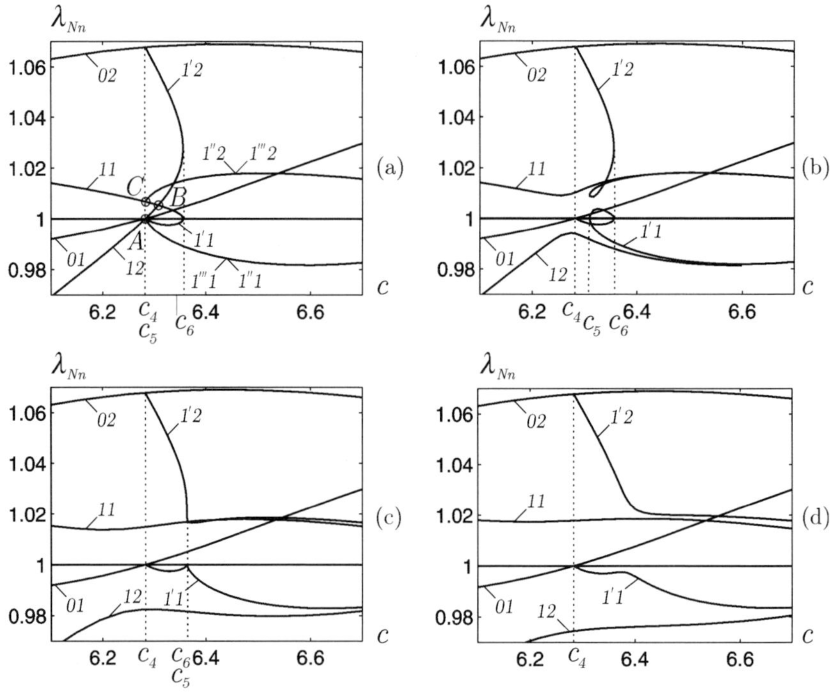

Figure A.10:

all curves become smooth. Only two solutions with $N = 1$ exist at $c > c_4$.

Note that bifurcation points of the appearance and disappearance type exist only for some of the even functions $F(\xi)$. For instance, they exist only at $\beta < 2/3$ for functions $F(\xi) = 1 - \beta\,|\xi|$. At $\beta = 2/3$ we have $c_4 = c_5 = c_6$, and then only one branching point c_4 remains.

A.3 Modified Phase Problem. Discrete Case

A.3.1 Problem formulation for linear antenna array. Lagrange–Euler equations

The above results concerning the continuous Fourier transformation (A.20) are carried over into the discrete Fourier transformation describing, in particular, the radiation pattern of equidistant linear antenna arrays [12], [17]. In this case the operator A maps the complex vector $u = \{u_m\}$ (the currents on the array radiators) into the function $f(\xi) \in L_2(-\pi/c_0\,, \pi/c_0\,)$ (radiation pattern):

$$f(\xi) = Au \equiv \sqrt{\frac{c_0}{2\pi}} \sum_{m=-M}^{M} u_m e^{imc_0\xi}. \tag{A.103}$$

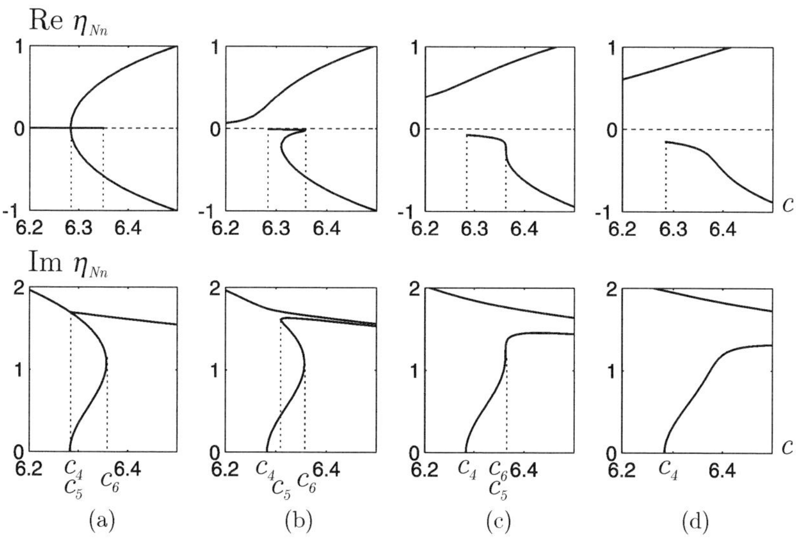

Figure A.11:

Here $c_0 = kd \sin \alpha \leq \pi$, d is the distance between radiators, α and ξ have the same sense as in Subsection A.1.4. For simplicity, we take the dimension of the vector space as an odd number $T = 2M + 1$. In this space the usual Euclidean norm

$$||u||_1^2 = \sum_{m=-M}^{M} |u_m|^2 \tag{A.104}$$

is used. The right-hand side of (A.103) is a periodical function of $\xi : -\infty < \xi < \infty$ with period $2\pi/c_0$. We consider only one period as a values domain of A and introduce the norm

$$||f||_2^2 = \int_{-\pi/c_0}^{\pi/c_0} |f(\xi)|^2 \, d\xi. \tag{A.105}$$

In these notations the same Parseval identity (A.21) holds.

The functional to be minimized is of the form

$$\sigma(u) = \int_{-\pi/c_0}^{\pi/c_0} [|f(\xi)| - F(\xi)]^2 \, d\xi, \tag{A.106}$$

where $F(\xi)$ is, as earlier, the given real nonnegative compactly supported function: $F(\xi) \equiv 0$ at $1 < |\xi| < \pi/c_0$. The Lagrange–Euler equations for this functional, analogous to (A.24), (A.25) are [37], [6]

$$u_m = \sqrt{\frac{2\pi}{c_0}} \int_{-1}^{1} F(\xi) e^{i(\arg f(\xi) - c_0 m \xi)} d\xi, \tag{A.107}$$

$$f(\xi) = \int_{-1}^{1} K(\xi - \xi')F(\xi')e^{i(\arg f(\xi') - c_0 m\xi')}d\xi', \tag{A.108}$$

where

$$K(t) = \frac{c_0 \sin(Tc_0 t/2)}{2\pi \sin(c_0 t/2)}. \tag{A.109}$$

Solutions to equation (A.108) have properties similar to properties 1–3, noted at the end of Section A.1.

Note the obvious analogue between the problem formulations for the continuous and discrete Fourier transformations and, as a result, between the Lagrange–Euler equations (A.24), (A.25) and (A.108), (A.109), respectively. Namely, all key formulas for the discrete case (with simple modifications) transform into those for the continuous case after passing to the limit: $T \to \infty$, $d \to 0$, $Td \to 2a$ and denoting $c = Tc_0/2$.

A.3.2 Theoretical results

Solutions to the nonlinear equation (A.108) can be expressed in the explicit form depending on a finite number of complex parameters. The results are formulated in the form of three propositions, which are proved in a similar way to those for equation (A.25). The following lemma allows to bring the problem into form convenient for application of the above technique:

Lemma 6 *Any function $f(\xi)$, $f(0) \neq 0$ of the form (A.103) can be written as*

$$f(\xi) = \frac{\alpha P_N(\tau) Q_{T-N-1}(\tau)}{(1+\tau^2)^M}, \tag{A.110}$$

where $|\alpha| = 1$,

$$\tau = \tan(c_0\xi/2), \tag{A.111}$$

the polynomial

$$P_N(\tau) = \prod_{n=1}^{N}(1 - \eta_n \tau) \tag{A.112}$$

describes all complex, nonconjugated zeros of $f(\xi)$:

$$\eta_n - \bar{\eta}_m \neq 0, \quad n, m = 1, 2, ...N, \tag{A.113}$$

$T = 2M + 1$, $Q_{T-N-1}(\tau)$ with real coefficients describes all real and complex conjugated zeros of $f(\xi)$.

(For simplicity, we write η_n instead of η_{Nn} omitting the first lower index N.)

Proof. It is easy to check that

$$e^{\pm imc_0\xi} = \frac{(1 \pm i\tau)^{2m}}{(1+\tau^2)^m}. \tag{A.114}$$

Substituting (A.114) into (A.103), we obtain

$$f(\xi) = \sqrt{\frac{c_0}{2\pi}} \frac{1}{(1+\tau^2)^M} \left[u_0 + \sum_{m=1}^{M} u_{-m}(1-i\tau)^{2m} + u_m(1+i\tau)^{2m}(1+\tau^2)^{M-m} \right]. \tag{A.115}$$

The expression in the square brackets is a polynomial of τ of power $2M$. Its general form is

$$f(0) \prod_{n=1}^{2M} (1 - \eta_n \tau), \tag{A.116}$$

where η_n^{-1}, $n = 1, 2, ...2M$ are all its zeros. Extracting the factor $\alpha = \exp(i \arg f(0))$ and splitting its zeros in an obvious way into two mentioned groups, we obtain (A.110). □

The following propositions are analogous to Propositions 2–4. Their proofs have the same schemes and are omitted here. Proposition 5 does not need its analogy in the discrete case, because, according to Lemma 6, all the solutions to (A.103) have a finite number of zeros in the complex plane τ.

Proposition 7 *Let $c_0 < \pi$ be a given positive number, $F(\xi) \in L_2(-1,1)$ be a given positive function. Then the function $f(\xi)$ having no zeros at $\xi \in (-1,1)$ is a solution to equation (A.103) if and only if it can be presented in the form*

$$f(\xi) = \alpha \frac{P_N(\tau)}{|P_N(\tau)|} \left| \int_{-1}^{1} K(\xi - \xi') F(\xi - \xi') \frac{P_N(\tau')}{|P_N(\tau')|} d\xi' \right|, \tag{A.117}$$

where α is an arbitrary complex constant with $|\alpha| = 1$, τ is given by (A.111), $P_N(\tau)$ is a polynomial of the form (A.112) with complex nonconjugated pairwise zeros η_n^{-1} (see (A.113)) satisfying the transcendental equation set

$$\int_{-1}^{1} \frac{\tau^{n-1} \sin(T c_0 \xi/2) F(\xi)}{\cos(c_0 \xi/2) |P_N(\xi)|} d\xi = 0, \tag{A.118a}$$

$$\int_{-1}^{1} \frac{\tau^{n-1} \cos(T c_0 \xi/2) F(\xi)}{\cos(c_0 \xi/2) |P_N(\xi)|} d\xi = 0, \tag{A.118b}$$

$$n = 1, 2, \ldots, N.$$

Proposition 8 *The equation set can have solutions only at*

$$N < \frac{T c_0}{\pi}. \tag{A.119}$$

Proposition 9 *If the function*

$$f(\xi) = \hat{f}(\xi)P_N(\tau), \tag{A.120}$$

where $\tilde{f}(\xi)$ *is a real positive function at* $\xi \in (-1,1)$ *and* $P_N(\tau)$ *is given by (A.112), (A.113), solves equation (A.108), then*

$$f_n(\xi) = f(\xi)\frac{1-\bar{\eta}_n\tau}{1-\eta_n\tau}, \quad n = 1,2,...,N \tag{A.121}$$

also solves equation (A.108).

For proofs of these propositions we refer the reader to [17].

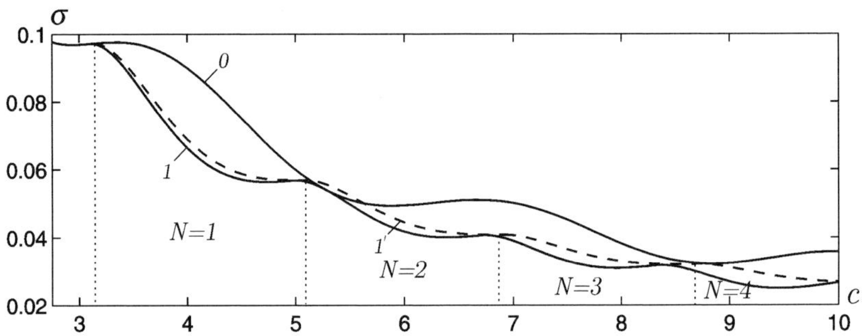

Figure A.12:

Table A.1:

m	$\lvert u^{(1)}\rvert$	$\arg u^{(1)}$	$\lvert u^{(2)}\rvert$	$\arg u^{(2)}$	$\lvert u^{(3)}\rvert$	$\arg u^{(3)}$
–4	1.513	0	0.689	1.679	1.394	1.792
–3	0.885	3.142	0.673	–1.440	1.569	–1.184
–2	0.884	3.142	0.442	2.154	1.631	2.342
–1	3.999	0	0.445	0.942	1.662	–0.118
0	2.995	0	5.615	0	4.681	–0.159
1	0.132	3.142	0.445	0.942	0.886	0.724
2	0.151	0	0.442	2.154	0.540	–3.133
3	0.213	3.142	0.673	–1.440	0.375	–0.732
4	0.145	0	0.689	1.679	0.352	2.064

A.3.3 Numerical results

Numerical results obtained for the same given function $F(\xi) = 1/\sqrt{2}$ and different values of M show that the qualitative behavior of solutions to equations (A.109) and (A.25) is similar. In particular, all bifurcation points of different type, existing in the continuous case, are observed in the discrete case as well. The same is also true for their transformation by nonsymmetrical perturbations of $F(\xi)$. Similarly to the continuous case, the best approximation of the given $F(\xi)$ at fixed c and T is provided by solutions corresponding to the largest possible polynomial of degree N. The only distinction of the discrete case from the continuous one is the upper bound of the value $c = Tc_0/2 \leq T\pi/2$ caused by the limitation of c_0 assumed in Subsection A.3.1.

For the above reason we do not show here the numerical results concerning the eigenvalues of the appropriate homogeneous equation describing the bifurcation process, dependence of the parameters η_n on c, etc. In Fig.A.12 the values of $\sigma(u)$ obtained on the "trivial" (curve 0) and optimal (curve 1) solutions are shown for the case $T = 9$. For comparison, the dependence $\sigma(c)$ for the continuous case is shown by the dashed line.

Three optimal distributions of the current $u^{(j)} = \{u_m^{(j)}\}$, $j = 1, 2, 3$ on the radiators, corresponding to equivalent solutions of equation (A.109) with $N = 4$ at $c = 9.5$ are given in Table A.1. As in the continuous case, we show solutions of three different types: real nonsymmetrical (1), complex symmetrical (2), complex nonsymmetrical (3).

Bibliography for Appendix

[1] A. G. Ramm. Optimal solution of the antenna synthesis problem. *Doklady AN SSSR* **180** (1968), 1071–1074 (Russian).

[2] A. G. Ramm. Nonlinear antenna synthesis problem. *Doklady AN SSSR* **186** (1969), 1277–1280 (Russian).

[3] Yu. I. Choni. On antenna system synthesis according to given amplitude radiation pattern. *Radioelektronika* **11** (12) (1968), 1325–1327.

[4] L. D. Bakhrakh, S. D. Kremenecky. *Synthesis of Radiating Systems. Theory and Calculation Methods.* Sovetskoe Radio, Moscow 1974 (Russian).

[5] N. N. Voitovich. On antenna synthesis according to prescribed amplitude radiation pattern (V. V. Semenov's method). *Radiotekhnika i Elektronika* **17** (12) (1972), 2491–2497 (Russian).

[6] M. I. Andriychuk, N. N. Voitovich, P. A. Savenko, V. P. Tkachuk. *Synthesis of Antennas by Amplitude radiation pattern. Numerical Methods and Algorithms.* Naukova Dumka, Kiev 1993 (Russian).

[7] P. A. Savenko. *Nonlinear Problems of Radiation System Synthesis.* IAPMM, Lviv 2002 (Ukrainian).

[8] N. N. Voitovich, Yu. P. Topolyuk. antenna synthesis according to prescribed amplitude radiation pattern and the phase problem. In: *Direct and Inverse Problems of Electromagnetic and Acoustic Wave Theory (DIPED–97)*, Lviv 1997, 90–92.

[9] N. N. Voitovich, Yu. P. Topolyuk, O. O. Reshnyak, J. Jarkowski. New closed solution of the linear antenna synthesis problem according to amplitude radiation pattern. In: *Proc. of Int. Conf. on Microwave and Radar (MIKON–98), vol. 2*, Krakow 1998, 304–308.

[10] N. N. Voitovich, O. M. Gis, O. O. Reshnyak, Yu. P. Topolyuk. New rigorous solution of some antenna synthesis problems according to prescribed amplitude radiation pattern. In: *Proc. of Int. Conf. on Mathematical Methods in Electromagnetic Theory (MMET–98), vol. 1*, Kharkov 1998, 133–135.

[11] N. N. Voitovich, O. M. Gis, Yu. P. Topolyuk. Mean square approximation of compactly supported functions with free phase by functions with bounded spectrum. *Dopovidi NAN Ukrainy* **3** (1999), 7–10 (Ukrainian).

[12] N. N. Voitovich, O. O. Reshnyak. Solutions of nonlinear integral equation of synthesis of the linear antenna arrays. *BSUAE J. of Appl. Electromagn.* **2** (2) (1999), 43–52.

[13] N. N. Voitovich, Yu. P. Topolyuk, O. O. Reshnyak. Approximation of compactly supported functions with free phase by functions with bounded spectrum. *Fields Institute Communications* AMS **25** (2000), 531–541.

[14] P. O. Savenko, Yu. P. Topolyuk, N. N. Voitovich. On differentiability of a functional arisen in antenna synthesis theory. In: *Direct and Inverse Problems of Electromagnetic and Acoustic Wave Theory (DIPED–2000)*, Lviv–Tbilisi 2000, 57–59.

[15] Yu. P. Topolyuk, N. N. Voitovich. Convergence of iterative method for problem with free phase in case of isometric operator. In: *Direct and Inverse Problems of Electromagnetic and Acoustic Wave Theory (DIPED–2000)*, Lviv–Tbilisi 2000, 52–56.

[16] O. M. Gis, Yu. P. Topolyuk. Mean-square approximation of nonnegative compactly supported functions by modulus of functions with finite spectrum (case of diminishing the approximant on the support). *Matematychni Metody i Fizyko–Mekhanichni Pola* **44** (2) (2001), 63–69 (Ukrainian).

[17] O. O. Bulatsyk, N. N. Voitovich. Analytical solutions to the nonlinear integral equation of approximation problem of compactly supported functions with free phase by functions with bounded discrete spectrum. *Matematychni Metody i Fizyko–Mekhanichni Pola* **44** (2) (2001), 70–78 (Ukrainian).

[18] O. O. Bulatsyk, O. M. Gis, N. N. Voitovich. Some peculiar properties of solutions of the generalized phase problem. In: *Direct and Inverse Problems of Electromagnetic and Acoustic Wave Theory (DIPED–2001)*, Lviv 2001, 38–41.

[19] O. O. Bulatsyk, N. N. Voitovich. Modified phase problem with nonsymmetrical data. In: *Direct and Inverse Problems of Electromagnetic and Acoustic Wave Theory (DIPED–2002)*, Lviv–Tbilisi 2002, 137–140.

[20] O. O. Bulatsyk. New properties of solutions to modified phase problem. In: *Mathematical Problems of Mechanics of Nonhomogeneous Structures*, Lviv 2003, 513–514 (Ukrainian).

[21] H. A. Ferwerda. Problem of the wave front phase reconstruction according to amplitude distribution and coherent functions. In: *Inverse Scattering Problems in Optics. Ed. H. P. Baltes*. Springer, Berlin 1978.

[22] V. K. Ablekov, S. A. Kolyadin, A. V. Frolov. *High-Resolved Optical Systems*. Mashinostroenie, Moscow 1985 (Russian).

[23] M. V. Klibanov. On uniqueness of definition of a finite function by modulus of its Fourier transform. *Doklady Akademii Nauk SSSR* **285** (1985), 278–280 (Russian).

[24] T. I. Kuznetsova. On the phase problem in optics. *Uspekhi Fizicheskikh Nauk* **154** (4) (1988), 677–690 (Russian).

[25] O. V. Ovcharenko. Relation between amplitude and phase of signal with finite analytical spectrum. *Radiotekhnika i Elektronika* **35** (1) (1990), 106–114 (Russian).

[26] H. A. Hauptman. The phase problem of X-ray Crystallography. In: *Signal Processing, Part II: Control Theory and Applications. Eds. F. A. Grunbaum, J. W. Helton, P. Khargonecar*. Springer, New York 1990, 257–273.

[27] A. G. Ramm. *Multidimensional Inverse Scattering Problems*. Longman/Wiley, New York 1992 (exp. Russian edition: Mir, Moscow 1994).

[28] A. G. Kurosh. *Course of Higher Algebra*. Gostekhizdat, Nauka, Moscow 1975 (Russian).

[29] B. M. Minkovich, V. P. Yakovlev. *Theory of Antenna Synthesis*. Nauka, Moscow 1969 (Russian).

[30] P. O. Savenko. On existence of solutions of a class of nonlinear inverse problems of mathematical physics concerning synthesis of radiating systems. *Dopovidi NAN Ukrainy* **8** (1994), 48–52 (Ukrainian).

[31] W. Rudin. *Real and Complex Analysis*. McGraw-Hill Book Company, New York 1966.

[32] M. M. Vainberg, V. A. Trenogin. *Theory of Bifurcation of Solutions of nonlinear equation*. Nauka, Moscow 1969 (Russian).

[33] A. N. Tikhonov and V. Y. Arsenin. *Solutions of Ill-Posed Problems*. Winston and Sons, Washington 1977.

[34] F. J. Lander. The Bezoitiant and inversion of Hankel and Toeplitz matrices. *Matematicheskie Issledovaniya, Kishinev* **2** (32) (1974), 69–87 (Russian).

[35] M. Fiedler. *Special Matrices and their Applications in Numerical Mathematics*. Martinus Nijhoff Publishers, Dordrecht/Boston/Lancaster 1986.

[36] B. A. Ilyin, V. A. Sadovnichiy, B. X. Sendov. *Mathematical Analysis*. Nauka, Moscow 1979 (Russian).

[37] P. A. Savenko. Synthesis of linear antenna arrays by the given amplitude pattern. *Izvestiya Vuzov, Radiofizika* **22** (12) (1979) 1498–1504 (Russian).

Index